The Mathematics of Various Entertaining Subjects

THE MATHEMATICS OF VARIOUS ENTERTAINING SUBJECTS

Volume 3

THE MAGIC OF MATHEMATICS

EDITED BY

Jennifer Beineke & Jason Rosenhouse

WITH A FOREWORD BY MANJUL BHARGAVA

National Museum of Mathematics, *New York* • Princeton University Press, *Princeton and Oxford*

Published by Princeton University Press,
41 William Street, Princeton, New Jersey 08540
6 Oxford Street, Woodstock, Oxfordshire OX20 1TR

press.princeton.edu

In Association with the National Museum of Mathematics,
11 East 26th Street, New York, New York 10010

 TWO SIGMA

LCCN: 2019936872
ISBN: 978-0-691-18257-5
ISBN (pbk.): 978-0-691-18258-2

British Library Cataloging-in-Publication Data is available

Editorial: Vickie Kearn, Susannah Shoemaker, and Lauren Bucca
Production Editorial: Nathan Carr
Text Design: Lorraine Doneker
Jacket/Cover Design: Lorraine Doneker
Production: Jacquie Poirier
Publicity: Matthew Taylor and Katie Lewis
Copyeditor: Cyd Westmoreland

Jacket/Cover Credit: Top row (left to right): Fig. 1. Possible sequence of outcomes
to a Smullyan puzzle. Fig. 2 Checkers game configuration. Fig. 3. Coin row
transformation puzzle. Middle row (left to right): Fig. 4: First scoring in a game
of Topology. Fig. 5. Octiles. Courtesy of Kadon Enterprises. Fig. 6: Spades dealt
into two piles. Bottom row (left to right): Fig. 7. Tarot cards. Fig. 8. Knotted bands.
Fig. 9. LEGO building made from sixteen jumper plates.

This book has been composed in Minion Pro

Printed on acid-free paper. ∞

Printed in the United States of America

1   3   5   7   9   10   8   6   4   2

Contents

Foreword

Manjul Bhargava

What exactly is recreational mathematics, and what distinguishes it from other types of mathematics? This is a question oft asked by amateur and professional mathematicians alike, though without any clear answers. Literally, recreational mathematics refers to mathematics that is pursued for enjoyment, for recreation, for fun. But by that definition, is not most—if not all—mathematics, recreational mathematics?

As becomes clearer every day, the dividing boundaries among pure, applied, and recreational mathematics are quite blurry, if not altogether nonexistent. Pure mathematics is used to solve important applied, real-life problems every day; conversely, applied mathematics research continuously generates ideas that advance theoretical and pure mathematics. Furthermore, both pure and applied mathematics are regularly pursued by mathematicians for their enjoyment and for the pursuit of beauty and aesthetics; indeed, much of the world's great research mathematics, both pure and applied, has arisen throughout history due to investigations primarily driven by play and curiosity. Therefore, whether given mathematical results, problems, papers, theorems, proofs, and so forth, should be classified as pure, applied, or recreational can often be very unclear—and can become increasingly unclear as time passes.

Despite this uncertainty of classification, the distinguishing feature of recreational mathematics—in the literature and in its practice—is that its explicit primary purpose is recreation (e.g., through its relation to puzzles, games, tricks, art, etc.). As a consequence, a further distinguishing feature of recreational mathematics is its wide accessibility: It can be enjoyed and pursued equally by all persons, including those without any particular training in mathematics. Of course, many problems in pure and applied mathematics can also be recreational by these definitions (especially problems in, say, number theory or combinatorics), which is why recreational mathematics cannot be considered as a category that is disjoint and separated from pure and applied research mathematics.

It is because of its playfulness and wide accessibility, together with its natural connections to pure and applied research, that recreational mathematics has helped attract numerous young people to mathematics and related fields. I, too, consider myself a product of this natural path to mathematical research.

In this regard, one can hardly forget to mention Martin Gardner and his role in developing and proving the value of this path. Ronald Graham famously observed that, through recreational mathematics, Gardner "turned thousands of children into mathematicians, and thousands of mathematicians into children." Gardner firmly established the power of recreational mathematics in inspiring mathematicians and nonmathematicians alike to pursue the enjoyment of mathematics. I, too, found myself as both the child and then the mathematician in Graham's quote, having read many of Gardner's writings both as a child in school and also as a "grown-up" mathematician.

It should therefore be clear that—with the increasing criticality of STEM for the advancement of society and humanity—recreational mathematics, besides being fun, can also play a critical role in encouraging youth to pursue mathematics and related fields. It is with the latter goal in mind that the MOVES (Mathematics Of Various Entertaining Subjects) conference was created in 2013. MOVES brings together teachers, students, amateurs, and professionals from around the country and the world to celebrate and share ideas and advances in recreational mathematics.

In recent years, each MOVES conference has had a theme. In 2017, the theme of MOVES was "The Magic of Math." It was a great pleasure to participate in MOVES 2017, and a truly humbling honor to be one of its two keynote speakers in conjunction with my teacher, and close friend, Persi Diaconis.

The very structure and nature of the MOVES conference inspires its participants to actively think, speak, and write about recreational aspects of mathematics, and in turn bring this material to classrooms, daily conversations, and research. Persi's participation at MOVES 2017 is a wonderful case in point. Persi related to us how he had been thinking for many years about the writings of Charles Sanders Peirce, the versatile nineteenth-century philosopher, scientist, and intellectual who also invented, in Persi's words, some "highly original and unperformable mathematical card tricks." The MOVES 2017 conference provided Persi with the inspiration and opportunity to collect his thoughts on and to analyze Peirce's work through the medium of his magical writings. The result was a beautiful lecture at MOVES, and a corresponding chapter in this volume—written jointly with long-time collaborator Ronald Graham— which in particular gets to the mathematical bottom of several of Peirce's tricks and provides valuable ideas on how to make some of his tricks more performable. The reader is highly encouraged to read Persi's chapter in this volume!

My own participation in MOVES was also a great motivation for me, personally, to look back on one of my very favorite numbers from childhood— 142,857—and to trace its recreational origins and connect it both with modern mathematics and with concepts from magic. As is well known—but perhaps not as well as it should be!—the number 142,857 has a remarkable property:

$$142857 \times 1 = 142857,$$

$$142857 \times 2 = 285714,$$

$$142857 \times 3 = 428571,$$

$$142857 \times 4 = 571428,$$

$$142857 \times 5 = 714285,$$

$$142857 \times 6 = 857142.$$

In other words, the first six multiples of 142,857 are obtained simply by cyclically permuting its digits. The number 142,857 is thus known as a *cyclic number*, that is, an n-digit number (possibly having some initial digits 0) whose first n multiples are given by cyclically permuting the original number in all possible ways.

The existence of such a remarkable number immediately raises further questions. First, do there exist other cyclic numbers? The answer is yes: 0,588,235, 294,117,647 is the next smallest cyclic number, having 16 digits; multiplying it by any number from 1 to 16 simply cycles its digits!

The next question that arises, then, is: Are there *infinitely many* cyclic numbers? This, remarkably, is an unsolved problem and connects to deep issues in number theory that I and many number theorists continue to think about in our research.

It turns out that every cyclic number in base 10 is given by the repeating pattern in the decimal expansion of $1/p$ for a prime p such that 10 is a *primitive root modulo p* (i.e., the first $p - 1$ powers of ten—10^0, 10^1, ..., 10^{p-2}—have distinct remainders when divided by p). For example, that 142,857 is a cyclic number is related to the fact that 1, 10, 10^2, 10^3, 10^4, and 10^5 all yield distinct remainders when divided by 7, namely, 1, 3, 2, 6, 4, and 5, respectively.

Whether or not there are infinitely many primes for which 10 is a primitive root is addressed by what is known as "Artin's Primitive Root Conjecture," a centrally important and much studied conjecture in modern number theory. Artin's Conjecture, first presented in 1927, predicts not only that 10 is a primitive root modulo p for infinitely many primes p, but also that 10 is a primitive root modulo p for approximately 37.3% of primes p. For these primes p, the quantity $1/p$ has in its decimal expansion a repeating pattern of length $p - 1$, which then yields a cyclic number of length $p - 1$. Thus, Artin's Conjecture implies not only that there are infinitely many cyclic numbers, but it also estimates how many such cyclic numbers there are of fewer than X digits, as X tends to infinity! Hooley[1] proved that Artin's Primitive Root Conjecture follows from certain cases of the generalized Riemann hypothesis. Thus, the question as to whether there are infinitely many cyclic numbers is directly related to the generalized Riemann hypothesis, which of course is one of the most central and important subjects of research in number theory today.

I find this a truly remarkable story: A question, born of recreation, that I played with and appreciated as a child, turns out to connect to some of the deepest open problems in number theory that I now think about seriously as a research mathematician. The story illustrates, again, the fine line between recreational and research mathematics.

Of course, the number 142,857 has been a favorite number not only of recreational mathematicians and number theorists but also of magicians. A number of surprisingly beautiful effects have been created by magicians over the years using the cyclic nature of this number. The reader may consult Martin Gardner[2] and the references contained therein for several examples. Gardner also discusses other remarkable properties of the number 142,857 (e.g., the fact that $14 + 28 + 57 = 99, 142 + 857 = 999$, and various other unconventional sums involving 142,857).

In my talk, I discussed how many of the unproved assertions of Gardner's article could be proved, for general cyclic numbers, using modern mathematical ideas, such as the 10-adic numbers. I also suggested magical effects that could be based on larger cyclic numbers (or on their close relatives, the doubly- and triply-cyclic numbers), so that if digits of such a lesser-known number were to come up, it would not raise any suspicions in the audience. Cyclic numbers and their generalizations demonstrate yet another beautiful area of interaction between mathematics and magic. I hope to write more details about the number theory and magic behind cyclic numbers in a future volume.

During one of the evenings of the conference, Art Benjamin also gave an exciting and energetic performance of mental math and magic, leading the audience through magic squares, calendar dates and days of the week, and rapid mental calculations. Mark Mitton performed his brilliant and captivating mathematical and close-up sleight-of-hand magic for participants during some of the other enjoyable evening events and dinners.

Other impromptu happenings included a Q&A session with my teachers/heroes/close friends John Conway and Persi Diaconis—I was pulled up onto the stage with them as well, a very humbling but fun experience—and we were bombarded with various outside-the-box and offbeat questions for over an hour that related to recreational mathematics and beyond (Figure 1). On another note, my birthday also happened to fall during the conference, and the conference organizers arranged a little celebration with cake, food, and singing during one of the longer tea breaks, which was very touching and a good excuse for a party (Figure 2).

Other academic highlights of the event included talks about the mathematics of BINGO, juggling (Figure 3), shuffling, the card game bridge, checkers, dice, chutes and ladders, SET, Yahtzee, Nim, Sudoko, KenKen, LEGO bricks, board games, card tricks, braids, flexagons, Engel machines, (Figure 4), logic puzzles, codes, and more, all of which stimulated enjoyable conversations among the conference participants and have resulted in corresponding articles in the current volume.

Figure 1. Panel with Persi Diaconis, John Conway, and the author.

Figure 2. Birthday cake! Persi Diaconis, John Conway, and the author.

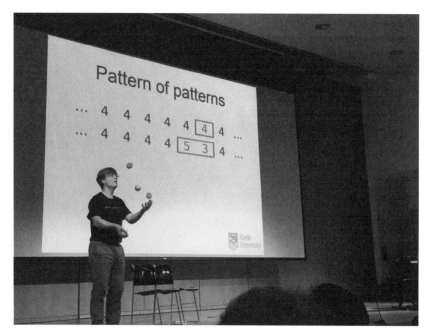

Figure 3. Colin Wright juggling.

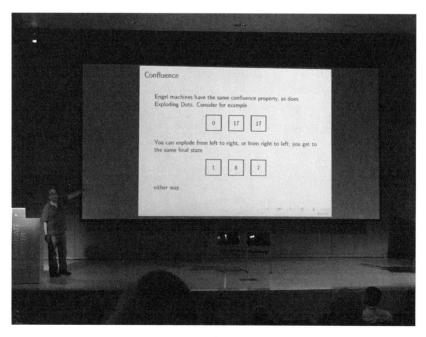

Figure 4. Jim Propp explaining Engel machines.

If you, fellow reader, enjoy this third MOVES volume even a third as much as we participants enjoyed the MOVES conference, we will all be doing pretty well! May MOVES continue to advance the pursuit of recreational (and pure and applied!) mathematics, and may it continue to inspire and attract more young people to the field of mathematics.

References

[1] Christopher Hooley. On Artin's conjecture. *J. Reine Angew. Math.* **1967** no. 225 (1967) 209–220.

[2] Martin Gardner. Cyclic numbers. In *Mathematical Circus: More Puzzles, Games, Paradoxes and Other Mathematical Entertainments from Scientific American.* Vintage Books, New York, 1981.

Here is a simple demonstration you can use to impress your friends: Remove 21 cards at random from a standard deck. We have just done that ourselves and ended up with this arrangement:

3♡	Q◇	2♣	3♣	A♣	K♠	10♣
J◇	A◇	7♠	10♠	6♣	4♠	2♡
4◇	5♣	9♡	Q♡	8♣	9♣	6♡

Ask your friend to remember any one of the cards, but not to tell you which one. Suppose she selects the 4♠. You now ask her for the row in which her card appears. She will tell you it is in the middle row. Form each row into a pile, and stack the piles one atop the other, so that what had been the middle row is now in the middle of the pile. Deal out the cards again into seven rows of three, but deal by going down the columns instead of across the rows. The result is this arrangement:

4◇	Q♡	6♡	7♠	4♠	Q◇	A♣
5♣	8♣	J◇	10♠	2♡	2♣	K♠
9♡	9♣	A◇	6♣	3♡	3♣	10♣

To clarify the correct dealing procedure, notice that the third row of the initial arrangement can now be found by reading down the leftmost columns of the new layout. Likewise, the remaining two rows of the initial arrangement are found reformed into the columns of the present arrangement.

Now ask your friend a second time for the row in which her card appears. This time, she will tell you it is in the first row. Repeat the procedure from the first step, by first forming the rows into piles and then stacking them so that what was formerly the top row becomes the middle of the pile. Once more deal them out into three rows of seven by dealing down the columns. Here is the result:

5♣	10♠	K♠	6♡	Q◇	9♣	3♡
8♣	2♡	4◇	7♠	A♣	A◇	3♣
J◇	2♣	Q♡	4♠	9♡	6♣	10♣

One last time you ask for the row in which her card appears, and she replies that it is in the third row. You now look to the middle card in that row, and identify the 4♠ as your friend's selection. The effect is all the more impressive if you maintain a steady stream of patter and devise some suitably dramatic way of revealing the card. It works every time, for reasons we shall explain at the end of this Preface.

This is the third volume of research papers in recreational mathematics inspired by the biennial MOVES conferences organized by the Museum of Mathematics in New York City. "MOVES" is an acronym for "The Mathematics of Various Entertaining Subjects." The conferences bring together mathematicians, educators, and students at all levels—from elementary to graduate school—to discuss mathematical problems arising from games, puzzles, and other "entertaining subjects." The first such conference was held in 2013, with subsequent editions in 2015 and 2017. With each conference more successful than the one previous, the MOVES conferences are today a major fixture on the American mathematical calendar.

The theme of the 2017 conference was "The Magic of Mathematics." Among the featured speakers were Persi Diaconis and Manjul Bhargava. Diaconis, himself a former professional magician, discussed the elaborate mathematics underlying various card tricks invented by one of America's greatest philosophers: Charles Sanders Peirce. Bhargava, a recipient of the Fields Medal, further developed this theme by showing how number theory can be used to astonishing effect in the performance of magic.

It is certainly true that you can appear to perform miracles simply by knowing some mathematics. An example is the effect we described above. Here is another: Hand your friend nine cards, and ask him to select one. Have him place that card on top of his pile of nine, and then place the remainder of the deck on top of that pile. You now explain that you are going to count out four piles of cards, but that you are going to do so in an unusual way. You are going to start counting backward from ten, each time simultaneously calling out the number and flipping the top card face up onto the table. If the number of the card matches the number you are saying, then you stop that pile and start a new one. If you count all the way down to one without getting a match, then the eleventh card is placed face down on top of the pile to "cap" it. You then move on to the next pile. Face cards (kings, queens, and jacks) count for ten, whereas aces count for one.

That seems very complicated, but an example will make it clear. We have a well-shuffled deck of cards on the table before us. Carrying out our procedure, the first pile comes up like this:

$$A♡ \quad 10♢ \quad 7♠ \quad J♢ \quad 6♣.$$

Because the six appeared just as we called out "Six!" we stop this pile and move on to the next one:

$$2\diamond \quad Q\clubsuit \quad 3\heartsuit \quad K\clubsuit \quad 8\heartsuit \quad 10\heartsuit \quad 7\clubsuit \quad 3\clubsuit.$$

This time the three appeared just as we called out "Three!" and we again stop the pile. The next pile is

$$5\spadesuit \quad 10\clubsuit \quad 8\diamond,$$

and the eight appeared just as we called out "Eight!" The final pile is

$$2\heartsuit \quad 6\heartsuit \quad 4\clubsuit \quad Q\heartsuit \quad 9\spadesuit \quad 3\diamond \quad A\diamond \quad J\spadesuit \quad 9\diamond \quad J\clubsuit.$$

This time we had no match, so we deal one further card face down on top of the pile. This "caps" it, which means that it plays no further role in the remainder of the trick.

The top cards of the piles remaining in play are 6, 3, and 8. This adds up to 17. If you count down 17 cards from those remaining in the deck, you will find that it is your friend's card. Really!

It seems astonishing, right up until you think about it mathematically. Your friend's card begins in the forty-fourth place in the deck, because it is on top of a pile of nine cards and therefore has 43 cards placed on top of it. Now look at each pile: In the first, we dealt five cards and stopped at six, which sums to 11. The "six" is telling you the number of additional cards you would need to have dealt to make 11. Likewise, the second pile stopped at three with eight cards dealt, and the third pile stopped at eight with three cards dealt. They sum to eleven each time. The final "capped" pile effectively had us deal out all eleven cards.

Thus, the sum of the top cards tells us how many cards short of 44 we are.

(If you followed that explanation, you will also know how to deal with the situation in which all four piles are capped. In this case, the card you would use to cap the final pile just is your friend's card, which you can then reveal with some appropriate flourish.)

The final keynote speaker was Arthur T. Benjamin, who shares with Diaconis a dual life as both a magician and a mathematician. His talk involved certain probabilistic paradoxes that arise in the venerable game of Bingo. Rather than use mathematics to create magic tricks, his talk revealed another connection between mathematics and magic: the ability of mathematical results to seem magical.

Magic is about showing someone an effect that appears to be impossible. For mathematicians, confrontations with the seemingly impossible are a regular occurrence. How can it be, for example, that objects as familiar as the counting numbers can do this:

$$1 = 1,$$

$$1 + 3 = 4,$$

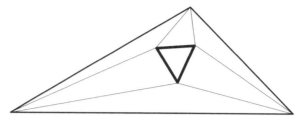

Figure 1. An illustration of Morley's theorem. When the angle at each vertex of any triangle is trisected, the points of intersection of adjacent trisectors form the vertices of an equilateral triangle.

$$1 + 3 + 5 = 9,$$
$$1 + 3 + 5 + 7 = 16,$$
$$1 + 3 + 5 + 7 + 9 = 25,$$

and so on? How can it be that consecutive odd numbers always add up to perfect squares?

Or how about this: In any triangle, the points of intersection of adjacent angle trisectors form the vertices of an equilateral triangle. This is known as Morley's theorem, after the Anglo-American mathematician who first proved it in 1899. The idea is illustrated in Figure 1. You can almost imagine Morley performing this as a magic trick. "Step right up, friends, and let me show you something. See this triangle I just drew? Well, I trisect each angle like so, and I connect these points of intersection, see, and right before your eyes, seemingly from nowhere, we get an equilateral triangle!" What else could the audience say but, "How did you do that?"

Just as with magic, sometimes mathematical tricks are easy to explain. That the sum of consecutive odd numbers is a perfect square seems surprising, until you consider the diagram shown in Figure 2. Now the connection between sums of odd numbers and perfect squares is not so mysterious. Other mathematical tricks, like Morley's theorem, are far more difficult to explain. (We will send you to the Internet for that one!)

As mathematicians, we routinely encounter amazing and seemingly impossible effects as we explore the relationships among abstract objects. But there is a difference between magic and mathematics as well. Whereas a magician's routine loses much of its appeal if the audience knows how it is done, mathematical reality only becomes more interesting when you come to understand its tricks!

As with the previous two volumes in this series, the chapters that follow vary widely with regard to the degree of mathematical sophistication required. Some chapters will appeal to anyone with a general interest in mathematics, while

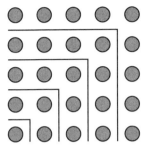

Figure 2. A visual explanation of why the sum of consecutive odd numbers is always a perfect square.

others might prove challenging to all but the most dedicated readers. Even in the most difficult chapters, however, it is our hope that much will be accessible even to those who choose not to work through all the equations.

We open with six chapters centered on clever puzzles and brainteasers. In Chapter 1, Peter Winkler opens the proceedings with some entertaining and stimulating problems in probability. Though they might seem challenging at first, all yield readily to some cleverness and clear thinking. Anany Levitin and Jason Rosenhouse (Chapter 2) discuss puzzles that are best solved by starting at the end and reasoning backward. Walker Anderson, Erik Demaine, and Martin Demaine devise "Spiral Galaxies" puzzles whose solutions form letters and numbers (Chapter 3). Steven Dougherty and Yusra Naqvi consider puzzles that are subtly related to coding and information theory (Chapter 4). James Stein and Leonard Wapner show in chapter 5 that sometimes it is possible to predict a coin flip more than half the time. Finally, Tanya Khovanova relates two classics of recreational mathematics (Chapter 6): coin-weighing puzzles (in which a balance scale must be used to find one counterfeit coin among a group of genuine coins) and knight/knave puzzles (in which knights always tell the truth and knaves always lie).

The second section focuses on problems inspired by games of various sorts. In Chapter 7, Barry Balof, Arthur Benjamin, Jay Cordes, and Joseph Kisenwether explore a seeming paradox in the game of Bingo: On any individual card, the probability of a vertical bingo is the same as the probability of a horizontal bingo, but in a game played with a large number of cards, the first bingo is far more likely to be horizontal than vertical. David Molnar (Chapter 8) uses abstract algebra to answer a combinatorial question arising from a consideration of wiggly games. You should absolutely read his chapter, if only to learn what a "wiggly game" is. Jeffrey Bosboom, Spencer Congero, Erik Demaine, Martin Demaine, and Jason Lynch apply the theory of computational complexity to questions related to checkers (Chapter 9). Where most people would see an amusing diversion for children in the classic game Chutes and

Ladders, Darren Glass, Stephen Lucas, and Jonathan Needleman see instead a challenging problem in probability (Chapter 10). The section concludes with Michael Allocca, Steven Dougherty, and Jennifer Vasquez relating the game of "Japanese ladders" to issues in both mathematics and biology (Chapter 11).

We then come to part III, with four chapters in which recreational topics lead to serious questions in algebra and number theory. In Chapter 12, Persi Diaconis and Ron Graham discuss the work of Charles Sanders Pierce. It seems that in addition to making fundamental contributions to philosophy, Peirce devised several card tricks that were both mathematically brilliant and hopelessly difficult to perform. Brian Hopkins finds his inspiration in a popular puzzle app for smartphones (Chapter 13). Max Alekseyev provides an elegant solution to a problem about Egyptian fractions, by which we mean fractions like $\frac{1}{2}$, $\frac{1}{3}$, and $\frac{1}{6}$, which sum to one and whose numerators are all one (Chapter 14). David Nacin rounds out part III by considering KenKen puzzles, which are a variant of the more familiar Sudoku (Chapter 15).

Part IV comprises three chapters that involve geometry and topology. Fair warning: These chapters are more challenging than most of what has come before, but we hope that will not discourage you from having a go at them. In Chapter 16, Yossi Elran and Ann Schwartz make a contribution to the theory of flexagons, which were first described in 1939 but are still going strong as an object of interest to mathematicians. Jei Mei and Edmund Lamagna explore Steiner trees on triangular grids (Chapter 17). Finally, we have Legos. No doubt we all played with them as kids, but David McClendon and Jonathan Wilson find deep mathematics underlying the manner in which they can be stacked and arranged (Chapter 18).

Assembling so much great mathematics into one book is itself something of a magic trick!

Have you figured out the original trick? Remember that each time we stack the cards in preparation for the next deal, the row containing your friend's card is placed in the middle of the pile. When you stack the 21 cards for the first time after the initial arrangement, your friend's card will therefore be in one of the positions from 8 to 14. When the cards are then dealt out for the second time, your friend's card will be either the third, fourth, or fifth card in its new row. It will therefore end up in one of the positions from 10 to 12 when the cards are dealt for the final time. But the cards in these positions are precisely the ones that comprise the fourth cards in the three rows in the final arrangement.

It only remains to thank the many people whose hard work and dedication made this book possible. Pride of place must surely go to Cindy Lawrence and Glen Whitney, without whom neither MoMath, nor the MOVES conferences, would exist. The entire mathematical community owes them a debt for their tireless efforts. The conference was organized by Joshua Laison and Jonathan

Needleman. When a conference runs as smoothly as this one, you can be sure there were superior organizers putting out fires behind the scenes. Particular thanks must go to Two Sigma, a New York–based technology and investment company, for their generous sponsorship. Finally, Vickie Kearn and her team at Princeton University Press fought hard for this project, for which they have our sincere thanks.

Enough! It is time to get on with the show.

Jennifer Beineke
Enfield, Connecticut

Jason Rosenhouse
Harrisonburg, Virginia

January 30, 2019

PART I

Puzzles and Brainteasers

1

◇◇◇

PROBABILITY IN YOUR HEAD

Peter Winkler

Probability theory is a well-developed science, and most probability puzzles, if translated into precise logical form, could in principle be solved by machine. But the theory also provides some remarkable problem-solving tools with which we humans, using our imagination, can often find delightful shortcuts.

Here are eight puzzles that you can try to solve by taking pencil in hand and "doing the math," but each can also be solved in your head, just by reasoning.

1 Problems

1.1 Flying Saucers

A fleet of saucers from planet Xylofon has been sent to bring back the inhabitants of a certain apartment building, for exhibition in the planet zoo. The earthlings therein constitute 11 men and 14 women.

Saucers arrive one at a time and randomly beam people up. However, owing to the Xylofonians' strict sex separation policy, a saucer cannot take off with humans of both sexes. Consequently, a saucer will continue beaming people up until it acquires a member of a second sex; that human is immediately beamed back down, and the saucer takes off with whomever it already has on board. Another saucer then swoops in, again beaming up people at random until it gets one of a new gender, and so forth, until the building is empty. What is the probability that the last person beamed up is a woman?

1.2 Points on a Circle

Three points are chosen at random on a circle. What is the probability that there is a semicircle of that circle containing all three?

1.3 Meet the Williams Sisters

Some tennis fans get excited when Venus and Serena Williams meet in a tournament. The likelihood of that happening normally depends on seeding and talent, so let us instead assume an idealized elimination tournament of 64 players, each as likely to win as to lose any given match, with bracketing chosen uniformly at random. What is the probability that the Williams sisters get to play each other?

1.4 Service Options

You are challenged to a short tennis match, with the winner to be the first player to win four games. You get to serve first. But there are options for determining the sequence in which the two of you serve:

1. Standard: Serve alternates (you, her, you, her, you, her, you).
2. Volleyball style: The winner of the previous game serves the next one.
3. Reverse volleyball style: The winner of the previous game receives in the next one.

Which option should you choose? You may assume it is to your advantage to serve. You may also assume that the outcome of any game is independent of when the game is played and of the outcome of any previous game.

1.5 Who Won the Series?

Two evenly-matched teams meet to play a best-of-seven World Series of baseball games. Each team has the same small advantage when playing at home. As usual, one team (say, Team A) plays games 1 and 2 at home, and, if necessary, plays games 6 and 7 at home. Team B plays games 3, 4 and, if needed, 5 at home.

You go to a conference in Europe and return to find that the series is over, and six games were played. Which team is more likely to have won the series?

1.6 Random Rice

You go to the grocery store needing 1 cup of rice. When you push the button on the machine, it dispenses a uniformly random amount of rice between nothing and 1 cup. On average, how many times do you have to push the button to get (at least) a cupful?

1.7 Six with No Odds

On average, how many times do you need to roll a die to get a 6, given that you do not roll any odd numbers *en route*? (Hint: The answer is not 3.)

1.8 Getting the Benz

Your rich aunt has died and left her beloved 1955 Mercedes-Benz 300 SL Gullwing to either you or one of your four siblings, according to the following stipulations. Each of the five of you will privately write "1," "2," or "3" on a slip of paper. The slips are put into a bowl to be examined by the estate lawyers, who will award the car to the heir whose number was not entered by anyone else. (If there is no such heir, or more than one, the procedure is repeated.)

For example, if the bowl contains one 1, two 2s and two 3s, the one who put in the 1 gets the car.

After stewing and then shrugging your shoulders, you write a 1 on your slip and put it in the bowl. What the heck, a 1/5 chance at this magnificent vehicle is not to be sneezed at! But just before the bowl is passed to the lawyers, you get a sneak peek and can just make out, among the five slips of paper, one 1 (which may or may not be yours), one 2, and one 3.

Should you be happy, unhappy, or indifferent to this information?

2 Solutions

2.1 Flying Saucers

This puzzle—and many others—is more easily tackled if we rephrase the question, perhaps by putting the randomness up front. Suppose we imagine that the building occupants are first lined up uniformly at random, then picked up by the flying saucers starting from the left end. Then the gender of the rightmost person in line would determine whether the last person picked up is a woman, and of course that gender is female with probability $14/(14 + 11) = 56\%$.

But there's a problem: This model is not correct, because the next saucer's beam-ups are newly randomized and might not begin with the person who was last rejected. You can see the difference by examining the case where there are just two women and one man.

So, let us imagine a different model, where the remaining occupants are re-lined up randomly each time a saucer arrives. Then the *next-to-last* saucer must be facing a line consisting entirely of one or more men followed by one or more women; or, one or more women followed by one or more men. Because these possibilities are equally likely (reversing the line, for example, transforms one set of possibilities to the other), the probability that the last saucer will be greeted by women is 1/2.

2.2 Points on a Circle

This puzzle, a classic, was suggested to me by combinatorics legend Richard Stanley of the Massachusetts Institute of Technology and the University of Miami.

Let us pick the three points in a funny way: Choose three random diameters, that is, three lines through the center of the circle at uniformly random angles. Each diameter has two endpoints on the circle, giving us six points, which we can label A, B, C, D, E, F clockwise around the circle, beginning anywhere. Then

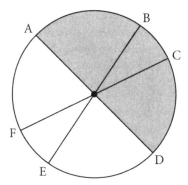

Figure 1.1. Three diameters with their endpoints labeled sequentially. Any three consecutive points, such as A, B, C or B, C, D, are easily seen to lie on a common semicircle. Nonconsecutive points that include one endpoint from each of the three diameters, such as A, C, E, do not lie on a common semicircle.

we will use a coin flip (three times) to decide, for each line, which of these two intersections becomes one of our three points.

It not hard to see that if the chosen points are consecutive (i.e., A, B, C, or B, C, D, etc.), up to F, A, B (six possibilities), then they are contained in a semicircle; otherwise not. This is illustrated in Figure 1.1.

Since there are eight possible outcomes for the coin flips, the probability that the desired semicircle exists is a whopping 3/4. (If you ask your friend to pick three random points on a circle, I expect they are less likely than not to be contained in a semicircle.)

Similar reasoning shows that if n points are chosen, $n \geq 2$, the probability that they are contained in some semicircle is $2n/2^n$. In fact, you can even apply a version of this argument in higher dimensions; for example, to deduce that the probability that four random points on a sphere are contained in some hemisphere is $1/8$.

2.3 Meet the Williams Sisters

This puzzle appears in Frederick Mosteller's wonderful little book, *50 Challenging Problems in Probability* [2], but the solution he offers involves working out examples, then guessing a general solution and proving its correctness by induction. Here, instead, is a solution you can work out in your head.

Given the problem's symmetry conditions, each of the $\binom{64}{2} = 63 \times 32$ pairs of players has the same probability of meeting. Since 63 matches are played (remember that all but one of the 64 players needs to be eliminated to arrive at the winner), the probability that the Williams sisters meet is $63/(63 \times 32) = 1/32$.

Here is another approach that almost works. It takes two coin flips on average to get a head and thus two matches in our random tournament to get a loss. Therefore, a player will meet on average two other players, so Serena's probability of meeting Venus is 2/63. Wait, that is not quite right. Can you find the flaw?

2.4 Service Options

This puzzle was inspired by one I heard from Dick Hess, author of *Golf on the Moon* [1] and other delightful puzzle books.

Assume that you play lots of games (maybe more than is needed to determine the match winner), and let *A* be the event that of the first four served by you and the first three served by your opponent, at least four are won by you. Then it is easily checked that no matter which service option you choose, you will win if *A* occurs and lose otherwise. Thus, your choice makes no difference. Notice that the independence assumptions mean that the probability of the event *A*, since it always involves four particular service games and three particular returning games, does not depend on when the games are played or in what order.

If the game outcomes were not independent, the service option could make a difference. For example, if your opponent is easily discouraged when losing, you might benefit by using the volleyball scheme, in which you keep serving if you win.

The idea that playing extra games may be useful for analytic purposes (despite—no, because!—they do not affect the outcome) will be even more critical in the next solution.

2.5 Who Won the Series?

This nice question came to me from Pradeep Mutalik, who writes the math column for the excellent online science magazine *QUANTA*. The solution is my own.

The key here is that *potential* games are as important as actual ones when it comes to computing the odds. It is tempting to think, for example, that Team A's extra home game is less of a factor, because the series does not usually go to seven games. But this is false reasoning: You *may as well* assume all seven games are played (since it makes no difference to the outcome if they are). Thus, the 4-to-3 advantage in home games enjoyed by Team A is real and is unaffected by the order of the games.

(To see a more extreme example of this phenomenon, imagine the series winner is to be the first team to win 50 games, and that the first 49 are home games for Team B and the rest for Team A. Then, since the series is a big favorite to end before game 98, most games will *probably* be played on Team B's home field—yet, it is Team A that has the advantage. This, again, can be seen by imagining that 99 games are played regardless; perhaps tickets for all 99 have already been sold and the fans don't mind watching games played after the series outcome is decided.)

Similar reasoning shows that if you know at most six games were played, then because half these potential games were at each team's home field, you would correctly conclude that the probability that Team A won the series is exactly 1/2. If, however, you know that at most five games were played, then Team B, with its 3-to-2 home game advantage, is more likely to have won.

It follows that if *exactly* six games were played, Team A is more likely to have won!

Given that the answer to this puzzle does not depend on the degree of home-field advantage, you could arrive at this answer the following way. Assume that home-field advantage is overwhelming. Therefore, given that exactly six games were played, it is highly probable that there was only one upset—that is, only one game was won by the visiting team. Then that upset must have occurred in games 3, 4, 5 or 6, because otherwise the series would have ended at game 5. In only one of those cases (upset at game 6) would Team B have won the series, thus Team A has nearly a three-to-one advantage! This kind of "extremal reasoning" can be very useful in puzzle solving. As noted, however, it depends on the assumption that changing the parameter does not affect the solution.

2.6 Random Rice

This puzzle is an oldie but goodie. You'll need a bit of math background to be able to solve it in your head (or any other way, for that matter).

The volume of the ith squirt of rice is, by assumption, an independent uniformly random real number X_i between 0 and 1. We want to determine the expected value of the first j for which $X_1 + X_2 + \cdots + X_j$ exceeds 1. That number will be at least 2, since the probability of getting a full cup on the first squirt is 0. But it might be 3 or even more if you are unlucky and begin with small squirts.

The key is to consider the fractional parts $Y_1, Y_2, \ldots$ of the partial sums. These numbers are also independent random numbers between 0 and 1, as you can easily see by noting that given the first i squirts, each possible value of Y_{i+1} arises from just one value of X_{i+1}.

We can assume the machine's output is never exactly 0 or 1, since those events have probability 0 and hence do not affect the expected number of squirts. Then, as you squirt rice, the value of Y_i keeps going up until your rice total exceeds 1, at which point Y_i goes down. Thus, the probability that you'll need more than i squirts is exactly the probability that $Y_1 < Y_2 < \cdots < Y_i$. This is just one of the $i!$ ways to order i numbers, so that probability is $1/i!$.

The expected value of any "counting" random variable is just the sum of the probabilities that that variable exceeds i, for all $i \geq 0$, so the expected value of the critical j is

$$\sum_{i=0}^{\infty} 1/i!,$$

which is the famous constant e. So on average, it takes exactly e (about 2.718281828459045) squirts to fill that rice cup.

2.7 Six with No Odds

This puzzle was communicated to me by MIT probabilist Elchanan Mossel, who came up with it as a problem for his undergraduate probability students before realizing that it was more subtle than he thought. The solution below is my own.

The first issue to be tackled is perhaps: Why is the answer not 3? Is this any different from rolling a die whose six faces are labeled with two 2s, two 4s, and two 6s? In that case, the answer would surely be 3, since the probability of "success" (rolling a 6) is $1/3$, and when doing independent trials with probability p of success, the expected number of trials to reach success is $1/p$.

But it *is* different. When rolling an ordinary die until a 6 is obtained, conditioning on no odd numbers favors short experiments—if it took you a long time to roll your 6, you would probably have rolled some odd number on the way. Thus the answer to the puzzle should be less than 3.

It might help you to think about a series of experiments. If you repeatedly roll until a 6 appears, but ignore odd numbers, you will find that on the average it takes 3 (non-odd) rolls to get that 6. But the correct experiment is to *throw out the current series of rolls* if an odd number appears, then start a new series with roll number 1. Thus only series with no odds will count in your experiment.

That *gedankenexperiment* might give you an idea. Each series of rolls would end with either a 6 or an odd number. Does it matter? The final number in the sequence (1, 3, 5, or 6) is independent of the length of the series. Thus, by *reciprocity of independence* if you like, the length of the series is independent of what number caused it to end.

If you simply roll a die until either a 1, 3, 5, or 6 appears, it takes $3/2$ rolls on average (since here the probability of "success" is $2/3$). Since ending in a 6 has no effect on the number of trials, the answer to the original puzzle is that same $3/2$.

2.8 Getting the Benz

You have only your author to blame for this last puzzle.

Your first thought, perhaps, is that since you have seen one of each number, all is fair and the odds have not changed.

But wait a second. The 1 you saw might have been yours, but the 2 and 3 were definitely submitted by your siblings. If you had guessed 2 or 3 instead of 1, you would now be plumb out of luck (for this round, anyway). So you did well to guess 1, thus your chances must be higher now.

Thinking of it another way, what is written on the two slips you did not see? There are $3^2 = 9$ possibilities, of which two—$(2, 3)$ and $(3, 2)$—get you the car, and three—$(1,1)$, $(2,2)$, and $(3,3)$—get you a rematch. You do not need a calculator to work out that this gives you quite a bit better than your original $1/5$ probability of getting the Benz.

Wonderful! You can already imagine your friends' jealous stares when you show up in your classic coupé. But something's nagging at your brain. It seems that your siblings, whatever they wrote on their slips, could all reason this way with the same peek that you got. Since you are all going for the same car, how can it be that you are all happy?

Acknowledgment

Research supported by NSF grant DMS–1600116.

References

[1] D. Hess. *Golf on the Moon*. Dover, Mineola, NY, 2014.
[2] C. F. Mosteller. *50 Challenging Problems in Probability*. Addison-Wesley, Reading, MA, 1965.

2

SOLVING PUZZLES BACKWARD

Anany Levitin and Jason Rosenhouse

In addition to the entertainment value of puzzles, some of them can be very effective in demonstrating general problem-solving strategies. In particular, puzzles can be used to illustrate the technique of solving problems backward, which is the subject of this chapter. The technique of solving backward, sometimes called *regressive reasoning*, has been recognized as an important strategy at least since the time of ancient Greece. Modern books on problem solving, starting with George Polya's classic *How to Solve It* [13], usually include a chapter devoted to this strategy. Examples used in such books do occasionally include puzzles, but the number of nontrivial examples is generally small. The reason may have something to do with the difficulty of regressive reasoning. Here is what Sherlock Holmes had to say on the subject, in the novel *A Study in Scarlet*:

> In the every-day affairs of life it is more useful to reason forwards, and so the other comes to be neglected. There are fifty who can reason synthetically for one who can reason analytically. ...Let me see if I can make it clearer. Most people, if you describe a train of events to them, will tell you what the result would be. They can put those events together in their minds, and argue from them that something will come to pass. There are few people, however, who, if you told them a result, would be able to evolve from their own inner consciousness what the steps were which led up to that result. This power is what I mean when I talk of reasoning backwards, or analytically.

The best-known puzzle solved by an insight based of regressive reasoning is the classic Tower of Hanoi puzzle invented by Édouard Lucas in 1883. There are n disks of different sizes and three pegs. Initially, all the disks are on the first peg in order of size, the largest on the bottom and the smallest on top. The objective is to transfer all the disks to the third peg, using the second one as an auxiliary, if necessary. Only one disk can be moved at a time, and it is forbidden to place a larger disk on top of a smaller one (see Figure 2.1).

For $n > 1$, consider the *last* move of the largest disk made by an algorithm that solves the puzzle in a minimum number of moves. All the smaller $n - 1$

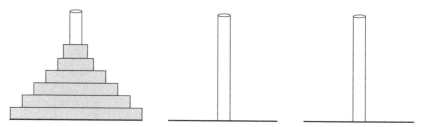

Figure 2.1. The Tower of Hanoi puzzle.

disks must then reside on the middle peg, making it possible to move the largest disk from the source peg to the destination peg in a single move. But moving the tower of $n - 1$ disks from the first peg to the second using the third peg as an auxiliary has the same structure as the original puzzle. This leads to the classic recursive algorithm for this puzzle: If $n = 1$, simply move the only disk from the first peg to the third. If $n > 1$, move recursively the smaller $n - 1$ disks from the first peg to the second peg using the third peg as an auxiliary, then move the largest disk from the first peg to the third one, and finally move recursively the $n - 1$ disks from the second peg to the third using the first peg as an auxiliary.

In the remainder of this chapter, we give eleven examples of puzzles illustrating regressive reasoning. Solutions are provided at the end. We did exclude from this short survey a few problem types related to working backward. Thus, Retrolife—the problem of finding a previous generation in John Conway's Game of Life—can be restated to avoid regressive reasoning altogether [6]. Retro analysis, which involves chess problems in which the solver must determine the moves played prior to reaching a given position (as presented, for example, by Smullyan [16]), requires at least some chess expertise. Finally, some simple two-person games, such as Wythoff's Nim [12, pp. 131–138], can also be solved by working backward, but games are not puzzles and are therefore not relevant here.

1 The Puzzles

Let us start with a puzzle from Martin Gardner's *My Best Mathematical and Logic Puzzles*, which illustrates the potential power and elegance of regressive reasoning in the most convincing way.

Interrupted Bridge. A bridge game starts with one of the four players dealing a deck of cards clockwise, starting with the player to the left of the dealer. Consider the following situation: After the dealer has dealt about half the cards, the telephone rings. The dealer puts down the undealt cards to answer the phone in

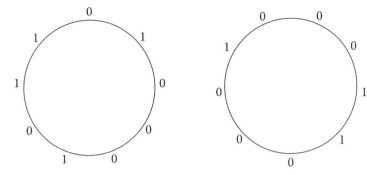

Figure 2.2. An example of the transformation rule in the Circle of Zeros and Ones puzzle.

another room. After the dealer returns, neither he nor anyone else can remember where the last card was dealt. No one has touched any of the dealt cards. Without counting the cards in any hand, or the number of cards yet to be dealt, how can the dealer finish the deal rapidly and accurately, giving each player exactly the same cards he would have received if the deal had not been interrupted? [7, p. 34]

This puzzle almost solves itself once you think to reason backward. Note the one-sentence solution we provide at the end of the chapter.

Here are two puzzles where regressive reasoning leads to easy solutions as well. The first one was provided by Dick Hess:

Trapping the Knight. A knight is placed on an infinite chessboard. What is the minimum number of moves needed for it to reach a position from which the knight can move only to a previously visited square? Note that every move of the knight must be made to a previously unvisited square. [9, p. 28]

The second puzzle is from a recent book by Marina Nogin titled *Strategies of Problem Solving*:

Circle of Zeros and Ones. Suppose four 1s and five 0s are written along a circle. Between two equal numbers we write a 1, and between two distinct numbers we write a 0. The original numbers are then wiped out, and this step is repeated. For example, a possible initial distribution of 1s and 0s is shown to the left in Figure 2.2. The result of applying the rule is then shown to the right in Figure 2.2. Is it ever possible to reach 9 ones by repeating this step a finite number of times? [11, p. 69]

Fred Schuh, a prominent Dutch researcher of puzzles, has also recognized the occasional usefulness of regressive reasoning:

Figure 2.3. The initial configuration (top) and the target configuration (bottom) for the Coin-Row Transformation puzzle.

> It frequently happens that a puzzle consists of obtaining a given final position from a given initial position by making the smallest possible number of moves. One can then start from the final position instead, and try to obtain the initial position by backward moves. If this has succeeded, all one has to do to get the solution of the original puzzle is to perform all moves in the reversed order and in the reversed directions. It is virtually certain that the new puzzle (the reversed one) will be simpler; otherwise the puzzle would surely have been presented in the reversed form. Hence, in most cases reversal of the puzzle will produce a simplification. [15, p. 17]

Schuh's comment certainly applies to most maze-traversal puzzles offered to the general public. As his example, however, Schuh gave a more interesting puzzle about transforming a row of coins:

Coin-Row Transformation. Transform in as few moves as possible a row of three dimes separated by two quarters to a row with all the dimes to the left of both quarters (see Figure 2.3). On every move, a dime and an adjacent quarter are to be moved as one whole to remain adjacent, but one is not allowed to reverse their order. The target row is not required to be in exactly the same place as the row given, but it must remain horizontal on the same line. [15, p. 17]

The advantage of solving this puzzle backward is the fact that while there are many possible moves from the starting position, there is just one pair of coins that can be moved starting from the target position. Martin Gardner called this problem "baffling and pretty" and included it in *My Best Mathematical and Logic Puzzles* [7, p. 28].

The same idea is exploited by the following puzzle invented by James Propp:

Penny Distribution Machine. The "machine" consists of a row of boxes. To start, one places n pennies in the leftmost box. The machine then redistributes the pennies as follows: On each iteration, it replaces a pair of pennies in one box with

6				
4	1			
2	2			
0	3			
0	1	1		

6				
4	1			
2	2			
2	0	1		
0	1	1		

Figure 2.4. Two possible ways of distributing six coins in the Penny Distribution Machine puzzle. Note that while the first two moves are forced, there are two ways of proceeding after the third line. However, each way of proceeding leads to the same final distribution.

a single penny in the next box to the right until there is no box with more than one coin. Two ways such a machine can distribute six coins are shown in Figure 2.4. Is it true or false that the final distribution of pennies never depends on the order in which the machine processes the coin pairs? [10, p. 62]

Our next puzzle is a version, due to W. Lloyd Milligan, of an old puzzle proposed by Henry Dudeney:

Crowning the Checkers. An even number of checkers, $n > 2$, are placed in a row on the table. The objective is to transform them into $n/2$ kings (stacks of two checkers) using $n/2$ moves. On the first move, a single checker should jump over its neighbor in either direction and land on another checker to make a king. On the second move, a single checker should jump exactly two checkers (which may be two single checkers side by side or two checkers forming a king). On the third move, a checker should jump three checkers (three singles or a single and a king), and so on, each time increasing by one the number of checkers to be passed over. You may disregard spacing between adjacent checkers. The procedure is illustrated for eight checkers in Figure 2.5. For what values of n is a solution possible? For values of n for which a solution exists, how do we find the solution? [8, p. 271]

This is a challenging puzzle, in which regressive reasoning not only helps determine the values of n for which a solution exists but also aids in designing an algorithm for finding the solution.

The next example is a famous puzzle invented by John Conway [2, p. 821]:

The Solitaire Army Problem. This peg solitaire game is played on an infinite two-dimensional board with a horizontal line separating the board into two halves. Some pegs are placed below the line. We imagine that these pegs represent soldiers

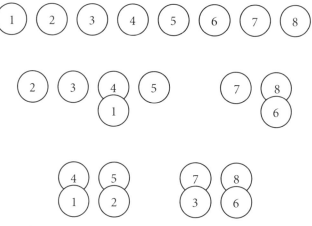

Figure 2.5. A solution to the Crowning the Checkers puzzle starting with eight checkers. The first row shows the starting arrangement. The middle row shows the first two moves: checker 6 jumps one checker to land on checker 8, while checker 1 jumps two checkers to land on checker 4. The bottom row shows the remaining two moves: checker 2 jumps over three checkers to land on checker 5, and then checker 3 jumps over four checkers to land on checker 7.

in an army, and that the other side of the line is enemy territory. The object is for the army to advance a scout as far into enemy territory as possible. On each move, a peg can jump over its immediate neighbor vertically or horizontally to land on an empty cell. After the jump, the jumped-over neighbor is removed from the board. How far above the enemy line can we send a scout?

In 1961, Conway gave an ingenious proof that no number of pegs, finite or infinite, can make it possible for one of the scouts to reach a cell at the fifth row above the line. A cell in the fourth row *can* be reached, as well as cells in rows one, two, and three. Ways to do this can be found by working backward.

Here is another well-known puzzle of a very different kind:

Pirates' Gold. Five pirates have 100 gold pieces to divide among them. A clear hierarchy exists among the pirates, so that there is a most powerful pirate, a second most powerful, and so forth. They come up with the following scheme: The most powerful pirate will propose a distribution. Then all of the pirates vote. If a majority support the distribution, then it is approved. If it is voted down, the proposing pirate is thrown overboard and the second pirate proposes a new distribution. We assume that every pirate wants to maximize his take of the gold, and will vote down any proposal that is not strictly an improvement over what he can get by taking his chances further down the line. We also assume that a tie vote, which can happen when an even number of pirates remain, implies that the distribution is adopted. Now the question is: What distribution should the first pirate propose? [14, p. 119]

Finally, for the sake of completeness, we should also mention several simple puzzles that can be solved either by working backward or working forward, with the former method leading to a solution much faster than the latter. These puzzles have appeared in numerous venues, but we have provided one citation for each:

Frog in the Well. In its search for water, a frog fell down a 30 foot well. Trying to get out, it managed to climb up 3 feet during each day, but then slipped back 2 feet during the ensuing night. How many days did it take the frog to get out of the well? [3, #75]

Water Lily. Water lilies double in area every 24 hours. At the beginning of summer, there is one water lily on the lake. It takes 60 days for the lake to become completely covered with water lilies. On which day is the lake half covered? [5]

The Dish of Potatoes. Three travelers stopped at a tavern and ordered a dish of potatoes for supper. When the landlord brought in the potatoes, the travelers were all asleep. The first of the travelers to awake ate a third of the potatoes and went back to sleep without disturbing the companions. The second traveler awoke, ate a third of the remaining potatoes and went back to sleep. A little later the third traveler did the same. When they were all sleeping again, the landlord came in to clear the table and found eight potatoes left. How many potatoes had the landlord provided initially? [4, #7]

2 Solutions

Interrupted Bridge. The dealer should continue the deal by taking the cards from the *bottom* of the packet of undealt cards, dealing first to himself, then counterclockwise around the table. (Note that this works because a standard deck has 52 cards, which is a multiple of four. Therefore, the dealer gets the last card.)

Trapping the Knight. The answer is 15 moves. Consider the knight in a position with all 8 squares reachable from it already visited. Since the knight needs at least 2 moves to go from one of these 8 positions to another, it will need to make at least 14 moves to do this before reaching its trapped position on the fifteenth move. An example of such a path is shown in Figure 2.6.

Circle of Zeros and Ones. Suppose the aim is attainable. Then consider the first time we have nine 1s, which cannot be an initial setup. Then one step prior to this, we must have nine equal numbers, which must all be 0s and

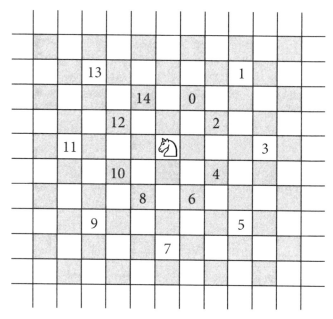

Figure 2.6. Starting from the square marked 0, the knight takes 15 moves to arrive on its present square. All squares to which it can move from its present square have already been occupied.

hence also cannot be an initial setup. But then one step before having all 0s, we would have nine alternating numbers. That is, we would have one of these two configurations around the circle:

$$0\!-\!1\!-\!0\!-\!1\!-\!0\!-\!1\!-\!0\!-\!1\!-\!0 \qquad 1\!-\!0\!-\!1\!-\!0\!-\!1\!-\!0\!-\!1\!-\!0\!-\!1$$

When we move from either of these states to the next step, we would produce a 1 between the last and first numbers, contradicting our assumption. Hence, the aim in question is not attainable.

Coin-Row Transformation. A transformation from the target row to the given one is shown below; reversing it solves the puzzle in four moves:

					d_1	d_2	d_3	Q_1	Q_2
		d_3	Q_1	d_1	d_2				Q_2
		d_3			d_2	Q_1	d_1	Q_2	
	d_1	Q_2	d_3		d_2	Q_1			
d_2	Q_1	d_1	Q_2	d_3					

Penny Distribution Machine. Number the boxes left to right starting with 0 for the leftmost box. Let $b_0 b_1 \ldots b_k$ be a bit string representing a result of the machine's distribution of n pennies, where b_i is equal to 1 or 0, depending on whether the ith box contains a coin or not. In particular, $b_k = 1$, where k is the number of the last box with a penny. That coin was obtained by replacing two coins in box $k-1$, which in turn replaced four coins in box $k-2$, and so on. Applying the same reasoning to any box i containing a coin in the final distribution, we obtain the formula

$$n = \sum_{i=0}^{k} b_i 2^i.$$

We see that the bit string of the final distribution represents the binary expansion of the initial number of pennies n, but with the digits reversed. Since the binary expansion of any natural number is unique, the machine always ends up with the same coin distribution for a given n, regardless of the order in which coin pairs have been processed.

Crowning the Checkers. To begin with, n must be a multiple of 4. To prove this, consider the state of the puzzle before the last move: $n-2$ coins are already paired. Therefore, one of the two remaining single coins should make a jump over an even number of coins to land on the other single coin. Since a coin is required to jump an even number of coins on move i if and only if i is even, the final move, numbered $\frac{n}{2}$, must be even, which implies that n is a multiple of 4.

One can devise an algorithm for solving the problem by backward thinking as follows. Consider the puzzle in its final state in which n coins are paired, with the pairs numbered left to right from 1 to $\frac{n}{2}$. To get back to the initial state of the puzzle with n single coins in a row, take the top coin of the pair numbered $\frac{n}{4}+1$, and move it to the left over all $\frac{n}{2}$ coins. Then move the top coin of the pair numbered $\frac{n}{4}$ to the left over all $\frac{n}{2}-1$ coins. Continue this process of moving the top coins of the pairs over all the coins to the left of them until the top coin of the leftmost pair is moved to the left over $\frac{n}{4}$ coins. Then, starting with the leftmost remaining pair (initially numbered $\frac{n}{4}+2$) and ending with the rightmost pair (initially numbered $\frac{n}{2}$), move the top coin of the pair to the left over $\frac{n}{4}-1, \ldots, 1$ coins, respectively. The top coins should be placed as single coins as just described—recall that we are supposed to disregard spacing between adjacent coins.

Reversing the above, we obtain the following algorithm for pairing n coins placed in a row and numbered left to right from 1 to n. First, perform the following operation for $i = 1, 2, \ldots, \frac{n}{4}-1$: For the rightmost single coin, find the coin to its left to have i coins between them and place that coin on top of

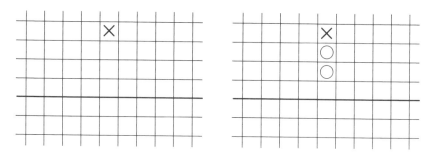

Figure 2.7. To reach the target cell (labeled ×), we need two pegs just below it on the previous move (labeled ◯).

the rightmost single coin. Then repeat the following operation for $i = \frac{n}{4}, \frac{n}{4} + 1, \ldots, \frac{n}{2}$: Take the leftmost single coin and jump it over i coins to the right.

This algorithm obviously makes the minimum number of moves possible, because it forms a new pair of coins on each move.

An illustration of this algorithm for 8 coins is shown in Figure 2.5. For another example, suppose we start with $n = 12$ coins numbered left to right. (Each coin retains its original number throughout the solution, even as empty spaces appear in the line of coins.) In this case, we have that $\frac{n}{4} - 1 = 2$. This tells us that the first two moves must be: Move coin 10 onto coin 12, then move coin 7 onto coin 11. For the remaining moves, we start from the left side, each time making the only legal move with the leftmost unpaired coin: Move coin 1 onto coin 5, move coin 2 onto coin 6, move coin 3 onto coin 8, and, finally, move coin 4 onto coin 9. Try it yourself for 16 coins. It works every time!

The Solitaire Army Problem. Consider a cell in the fourth row above the line we want to reach. For the scout to reach this cell, we must have two pegs just below it on the previous move. This is shown in Figure 2.7.

To get the two pegs below the target from a configuration one row closer to the enemy line requires four pegs arranged as shown in Figure 2.8 (or possibly a symmetric configuration, which is not shown). The two pegs numbered 1 on the right produce the sole peg numbered 1 on the left. After this, the two pegs numbered 2 produce the peg numbered 2 on the left.

We now go back one more move. To obtain the configuration on the right in Figure 2.8, we can have pegs arranged as shown to the right in Figure 2.9. The pegs numbered 1, 2, and 3 in the figure on the right produce, respectively, the pegs numbered 1, 2, and 3 in the target figure on the left, after which two pegs numbered 4 on the right produce the peg numbered 4 on the left.

Finally, a possible configuration to get to the image on the right in Figure 2.9 is shown to the right in Figure 2.10. Each numbered grouping on the right, after suitable jumps, produces the correspondingly numbered grouping on the left.

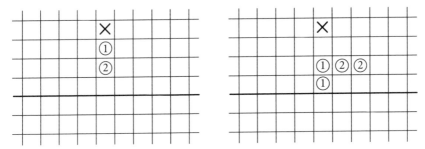

Figure 2.8. To attain the arrangement of two pegs on the left, we must have four pegs on the previous move. A possible arrangement of those pegs is shown on the right.

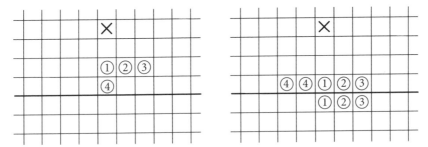

Figure 2.9. To attain the arrangement of pegs on the left, we must have eight cells on the previous move. A possible arrangement of those pegs is shown on the right.

The 20-peg solution shown in Figure 2.10 is not unique. Beasley mentioned two alternatives with the same number of pegs [1, p. 212]. He also proved that 20 is the minimum number of pegs needed to send a scout four lines beyond the enemy line by using what is known as a *pagoda function*. This technique was invented by Conway to prove the impossibility of a scout reaching line 5 [2, pp. 822–823].

Pirates' Gold. The key is to work backward! If only pirates 4 and 5 remain, then 4 will just keep all the gold. The vote will be a tie, and then he wins. So when 3, 4, and 5 remain, pirate 3 should keep 99 for himself, give 0 to number 4, and give 1 to number 5. Pirate 5 knows he will get nothing if he votes down the proposal, so he will vote yes. Thus, when 2, 3, 4, and 5 remain, pirate 2 will keep 99 for himself, give 0 to pirate 3, give 1 to number 4, and give 0 to number 5. This will then lead to a tie vote and adopting the proposal. Now we see what pirate 1 should do. He will keep 98 gold pieces for himself, give 0 to pirate 2, 1 piece to

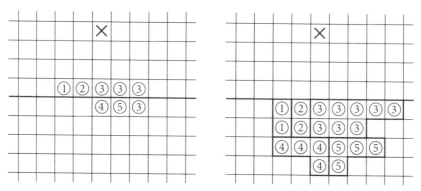

Figure 2.10. A possible starting configuration for advancing a scout four rows above the enemy line is shown on the right. After suitable jumps, the numbered groupings on the right produce the correspondingly numbered groupings on the left. Although this arrangement is not unique, it does employ the minimal number of pegs.

pirate 3, 0 to pirate 4, and 1 piece to pirate 5. Pirates 3 and 5 do strictly better with this proposal than they would if the next pirate proposes the distribution.

Frog in the Well. Since the frog advances 1 foot per day, the seemingly obvious answer is 30 days. But this is not correct! Thinking backward immediately shows that the frog will get out of the well during the last day's up move of 3 feet. It will start the last day in the well 27 feet above the bottom, which will take the frog 27 days to reach. Hence, the frog will get out of the well on the twenty-eighth day.

Water Lily. Thinking backward immediately reveals that the lilies will cover half of the pool on day 59. The lilies then double in size the next day to cover the whole pond.

The Dish of Potatoes. After each of the three travelers ate 1/3 of the available potatoes, 2/3 of the available potatoes remained. Since there were 8 potatoes left after the third traveler ate, there must have been 12 potatoes before he ate. It follows that there were 18 potatoes before the second traveler ate, and 27 potatoes before the first traveler ate.

References

[1] J. D. Beasley. *The Ins and Outs of Peg Solitaire.* Oxford University Press, Oxford, 1992.

[2] E. R. Berlekamp, J. H. Conway, and R. K. Guy. *Winning Ways for Your Mathematical Plays*, volume 4, second edition. A K Peters, Boca Raton, FL, 2004.

[3] B. Bolt. *The Amazing Mathematical Amusement Arcade*. Cambridge University Press, 1984.

[4] G. Brandreth. *Classic Puzzles*. Barnes & Noble, New York, 1998.

[5] G. T. Dow and R. E. Mayer. Teaching students to solve insight problems: Evidence for domain specificity in creativity training. *Creativity Res. J.* **16** no. 4 (2004), 389–402.

[6] Y. Elran. Retrolife and the pawns neighbors. *College Math. J.* **43** no. 2 (March 2012) 147–151.

[7] M. Gardner. *My Best Mathematical and Logic Puzzles*. Dover, Mineola, 1994.

[8] M. Gardner. *The Colossal Book of Short Puzzles and Problems*, edited by Dana Richards. W. W. Norton, New York, 2006.

[9] D. Hess. *All-Star Mathlete Puzzles*. Sterling, New York, 2009.

[10] A. Levitin and M. Levitin. *Algorithmic Puzzles*. Oxford University Press, Oxford, 2011.

[11] M. Nogin. *Strategies of Problem Solving*, 2nd ed. California State University, Fresno, 2014.

[12] T. H. O'Beirne. *Puzzles and Paradoxes*. Oxford University Press, Oxford, 1965.

[13] G. Polya. *How to Solve It: A New Aspect of Mathematical Method*, 2nd ed. Princeton University Press, Princeton, NJ, 1957.

[14] W. Poundstone. *How Would You Move Mount Fuji? Microsoft's Cult of the Puzzle: How the World's Smartest Companies Select the Most Creative Thinkers*. Little-Brown, Boston, 2003.

[15] F. Schuh. *The Master Book of Mathematical Recreations*. Dover, Mineola, NY, 1968.

[16] R. M. Smullyan. *The Chess Mysteries of Sherlock Holmes*. Dover, Mineola, NY, 2011.

3

SPIRAL GALAXIES FONT

Walker Anderson, Erik D. Demaine, and Martin L. Demaine

In a Spiral Galaxies puzzle, you are given a grid of unit squares and a collection of *centers* (points, drawn as dots, typically at square centers or edge midpoints). The goal is to decompose the grid into regions called *galaxies*, one containing each center, so that each galaxy is a polyomino (an edge-to-edge joining of unit squares) and is 180° rotationally symmetric about its center. In general, the solution may not be unique, as shown in Figure 3.1, but most puzzles are designed to have a unique solution.

Spiral Galaxies or Tentai Show is one of the many pencil-and-paper puzzles designed for the Japanese puzzle magazine and publisher Nikoli [4]. Like most Nikoli puzzles, Spiral Galaxies is *NP-complete* [2], meaning that there is no efficient algorithm to solve them, assuming a famous conjecture in computational complexity theory (P ≠ NP). See works by Demaine and Hearn [1, 3] for related complexity results about other puzzles.

In this chapter, we design a typeface featuring 36 Spiral Galaxies puzzles, one for each numeral and letter of the alphabet. Figure 3.2 presents the unsolved font, which shows just the grid and centers. Like the Spiral Galaxies puzzles published by Nikoli, the centers are drawn with dots of two different colors, black and white, with the intent that the corresponding galaxy polyomino should be filled the same color. In our puzzles, the black galaxies will form a letter or numeral. Figure 3.5 later in the chapter illustrates these (unique)

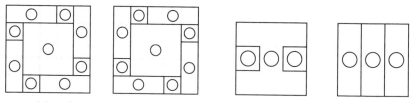

(a) Reflectional symmetry   (b) No reflectional symmetry

Figure 3.1. Examples of Spiral Galaxies puzzles with multiple solutions.

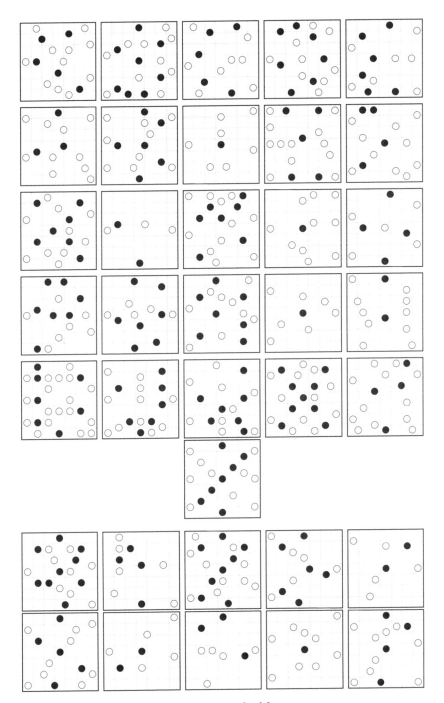

Figure 3.2. Unsolved font.

solutions, so do not look if you want to solve the puzzles on your own. You can write custom messages in both fonts using a free web app: http://erikdemaine.org/fonts/spiralgalaxies/.

This mathematical puzzle typeface is part of a series of fonts by the second and third authors and their collaborators. These fonts aim to illustrate mathematical ideas, theorems, and open problems for the general public, inviting the reader to engage with mathematics through solving puzzles to read the letters. See http://erikdemaine.org/fonts/ for more examples.

1 Solving Spiral Galaxies Puzzles

Before we describe how we designed our Spiral Galaxies font puzzles, we describe in general how you might solve a Spiral Galaxies puzzle. This approach will let you try solving the font puzzles yourself, and it is also central to how we design puzzles with guaranteed unique solutions.

The general approach is to follow logical deductions from the basic rules of the puzzle and the resulting properties of the solution:

1. **Membership**: Every grid square belongs to exactly one galaxy.
2. **Centers**: Every center is contained in one galaxy, so its incident grid squares belong to that galaxy.
3. **Symmetry**: Every galaxy is 180° rotationally symmetric about its center. This symmetry applies to both the constituent grid squares and the boundary of the galaxy.

Figure 3.3 walks through a sample 5 × 5 Spiral Galaxies puzzle. We start with the puzzle centers in Figure 3.3a.

The first rule we apply (Figure 3.3b) is the centers rule, which tells us that when a center touches multiple grid squares, those squares must all be part of the same galaxy. We draw these same-galaxy relations with gray lines. This rule helps us get started, but we will not need it any further.

It is also helpful to keep track of the boundaries of the galaxies (Figure 3.3c). When two adjacent grid squares must belong to different galaxies (being connected to different centers via gray edges, say), we can draw a black boundary line between them.

Next we apply the symmetry rule to the galaxy marked with a star in Figure 3.3d. This galaxy's upper-left square has a boundary edge to its left. Because the region is rotationally symmetric, its lower-right square must have a boundary edge to its right. Can you see how similar logic can be used to place all of the dotted-line borders in Figure 3.3e?

Then we apply the membership rule to the starred grid square in Figure 3.3f. This square must belong to some galaxy, but it is surrounded by three boundary edges. Therefore, we can draw a gray edge in the fourth (downward) direction.

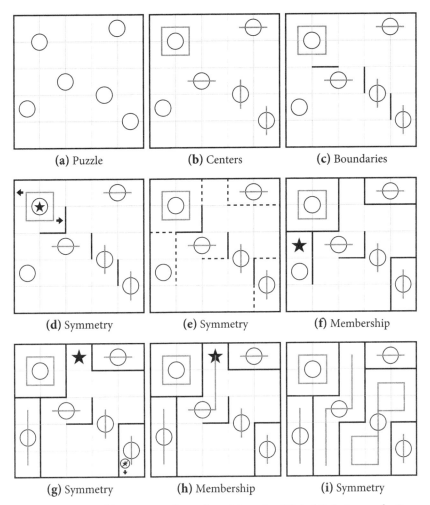

Figure 3.3. A possible sequence of steps for solving a 5 × 5 Spiral Galaxies puzzle. Gray lines connect grid squares that must be in the same galaxy; black lines separate grid squares that must be in different galaxies.

Applying the symmetry rule again (Figure 3.3g), we can complete the lower-left polyomino.

Now we would like to apply the membership rule to the starred grid square in Figure 3.3g. Which of the two remaining incomplete galaxies does it belong to? By the symmetry rule, the square cannot belong to the right galaxy, because its then-symmetric square is located outside of the grid. By the membership rule, it must then belong to the left galaxy, as in Figure 3.3h.

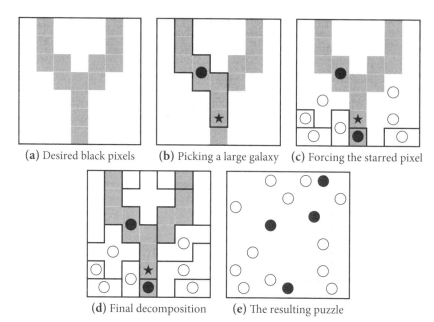

(a) Desired black pixels   (b) Picking a large galaxy   (c) Forcing the starred pixel

(d) Final decomposition   (e) The resulting puzzle

Figure 3.4. Design walkthrough for the Y puzzle.

Applying the symmetry rule one last time to the left polyomino, we solve the puzzle (Figure 3.3i).

2 Designing a Spiral Galaxies Font

Each puzzle was designed by hand. Here we describe the process, using the Y puzzle as an example; refer to Figure 3.4.

First, we designed a black/white pixel rendering of the letter in question; see Figure 3.4a. Our goal is to make the black pixels decompose into the black galaxy polyominoes, and the white pixels decompose into the white galaxy polyominoes.

In each successive step, we remove a rotationally symmetric polyomino from the not-yet-decomposed black or white region. We tried to choose polyominoes that would be fun to discover, either with an unusual shape or having a large size. In Figure 3.4b, we chose a large polyomino.

Before proceeding to the next step, we try to make sure that this polyomino is necessary/forced. One simple way to do this is to pick a square and set up clues, so that the square can only be part of the desired polyomino. For example, consider the starred square in Figure 3.4b. Once we add the white centers in Figure 3.4c, the starred square must be part of the polyomino for the

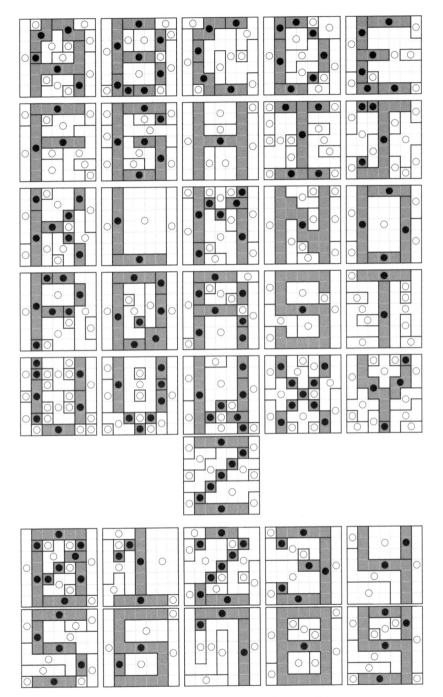

Figure 3.5. Solved font for uppercase letters and digits.

already-placed black center, because including it in any other polyomino would force that (rotationally symmetric) polyomino to extend past the border of the grid or into squares already occupied by a different polyomino.

Once we are near the end, we simply fill in the remaining regions with rotationally symmetric polyominoes (Figure 3.4d) and place the corresponding centers (Figure 3.4e).

Finally, we need to test the puzzle to ensure that the solution is unique. To do so, we solved the puzzle in the traditional way, only writing something into the grid if we could logically justify that it had to be the case. If the solution ended up being not unique, we would go back and tweak a few of the polyominoes to ensure uniqueness. The result is shown in Figure 3.5.

We encourage the reader to design their own puzzles—for example, another font of a different grid size, or lowercase letters, or a different language—and send them to us!

Acknowledgments

We thank Jason Ku for helpful discussions, in particular about puzzles with multiple solutions.

References

[1] E. D. Demaine and R. A. Hearn. Playing games with algorithms: Algorithmic combinatorial game theory. In *Games of No Chance 3*, M. H. Albert and R. J. Nowakowski, eds. Cambridge University Press, Cambridge, 2009, 3–56.

[2] E. Friedman. Spiral Galaxies puzzles are NP-complete. Online at http://www.stetson.edu/~efriedma/papers/spiral/spiral.html. Last accessed June 2018.

[3] R. A. Hearn and E. D. Demaine. *Games, Puzzles, and Computation.* A. K. Peters, Boca Raton, FL, 2009.

[4] Nikoli Co., Ltd. Tentai Show. Online at http://nikoli.co.jp/en/puzzles/astronomical_show.html. Last accessed June 2018.

4

KINGS, SAGES, HATS, AND CODES

Steven T. Dougherty and Yusra Naqvi

In the Kingdom of Noether, a capricious and mathematically inclined king ruled the land. He was known for posing difficult mathematical challenges to those over whom he ruled. He used these challenges to help make royal decisions. We describe two such occurrences together with the clever solutions of those who were able to defeat the king's challenges. We then show how these challenging problems are related to modern information theory.

1 Problems

1.1 Hiring Royal Assistants

The king wishes to hire two sages to assist him in his royal duties and intends to pay them generously for their services. However, to ensure that they are not frauds, he wants to test their abilities. If they prove themselves worthy, they will be given a place in court. If not, they shall be considered impostors and thrown in jail. The test that the king proposes is as follows.

Question 1. *The sages will be put in two separate waiting rooms. The king will lay out an 8 by 8 chess board in the throne room and place some number of stones on different squares of the chessboard. There can be any number of stones from 0 to 64. Then the first sage will be brought in, and the king will point to an arbitrary square on the board (the square can either have a stone or not have a stone). The first sage then has the choice of one of the following three options:*

1. *place one stone on an unoccupied square on the board,*
2. *remove one stone that is already on the board, or*
3. *do nothing.*

The first sage will then be taken back to the first waiting room, and the second sage will be brought in to the throne room. This second sage must now look at the board and determine to which square the king has pointed. Of course, the two sages are informed of this test when they apply for the job, and they can discuss a strategy before they are taken to their respective waiting rooms, but they may not communicate at all after that. What strategy do they employ to determine to which square the king has pointed?

1.2 Prisoners

It should not be surprising that such a temperamental and demanding king has had several occasions to imprison people on various insignificant charges. However, continuing his tradition of enjoyment of mathematical challenges, he would often allow them a challenge to free themselves. Here we describe one such challenge.

Question 2. *A group of seven prisoners will be taken into separate cells. They will each be given either a red hat or a yellow hat. The hats are chosen from a sack containing thousands, and it is safe to assume that each prisoner has equal likelihood of getting either color of hat. They will then all be brought into the same room. No one can see the color of his own hat, but they can, of course, see the color of the other hats. When prompted, they must immediately choose from one of the following two options:*

    1. guess the color of their own hat, or
    2. pass on the question.

    Notice that all prisoners give their answer at the same time, either by guessing a color or by signaling that they pass. If someone guesses wrongly or all of them pass, they will all have their prison sentences extended, but if any one prisoner guesses correctly and no one guesses wrongly, they will all be freed. They are allowed to communicate ahead of time, but they are not allowed to communicate in any way after the hats are handed out, nor are they allowed to delay in responding. What should their strategy be so that their answer has a better chance of succeeding than simply guessing the color of a random hat with probability $\frac{1}{2}$?

2 Hints

Before providing solutions, we make a few observations about these two problems.

> **Observation 1.** Both challenges involve a number of objects (chessboard squares or hats) that take on exactly two modes. The chessboard squares can either be *covered* or *uncovered*. The hats can be either *red* or *yellow*. Therefore, all possible configurations of these objects can be represented in binary, as a sequence of 1s and 0s.

Observation 2. The second sage in Question 1 does not need to determine which stone, if any, was changed by the first sage, only to which square the king pointed. The second sage answers based entirely on the configuration of stones at hand.

Observation 3. Since the prisoners in Question 2 do not know in which order they will be called, they do not need to use previous passes, if any, as part of their strategy. They each determine how to answer based on the configuration of hats that they see.

The King of Noether would give those who were questioned one hour to think about a solution before being put to the test. The interested reader may wish to pause at this point to come to a solution on her own!

3 Solutions

3.1 The Chessboard Question

At first glance, the question posed to the sages seems to be quite difficult. If they were allowed to move more than one stone, it would seem that they could arrange them in a manner to indicate the row and the column, but being able to change only one location seems to make the problem very difficult.

Let's examine a small case first and then examine the general solution. Consider a 2×2 chessboard. Label each square with a different number from 0 to 3:

$$\begin{array}{|c|c|} \hline 0 & 1 \\ \hline 2 & 3 \\ \hline \end{array}$$

The king places stones on the board. Construct a vector $\mathbf{v}$ of length 4 with coordinates labeled 0, 1, 2, 3 as follows: If he has placed a stone on a square, then the coordinate corresponding to that square has a 1 in it, and all other coordinates are 0. For convenience, we think of this vector as a column vector, that is,

$$\mathbf{v} = \begin{pmatrix} v_0 \\ v_1 \\ v_2 \\ v_3 \end{pmatrix}.$$

Thus, if the king places stones on squares 0 and 3:

then our vector $\mathbf{v}$ is

$$\mathbf{v} = \begin{pmatrix} 1 \\ 0 \\ 0 \\ 1 \end{pmatrix}.$$

Next, construct a matrix M_4 by writing each possible length-2 vector consisting of zeros and ones as a column:

$$M_4 = \begin{pmatrix} 0 & 0 & 1 & 1 \\ 0 & 1 & 0 & 1 \end{pmatrix}.$$

Then $\mathbf{w} = M_4 \mathbf{v}$ is a two-coordinate vector. (Note that all computations are done over the finite field $\mathbb{F}_2$; that is, everything is done modulo 2. In other words, we have only two elements, 0 and 1. Arithmetic works as usual for multiplication and addition, except that $1 + 1 = 0$.) In our example, this vector would be

$$\mathbf{w} = M_4 \mathbf{v} = \begin{pmatrix} 0 & 0 & 1 & 1 \\ 0 & 1 & 0 & 1 \end{pmatrix} \begin{pmatrix} 1 \\ 0 \\ 0 \\ 1 \end{pmatrix} = \begin{pmatrix} 1 \\ 1 \end{pmatrix}.$$

This vector $\mathbf{w}$ is called the *syndrome*. One might notice that many different vectors $\mathbf{v}$ give the same syndrome, including the vector that has a 0 in every place except the coordinate that corresponds to the column in M_4 that is the syndrome. In this example, that would be the vector

$$\begin{pmatrix} 0 \\ 0 \\ 0 \\ 1 \end{pmatrix}.$$

Each number from 0 to 3 can be represented as a two-digit binary number $a_1 a_0$, so given a vector $\begin{pmatrix} a_1 \\ a_0 \end{pmatrix}$, it corresponds to the number $2a_1 + a_0$. Therefore, in our example, $\mathbf{w} = \begin{pmatrix} 1 \\ 1 \end{pmatrix}$ would correspond to the number 3. Notice that the number 3 is in the coordinate of the array corresponding to where we put the 1 in the vector corresponding to the syndrome.

Now, the king may point to any one of the squares on this board. Thus, our strategy is to alter the vector $\mathbf{v}$ in one coordinate to form a new vector $\mathbf{v}'$ so that $\mathbf{w}' = M_4 \mathbf{v}'$ gives the two-coordinate vector that represents the number of the square to which the king pointed.

If $\mathbf{w}$ is the two-coordinate vector given by the original configuration of stones, and $\mathbf{w}'$ is the actual vector we wish to convey to the second sage, we need to consider the vector $\mathbf{w} + \mathbf{w}'$. (Again, this computation is done modulo 2.) The result is a length-2 vector, which must appear as one of the four columns in the matrix M_4. Construct a length-4 column vector $\mathbf{u}$ that only has a 1 in the coordinate corresponding to that column of M_4 (and zeros everywhere else). Then

$$M_4(\mathbf{v} + \mathbf{u}) = \mathbf{w} + (\mathbf{w} + \mathbf{w}') = \mathbf{w}'.$$

Hence, by changing the square corresponding to this nonzero coordinate of $\mathbf{u}$, we effect the desired change, so that the second sage can determine the answer. More specifically, if there is a stone already on the square indicated by $\mathbf{u}$, the sage removes it, and if there is no stone there, then the sage adds one. In other words, the sage adds a 1 (modulo 2) to that coordinate. If $\mathbf{w} = \mathbf{w}'$, then the sage does nothing.

Let us return to our example, and suppose that the king points to square 2. Then $\mathbf{w}' = \left(\begin{smallmatrix}1\\0\end{smallmatrix}\right)$. Since $\mathbf{w}$ in our example is $\left(\begin{smallmatrix}1\\1\end{smallmatrix}\right)$, the vector $\mathbf{w} + \mathbf{w}'$ (modulo 2) is $\left(\begin{smallmatrix}0\\1\end{smallmatrix}\right)$. This vector is the second column of of M_4, and therefore

$$\mathbf{u} = \begin{pmatrix} 0 \\ 1 \\ 0 \\ 0 \end{pmatrix},$$

and so the first sage must add a new stone to the square labeled 1, since this square was empty before:

The second sage, on entering the room, will see this new configuration,

$$\mathbf{v}' = \mathbf{v} + \mathbf{u} = \begin{pmatrix} 1 \\ 0 \\ 0 \\ 1 \end{pmatrix} + \begin{pmatrix} 0 \\ 1 \\ 0 \\ 0 \end{pmatrix} = \begin{pmatrix} 1 \\ 1 \\ 0 \\ 1 \end{pmatrix},$$

and compute $M_4 \mathbf{v}' = \left(\begin{smallmatrix}1\\0\end{smallmatrix}\right)$, and will correctly declare that the king pointed to the square labeled 2.

The question however, was for an 8×8 chessboard. Let us see to which boards we can extend this solution. We need to construct a matrix with all possible binary columns of length k. There are 2^k such vectors. So for an $n \times n$ chessboard, we need $n^2 = 2^k$, which requires n to also be a power of 2.

The general algorithm for solving this problem can be described as follows:

1. Label the squares from 0 to $2^k - 1$.
2. Construct M_{2^k} consisting of all possible binary columns of length k.
3. Construct $\mathbf{v}$ where $\mathbf{v}_i = 1$ if the square corresponding to i has a stone on it, and $\mathbf{v}_i = 0$ if it does not.
4. Compute $\mathbf{w} = M_{2^k}\mathbf{v}$ and $\mathbf{w}'$ as the vector corresponding to the point to which the king pointed.
5. Change the value in the square corresponding to the column $\mathbf{w} + \mathbf{w}'$ by adding or removing a stone.
6. Compute $M_{2^k}(\mathbf{w} + \mathbf{w}')$.
7. Receive employment, and accompanying riches, from the king of Noether.

3.2 The Hat Question

For this question, we once again start by examining a smaller case. Consider the same problem, but with only three prisoners. Denote a red hat as R and a yellow hat as Y.

The possibilities for three people wearing these hats are as follows:

RRR	YYY
RRY	YYR
RYR	YRY
RYY	YRR

The standard solution is as follows: The probability that all three hats have the same color is $\frac{2}{8} = \frac{1}{4}$. Hence, it is much more probable that they are not all the same color. So here is how you proceed. If you look at the other two people and they both have the same color, then you say that you have the opposite color. If you see the other two people have different colors, then you pass.

With this strategy, you will win with probability $\frac{3}{4}$, which is better than just guessing, in which case you win with probability $\frac{1}{2}$.

To generalize this solution, we must first examine the question in a more mathematical setting. Let R be associated with the number 0, and Y be associated with the number 1, and look at the following matrix:

$$H_3 = \begin{pmatrix} 0 & 1 & 1 \\ 1 & 0 & 1 \end{pmatrix}.$$

(Note that H_3 is composed of all possible nonzero binary vectors of length 2 as its columns.) The set C_3 of vectors $\mathbf{v}$ such that $H_3\mathbf{v}^T = \mathbf{0}$ is $\{(000), (111)\}$. These vectors precisely correspond to the case of all red hats or all yellow hats.

However, it is much more likely that the correct configuration lies outside the set C_3 rather than inside it, and so our strategy was based on this assumption.

Now consider the case for seven people. In this case, we must look at the matrix

$$H_7 = \begin{pmatrix} 0\ 0\ 0\ 1\ 1\ 1\ 1 \\ 0\ 1\ 1\ 0\ 0\ 1\ 1 \\ 1\ 0\ 1\ 0\ 1\ 0\ 1 \end{pmatrix}.$$

Note that H_7 consists of all nonzero binary columns of length 3 (of which there are precisely seven).

Let C_7 be the set of all binary vectors $\mathbf{v}$ such that $H_7\mathbf{v}^T = \mathbf{0}$. Since H_7 consists of three independent rows, we know that C_7 must have dimension 4, and therefore C_7 contains the following 2^4 vectors:

0000000	1111111
1110000	0001111
1001100	0110011
0101010	1010101
0010110	1101001
1000011	0111100
0100101	1011010
0011001	1100110

Therefore, the probability that the configuation of hats is in C_7 is $\frac{2^4}{2^7} = \frac{1}{8}$, and so once again, it is much more probable that the actual configuration lies outside this set. This assumption allows us to design a strategy that will lead to freedom $\frac{7}{8}$ of the time.

To proceed with this strategy, we must first order the prisoners by assigning each one a unique number from 1 to 7. Then, we construct $\mathbf{v}$ by setting its ith coordinate to 0 if prisoner i has a red hat and 1 if yellow. For example, if the assignment of hats is

Number:	1	2	3	4	5	6	7
Hat:	Y	R	Y	R	R	R	Y

then $\mathbf{v} = (1, 0, 1, 0, 0, 0, 1)$.

Now as long as $\mathbf{v}$ is not in C_7, the vector $\mathbf{w} = H_7\mathbf{v}^T$ (computed again in the binary field) must be nonzero, and so in fact, it must appear as one of the columns of H_7. Notice that this vector $\mathbf{w}$ is the syndrome, just as in the previous problem. In the case of our example,

$$H_7 \mathbf{v}^T = \begin{pmatrix} 1 \\ 0 \\ 1 \end{pmatrix},$$

which is the same as column 5 of H_7. This column also corresponds to the binary representation of 5, which is 101.

If $\mathbf{w}$ matches column i of H_7, construct a length-7 vector $\mathbf{u}$ that only has a 1 in coordinate i and zeros everywhere else. Then $H_7 \mathbf{u}^T$ also equals $\mathbf{w}$, and so, modulo 2, we have

$$H_7 (\mathbf{v} + \mathbf{u})^T = \mathbf{w} + \mathbf{w} = \mathbf{0}.$$

This implies that if prisoner i were to have the opposite hat to the one currently assigned, the vector $\mathbf{v}$ would be transformed to a new vector $\mathbf{v}' = \mathbf{v} + \mathbf{u}$ that lies inside C_3. In our example, if prisoner 5 had a yellow hat instead of a red one, the resulting vector $\mathbf{v}'$ would be $(1, 0, 1, 0, 1, 0, 1)$, which is indeed in C_7.

Therefore, the prisoners use the following strategy. Each prisoner guesses $\mathbf{v}$ by looking at the hats of the other six prisoners and assuming that their own hat is red (i.e., they each use 0 for their own coordinate). They all compute $\mathbf{w} = H_7 \mathbf{v}^T$ based on their personal guess for $\mathbf{v}$. Then as long as the actual $\mathbf{v}$ is not in C_7, there is a unique prisoner who will get either $\mathbf{0}$ or the column in H_7 corresponding to their own number. If this prisoner gets $\mathbf{0}$, then the correct guess is "yellow;" otherwise, it is "red." Everyone else will get a column of H_7 that is different from the one corresponding to their number and should pass on their turn.

In our example, suppose that prisoner number 3 is asked to guess first. Prisoner 3 would guess $\mathbf{v}$ to be $(1, 0, 0, 0, 0, 0, 1)$, using a 0 for coordinate 3 based on the assumption of having received a red hat. Then $H_7 \mathbf{v}^T$ would be

$$\begin{pmatrix} 0 & 0 & 0 & 1 & 1 & 1 & 1 \\ 0 & 1 & 1 & 0 & 0 & 1 & 1 \\ 1 & 0 & 1 & 0 & 1 & 0 & 1 \end{pmatrix} \begin{pmatrix} 1 \\ 0 \\ 0 \\ 0 \\ 0 \\ 0 \\ 1 \end{pmatrix} = \begin{pmatrix} 1 \\ 1 \\ 0 \end{pmatrix},$$

which is nonzero and is also not column 3 of H_7. So prisoner 3 passes. Similar computations would show that all other prisoners besides prisoner 5 would also pass. However, prisoner 5 computes $\mathbf{v} = (1, 0, 1, 0, 0, 0, 1)$ and sees that $H_7 \mathbf{v}$ is in fact column 5 of H_7 in this case. Therefore, this prisoner would correctly guess "red."

We notice at this point that the matrices H_3 and H_7 are the same as M_4 and M_8 of the last section, with the all-zero columns removed. These solutions are thus easily generalized to H_{2^k-1}, which deals with the case of $2^k - 1$ prisoners.

The general algorithm for solving this problem can be described as follows:

1. Label the prisoners from 1 to $2^k - 1$.
2. Construct H_{2^k-1} consisting of all possible nonzero binary columns of length k.
3. Compute the set of all vectors $\mathbf{u}$ such that $H_{2^k-1}\mathbf{u} = \mathbf{0}$, and call this set C_{2^k-1}. The matrix H_{2^k-1} has k linearly independent rows and so by the Rank-Nullity Theorem, the dimension of C is $2^k - 1 - k$.
4. Construct $\mathbf{v}$, where $\mathbf{v}_i = 1$ if the person corresponding to i has a yellow hat, and $\mathbf{v}_i = 0$ if that person has a red hat, and using a 0 for your own coordinate.
5. If $H_{2^k-1}\mathbf{v}$ is $\mathbf{0}$, guess "yellow." If $H_{2^k-1}\mathbf{v}$ is the same as the column of H_{2^k-1} corresponding to your prisoner number, guess "red." If it is neither of these, pass.
6. Enjoy your freedom with probability $\frac{2^k-1}{2^k}$.

4 Codes

It may be clear to some readers at this point that the King of Noether has an intimate knowledge of coding theory, that branch of mathematics developed in the second half of the twentieth century to correct errors that occur in electronic communication.

The king proposed two seemingly different problems, which really have solutions that come from the same structure: specific matrices and the set of vectors that multiply by those matrices to get the zero vector. This set of vectors is actually a well-known code called the "Hamming code."

We can explain the coding theory that the king obviously knew when proposing such problems. Given a fixed length r, there are $\ell = 2^r - 1$ nonzero binary vectors of length r. Let H_r be the $r \times \ell$ matrix such that the columns of H_r consist of all possible nonzero binary vectors. This matrix H_r is known as the parity check matrix. The set of binary vectors $\mathbf{v}$ such that $H_r\mathbf{v}^T = \mathbf{0}$ forms a binary vector space C_r in $\mathbb{F}_2^\ell$. This set of vectors is known as a code. The matrix H_r has r linearly independent rows, and so by the well-known rank-nullity theorem, the dimension of C_r over $\mathbb{F}_2$ is $\ell - r$. This vector space C_r is known as the *Hamming code*.

The *Hamming distance* between two binary vectors is the number of coordinates by which they differ. For instance, the vectors 1001100 and 0101010 in $\mathbb{F}_2^7$ have a distance of 4, since they differ in the 4 positions indicated below:

$$\boxed{1}\,\boxed{0}\,0\,1\,\boxed{1}\,\boxed{0}\,0$$
$$\boxed{0}\,\boxed{1}\,0\,1\,\boxed{0}\,\boxed{1}\,0.$$

If the vector has length n, then there are $\binom{n}{s}$ vectors in $\mathbb{F}_2^n$ that are distance s from a given vector, each obtained by flipping s of the n coordinates in the vector. For example, since the vector 010 has length 3, there must be $\binom{3}{2} = 3$ vectors at a distance of 2 from it, obtained by changing 2 of the digits in each case:

$$\boxed{0}\boxed{1}\boxed{0} \longrightarrow \boxed{1}\boxed{0}\boxed{0}$$

$$\boxed{0}\boxed{1}\boxed{0} \longrightarrow \boxed{1}\boxed{1}\boxed{1}$$

$$0\boxed{1}0 \longrightarrow 0\boxed{0}\boxed{1}.$$

Around any vector, we define a sphere of a given radius r to be the set of all vectors that are distance r or less from the vector. So in a sphere of radius r, there are $\left(\binom{n}{0} + \binom{n}{1} + \binom{n}{2} + \cdots + \binom{n}{r}\right)$ vectors. If the minimum distance between vectors in a code (a set of vectors) is $2r + 1$, then the spheres around the vectors in the code are disjoint:

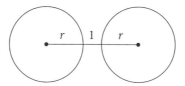

The minimum distance between any two vectors in C_r is 3, since no two columns of H_r are linearly dependent, but some three of them are. This tells us that any sphere of radius 1 must be disjoint from any other.

As we have seen above, each vector in the code C_r has length $\ell = 2^r - 1$, and C_r has dimension $\ell - r$ over $\mathbb{F}_2$. It follows that there are $2^{\ell - r}$ different vectors in C_r. Thus there must be $2^{\ell - r}$ disjoint spheres of radius 1 in $\mathbb{F}_2^n$, one for each vector in C_r. Since each sphere of radius 1 contains $\ell + 1$ vectors (i.e., the center vector that lies in C_r, along with the ℓ additional vectors that are obtained by changing exactly one coordinate), we see that in total, these spheres contain

$$2^{\ell - r} \cdot (\ell + 1) = 2^{\ell - r} \cdot 2^r = 2^\ell$$

vectors, which is exactly the total number of vectors in $\mathbb{F}_2^\ell$. Codes for which this occurs are known as *perfect codes*.

The construction of this perfect code is precisely the amazing fact that allows us to solve both puzzles! It implies that in the ambient space, any vector is either in the code (the set of vectors) or distance 1 from a unique vector in the code. So to solve the first puzzle, you are simply changing one position on the grid to get to the vector that represents the spot where the king points. In the second puzzle, each scenario either represents an element in the code or something distance 1 from the code. We assume then with ever greater probability that the element is not in the code, and we have our solution to the problem.

In terms of electronic communication, a similar technique helps us detect errors that occur during transmission. Suppose we receive a vector **v** of length $2^r - 1$ as a possible message. Then we can compute the *syndrome* $\mathbf{w} = H_r \mathbf{v}$. If the syndrome is the zero vector, then we assume that the vector **v** was correctly transmitted. If the syndrome is not the zero vector, then **w** is the same syndrome as for a vector **u** with only a single 1 in it, by the above argument. We assume that the error in transmission occurred in the coordinate where the 1 is in **u**, and this coordinate is corrected by adding **u** to **v**. The resulting vector $\mathbf{v} + \mathbf{u}$ has a syndrome of **0**, and so we take this to be the intended message.

Notice that in both problems posed by the King of Noether, the key to the solution is simply finding the coordinate where the *error* is. The same type of technique also allows for nearly error-free transmission of information in such diverse areas as space probes, cable television, phones, compact disc players, or Internet communication. So whether you want to free yourself from the dungeon of the King of Noether, win a position in his court, or send a picture from space back to earth, the strategy is really the same!

5 Further Reading

For those interested in reading Hamming's original paper, you can read his early work from 1950 [2]. For an undergraduate text describing algebraic coding theory see Hill's classic text [3]; for more advanced treatment of the subject, see the text by Huffman and Pless which has become the standard in classical coding theory [4]. For a very broad approach to coding theory as pure mathematics, see the recent text by one of the authors of this chapter [1].

References

[1] S. T. Dougherty. *Algebraic Coding Theory over Finite Commutative Rings*. Springer Briefs in Mathematics. Springer, Cham, Switzerland, 2017.
[2] R. W. Hamming. Error detecting and error correcting codes. *Bell Syst. Tech. J.* **29** (1950) 147–160.
[3] R. Hill. *A First Course in Coding Theory*. Oxford University Press, Oxford, 1990.
[4] W. C. Huffman and V. S. Pless. *Fundamentals of Error-Correcting Codes*. Cambridge University Press, Cambridge, 2003.

5

HOW TO PREDICT THE FLIP OF A COIN

James D. Stein and Leonard M. Wapner

It is widely believed—and may well be true—that you can only correctly predict the flip of a fair coin 50% of the time. If the coin is biased and lands heads 60% of the time, it is also widely believed—and also may well be true—that the best you can do is to predict that it will land heads all the time, and you will be right 60% of the time. If there were a strategy that would correctly predict the flips of the biased coin more than 60% of the time, then we would say that we can predict the outcome of that experiment better than chance. There *are* such experiments, some of which we describe in this chapter.

1 A Random Ride on a Railroad

Imagine you are a passenger on a train traveling on an east-west line consisting of infinitely many equally-spaced stations. You are not familiar with the names of any of the stations. The train moves from station to station according to a simple random walk, where the direction of departure at each station is determined by the engineer's flip of a coin. If the outcome is heads, the train proceeds one station to the east; if tails, it proceeds one station to the west. This random process is repeated indefinitely.

You board the train and take your seat. It has been a tiring day, and you fall asleep as the train departs, making an unknown number of stops while you sleep. At some point, you awaken while the train is stopped at some unknown station. After a few minutes, the conductor passes through, announcing "Next stop—Willoughby," indicating that the coin has been flipped and the direction of the next departure has been determined by the engineer. The train will depart again in a few minutes.

Before reading further, decide which of the following statements reflects how you would summarize the previous information with regard to your whereabouts:

1. You are equally likely to be at A or B:

$$(\longleftarrow \text{west}) \quad A\text{---Willoughby---}B \quad (\longrightarrow \text{east}).$$

2. Willoughby is equally likely to be at A or B:

$$(\longleftarrow \text{west}) \quad A\text{---Your present position---}B \quad (\longrightarrow \text{east}).$$

3. Both 1 and 2 are true.
4. Neither 1 nor 2 is true.

This may appear to be a relatively inconsequential question, but we shall show that there are surprising ramifications if alternative 1 is true—or even true just some of the time.

There are compelling arguments for choosing either 1 or 2. Option 1 could be true because Willoughby is the only place that has a clearly defined location (even if you do not know where that location is), and for all you know, you could be on either side of it. Option 2 could be true because even though you cannot specify your present position (it is most likely described by some sort of probability distribution), your destination is equally likely to be east or west of it.

Why Willoughby? "A Stop at Willoughby" is an episode from the television series *The Twilight Zone*, first aired in 1960. A harried Madison Avenue advertising executive is besieged on two fronts: an acerbic boss during the day and a gold-digging wife at night. One evening, on the train ride back from Manhattan to his suburban home, he falls asleep and then hears the conductor announce that the next stop is Willoughby. He finds that Willoughby is an idyllic town from the 1890s, where people lead unhurried, fulfilling lives and attend ice-cream socials.

We focus on some peculiar results that occur in a random walk when the destination has the somewhat ethereal characteristics of Willoughby. These results are connected by a technique that traces back (as far as we know) to the statistician David Blackwell [1]. A nontechnical exposition of the technique is described as "Blackwell's Bet" by Wapner [2].

2 Next Stop, Willoughby

Under the assumption that the train is equally likely to be one stop east or one stop west of Willoughby, a clever technique enables you to guess the direction in which the train will depart with a probability of success greater than $1/2$.

Let W denote Willoughby, and let S_1 and S_2 be the stations to the west and east of Willoughby, respectively. The three-station configuration looks like this:

$$(\longleftarrow \text{west}) \quad S_1 \text{—} W \text{—} S_2 \quad (\longrightarrow \text{east}).$$

Let x denote a random variable selected from a continuous distribution along the railroad track that is positive on every open interval. We think of x as a pointer indicating the direction in which the train will proceed; if it is west of the train's current location, you will predict that the train will head west. If it is east of the train's current location, you will predict that the train will head east. You receive the information about the relative position of the pointer from an application called "Pointer" on your cell phone.

Let p be the probability that x is west of S_1, and q the probability that x is east of S_2. The train is currently at S_1 with probability $1/2$; the pointer will correctly indicate the direction in which the train will proceed (which is toward W) with probability $1 - p$.

So the combined probability that the train is at S_1 and the pointer will indicate the correct direction is $\frac{1}{2}(1 - p)$. A similar argument shows that the probability that the train is at S_2 and the pointer will indicate the correct direction is $\frac{1}{2}(1 - q)$. The probability that the pointer will correctly indicate the direction in which the train will proceed is then

$$\frac{1}{2}(1 - p) + \frac{1}{2}(1 - q) = \frac{1}{2} + \frac{1}{2}(1 - (p + q)) > \frac{1}{2},$$

as we have specified that the pointer comes from a continuous distribution that is positive on every open interval. We can view this probability as $1/2$ plus half the probability that the pointer falls in the gap between the two possible stations of origin.

Notice that successfully guessing the direction in which the train will be headed is equivalent to successfully guessing the result of the coin flip that determined the direction in which the train would proceed. This is not a prediction of the result of the coin flip. The coin has already landed, and we have received information about the result of the flip, since the destination—Willoughby—was determined by it.

3 Correctly Predicting the Result of a Coin Flip More Than Half the Time

Now imagine that the setup remains the same, but a second passenger, sitting across the aisle from you, also has the application "Pointer" on her cell phone. It returns the identical information to the second passenger that it does to you.

The second passenger, however, makes her guess as to the direction in which the train will proceed before the coin is flipped.

The second passenger is in the same station as you are, receives the same pointer that you do, and will therefore make the identical guess as you do. Her guess, however, is a prediction of the direction in which the train will proceed and is thus a prediction of the result of the coin flip.

It would seem there are now two possibilities. One is that both you and the second passenger have a probability of success that is greater than 1/2, and therefore the second passenger can predict the result of the coin flip in this particular situation with a probability greater than 1/2. The second possibility is that the probability of successfully guessing the result of the coin flip is actually less than or equal to 1/2. But how can this be? You were successfully guessing the result of the coin flip more than half the time prior to the appearance of the second passenger, and the only thing that has changed is that the second passenger is now making predictions. How can this possibly cause your rate of successful guessing to decrease from more than 1/2 to less than or equal to 1/2?

There is a third possibility, however, which is that the assumption that the destination has two equally probable stations of origin is invalid. However, a random walk on a linear track with two end stations is a Markov chain with reflecting barriers, and the steady state probabilities for all the stations in the middle are the same. Similarly, a random walk on a circular track with a finite number of stations has the same steady state probability for all stations. Also, one could simply mandate the assumption. So there is apparently justification for assuming that a destination can have two equally probable stations of origin.

Of the two alternatives listed above, we believe that the simple mathematical arguments presented here are correct, and that the first alternative is true.

It is certainly true that the second passenger is not predicting the result of a coin flip by itself; the coin flip here has a complicated structure interwoven with it. There might possibly be some "hidden variable" underlying this; we shall have more to say about hidden variables—and the structure in which the coin flip is interwoven—in subsequent sections.

It is worth observing that as long as there is a positive probability that the station of origin is equally likely to be on either side of Willoughby, then the second passenger can predict the result of the coin flip with a probability greater than 1/2.

4 Using the Pointer Technique to Reveal Hidden Variables

Hidden variables are those that are opaque to the experimenter but nonetheless affect the outcome of an experiment. As previously noted, one possible explanation of the ability of the second passenger to successfully predict the result of a coin flip more than half the time is that some such hidden variable underlies the experimental setting. We now present two different experiments that appear identical to the experimenter in all respects, but one involves hidden variables and the other does not. The pointer technique used in Section 3 enables the experimenter to discover the existence of a hidden variable. Both experiments involve independent trials with an outcome probability of 1/2, but the pointer technique enables the experimenter to correctly predict with probability greater than 1/2 the outcome of the experiment with the hidden variable.

The first experiment, which contains a hidden variable, is a situation in which an outcome with probability 1/2 is determined by using a randomizing device, and for which it is possible to predict the outcome correctly with probability greater than 1/2. The hidden variable actually makes it possible to make this prediction before the randomizing device is used. Repeated trials are independent of one another.

The second experiment is identical, from the standpoint of the experimenter, to the first, but no hidden variables are present. The pointer technique enables us to distinguish between the two.

4.1 Experiment 1: A Random Walk with a Hidden Variable

Once again, Willoughby will be located between the stations S_1 and S_2, and again we will assume the passenger is equiprobably at S_1 or S_2. Station A lies immediately to the west of S_1, and Station B lies immediately to the east of S_2. The five-station configuration, reading west to east looks like this:

$$(\longleftarrow \text{west}) \quad \text{—}A\text{—}S_1\text{—}W\text{—}S_2\text{—}B\text{—} \quad (\longrightarrow \text{east}).$$

As before, the passenger has the Pointer application on her cell phone. However, the direction of the train will now be determined by a spinner. Let r be a real number such that $\frac{1}{2} < r < 1$. The spinner is divided into red and blue regions; the probability that the spinner lands in the red region is r.

If the spinner lands in the red region, the train will proceed toward Willoughby. If the spinner lands in the blue region, the train will proceed away from Willoughby. So if the train is currently at station S_1, the train will proceed toward A, and if the train is at station S_2, the train will proceed toward B.

Notice that if the train is at station S_1 (which it is with probability 1/2), the train will proceed east with probability r. If the train is at station S_2 (which also occurs with probability 1/2), it will proceed east with probability $1 - r$. Therefore, the probability that the train will proceed east is

$$\frac{1}{2}r + \frac{1}{2}(1 - r) = \frac{1}{2}.$$

We now show that the pointer employed in the two-passenger scenario can be used to predict the direction in which the train will move and that this prediction will be correct with probability greater than 1/2. Moreover, this

TABLE 5.1.
Probabilities of the four scenarios for Experiment 1

Station	Spinner	Direction	Pointer Correct?	Combined Probability
$S_1 \left(\frac{1}{2}\right)$	Red (r)	East	$p_2 + p_3$	$\frac{1}{2}r\left(p_2 + p_3\right)$
$S_1 \left(\frac{1}{2}\right)$	Blue $(1-r)$	West	p_1	$\frac{1}{2}(1-r)p_1$
$S_2 \left(\frac{1}{2}\right)$	Red (r)	West	$p_1 + p_2$	$\frac{1}{2}r\left(p_1 + p_2\right)$
$S_2 \left(\frac{1}{2}\right)$	Blue $(1-r)$	East	p_3	$\frac{1}{2}(1-r)p_3$

Note: The first two columns respectively record your present direction and the spinner color. The third column records the train's direction on its way to the next station. The final two columns record the probability that the pointer makes the correct prediction, and the combined probability of the scenario in that row actually occurring, respectively.

prediction can be made before the spinner is used to determine the direction of movement of the train.

Let x be an independent, continuous random variable whose domain is the railroad track (which we can view as the real line). Let

$p_1 = $ the probability that x is west of station S_1,

$p_2 = $ the probability that x lies between stations S_1 and S_2, and

$p_3 = $ the probability that x is east of station S_2.

As in the two-passenger scenario, if the pointer is to the passenger's east, she will guess that the train will move east. If the pointer is to the passenger's west, she will guess that the train will move west. The pointer can be consulted prior to using the spinner that directs the train.

We have four separate cases, as there are two train locations (S_1 or S_2) and two spinner results (red or blue). In Table 5.1, we present certain probabilities relevant to these four scenarios. Note that since location, spinner, and pointer are independent, the relevant probabilities are simply obtained by taking products. In the table, the station and spinner probabilities are included parenthetically for convenience.

The overall probability that the pointer will correctly indicate the direction of travel is the sum of the probabilities in the last column:

$$\frac{1}{2}(1-r)(p_1 + p_3) + \frac{1}{2}r(p_1 + 2p_2 + p_3) = \frac{1}{2}(p_1 + p_3) + rp_2$$

$$= \frac{1}{2}(p_1 + p_2 + p_3) + \left(r - \frac{1}{2}\right)p_2$$

$$= \frac{1}{2} + \left(r - \frac{1}{2}\right)p_2.$$

TABLE 5.2.
Probabilities associated with the scenarios in Experiment 2

Coin	Direction	Pointer Correct?	Combined Probability
Heads $\left(\frac{1}{2}\right)$	East	q_2	$\frac{1}{2}q_2$
Tails $\left(\frac{1}{2}\right)$	West	q_1	$\frac{1}{2}q_1$

Since $r > \frac{1}{2}$, this last quantity is greater than $\frac{1}{2}$.

The restriction $r < 1$ has been chosen to ensure the possibility that the train can move in either direction from either station. It is intriguing that it is our lack of knowledge of whether the train is at S_1 or S_2, coupled with the fact that the pointer measurement is made after the train has arrived at a station, that enables us to obtain this curious result.

4.2 Experiment 2: A Random walk with No Hidden Variable

The setup is similar to that of Experiment 1, except that (1) there is no station S_2—the passenger is always at station S_1, and (2) the direction of travel is determined by the flip of a fair coin. If the coin lands heads, the train will go east, if it lands tails, the train will go west. The station configuration now looks like

$$(\longleftarrow \text{west}) \quad —A—S_1—B— \quad (\longrightarrow \text{east}) .$$

The pointer is obtained in the same way as before and is used by the passenger in the same way, but we define the probabilities to be

$$q_1 = \text{the probability that } x \text{ is west of station } S_1,$$

$$q_2 = \text{the probability that } x \text{ is east of station } S_2.$$

It is clear that the train will go east 50% of the time and west 50% of the time. There are only two scenarios this time, heads or tails, and the probabilities associated with these scenarios are displayed in Table 5.2. The sum of the probabilities in the last column is the overall probability that the pointer will correctly indicate the direction of travel. This sum is $\frac{1}{2}\left(q_1 + q_2\right) = \frac{1}{2}$.

4.3 A Comparison of Experiments 1 and 2

From the standpoint of the passenger on the train, the two experiments are identical. In each experiment, he simply uses a pointer, which he compares to his current location, to guess the direction of travel—and in each experiment, the train randomly goes east in independent trials 50% of the time.

Experiment 1 relies on the equiprobability hypothesis. Admittedly, in Experiment 1, the train travels east from station S_1 more often than it travels east from station S_2. But there is no way the passenger can ascertain this, because in both experiments, he is simply at an unknown station making a measurement, and in both experiments, the two outcomes east and west occur independently with probability $1/2$.

However, using the pointer in Experiment 1 enables him to guess correctly the direction in which he is traveling more than 50% of the time and also has the curious feature that he can guess this even before the direction of the train is determined. Of course, this is because the train travels east from station S_1 more often than it travels east from station S_2, and the pointer "detects" this, and thus "knows" in which way the train is likely to move before the train's direction is determined.

There is an intriguing contrast between Experiment 1 and loaded dice. In the case of loaded dice, empirical measurements would contradict the Principle of Indifference, which assumes that in the absence of additional information, all outcomes are equally probable. However, in Experiment 1, empirical measurements would confirm the results expected according to the Principal of Indifference, and only the pointer illuminates the hidden variable.

4.4 Hidden Variables Redux

We now compare Experiment 2 with the two-passenger scenario described in Section 3. The two-passenger scenario takes place under the hypothesis that there is one destination—Willoughby—with two equiprobable stations of origin. With this description, the use of the pointer enables the passengers to correctly guess the direction in which the train will proceed more than 50% of the time.

Experiment 2 could easily be conducted with another passenger using the same pointer on the same train, but it takes place under the hypothesis that there is one station of origin—S_1—with two equiprobable destinations. With this description, this passenger correctly guesses the direction in which the train will proceed only half the time.

This was the reason that we posed the question to you in Section 1. Both descriptions (one destination, two equiprobable stations of origin and one station of origin, two equiprobable destinations) seem reasonable. However, they cannot both be correct. The passengers using the pointer can discern which description is the correct one. If the use of the pointer results in guessing which way the train will go 50% of the time, the one station of origin with two equiprobable destinations description is correct. If the use of the pointer results in guessing which way the train will go more than 50% of the time, then at least some of the time the train was in the one destination with two equiprobable station-of-origins configuration.

4.5 Experiment 3: Pointers Can Yield Either Superior or Inferior Results Compared to Chance

The pointer always improves the probability of successful guessing in Experiment 1, and it never improves the probability of successful guessing in Experiment 2. We now examine a third experiment, in which the choice of pointers can result in better or poorer results than chance.

The station configuration is identical to that in Experiment 1, and the train is again equiprobably at either S_1 or S_2. However, this time the spinner directs the train to head east if and only if the spinner lands in the red region. The passenger's cell phone application has been modified so that if the pointer is on the Willoughby side of the station, it will direct the passenger to guess that the train will move toward Willoughby, and conversely.

If the train is at station S_1, then it will head east toward Willoughby with probability r, and if it is at station S_2, it will head west toward Willoughby with probability $1 - r$. Therefore, the probability that it will head toward Willoughby is once again seen to be $\frac{1}{2}r + \frac{1}{2}(1 - r) = \frac{1}{2}$. Table 5.3 records the relevant probabilities. The probability of a successful guess in this case is

$$\frac{1}{2}r(p_2 + 2p_3) + \frac{1}{2}(1 - r)(2p_1 + p_2) = \frac{1}{2}p_2 + rp_3 + (1 - r)p_1$$

$$= \frac{1}{2} + \left(r - \frac{1}{2}\right)p_3 + \left(\frac{1}{2} - r\right)p_1$$

$$= \frac{1}{2} + \left(r - \frac{1}{2}\right)(p_3 - p_1).$$

So the pointer improves the probability of successfully guessing whether the train will move toward or away from Willoughby if and only if $p_3 > p_1$.

5 Bernoulli Trials

Is Experiment 1 a Bernoulli trial? A Bernoulli trial has two outcomes to the experiment, often referred to as success and failure. The probability of success remains the same from trial to trial, and successive trials are independent. The probability of exactly k successes in n trials is the same for all Bernoulli trials with the same success probability. Because this formula is the same for all

TABLE 5.3.
Probabilities for the possible scenarios in Experiment 3

Station	Spinner	Direction	Pointer Correct?	Combined Probability
$S_1\left(\frac{1}{2}\right)$	Red (r)	Toward	$p_2 + p_3$	$\frac{1}{2}r\left(p_2 + p_3\right)$
$S_1\left(\frac{1}{2}\right)$	Blue ($1-r$)	Away	p_1	$\frac{1}{2}(1-r)p_1$
$S_2\left(\frac{1}{2}\right)$	Red (r)	Away	p_3	$\frac{1}{2}rp_3$
$S_2\left(\frac{1}{2}\right)$	Blue ($1-r$)	Toward	$p_1 + p_2$	$\frac{1}{2}(1-r)(p_1 + p_2)$

Note: The "Direction" column indicates whether the train is moving toward or away from Willoughby.

Bernoulli trials, it is generally assumed that when you have seen one Bernoulli trial with success probability p, you have seen them all.

Our contention is that there are different types of Bernoulli trials.

We start by looking at an experiment that has features in common with the two-passenger scenario and Experiment 1 of Section 4. A bag contains two coins, both of which are biased. One of the coins has a heads probability of $2/3$; the other has a heads probability of $1/3$. The coins are indistinguishable from the standpoint of the experimenter. If one of the two coins is chosen at random and the coin is flipped, the probability of the coin landing heads is $1/2$. Whether we compute the probability before the coin is chosen or after the coin is chosen, the probability of the coin landing heads is still $1/2$.

What happens in Experiment 1 is similar. The train goes east 50% of the time, no matter whether we compute this probability before the spinner determines the direction of the train or after it. We can modify the two coins in a bag experiment slightly to produce a situation similar to Experiment 1. Someone other than the experimenter places the two coins on a table with the one which has a heads probability of $2/3$ to the left of the other. He flips a fair coin to decide which of the two coins to remove from the table. The experimenter now enters the room and chooses a random location on the table as a pointer. If the pointer is to the left of the remaining coin, she guesses that when the coin is flipped, it will land tails. If the pointer is to the right of the remaining coin, she guesses heads. This guess will be correct more often than not.

There is an important common thread connecting the one destination, two equiprobable stations of origin Willoughby scenario (Experiment 1), and the two coins in a bag experiment outlined in this section. These experiments differ in significant respects from classic examples of Bernoulli trials, such as throwing a 6 with a die or flipping a heads with a coin. Throwing a die and flipping a coin are "one-stage" experiments; the die is thrown, or the coin is flipped, and the result is observed. The one destination, two equiprobable stations of origin scenario, Experiment 1, and the two coins in a bag experiment described in this section are two-stage experiments. In Experiment 1, for example, in the first stage the train is positioned at one of two alternative locations before the pointer observation is made. As we have noted, it is the ambiguity of location at which the experiment takes place, coupled with the fact that the pointer is used midway through an experiment, that helps enable the somewhat curious results of this chapter. It is almost as if the less one knows, the better one can guess—which this runs counter to what we generally think of as the value of knowledge.

What you do not know can actually help you—at least in some cases.

5.1 Compound Bernoulli Trials

We can fit all these observations into a general framework.

Let us define a compound (two-stage) Bernoulli trial as consisting of two independent experiments conducted in sequence. The first experiment has sample space S with outcomes $\{x_1, \ldots, x_N\}$ with probabilities $\{p_1, \ldots, p_N\}$, respectively. The second experiment has two outcomes, which we shall denote s (success) and f (failure). Define the success probability p for the composite experiment by

$$p = \sum_{k=1}^{N} p_k P(s|x_k).$$

The composite experiment looks like a Bernoulli trial with success probability p. We will assume $p \geq \frac{1}{2}$, so that success is no less likely than failure.

Let X be a probability space, and let $E_1, \ldots, E_N$ be subsets of X. Let v be an independent random variable defined on X; v is the pointer. If the outcome of the first experiment is x_k, we predict success for the compound Bernoulli trial if $v \in E_k$, and failure if $v \in X \backslash E_k$.

The probability PSP of successfully predicting whether a trial outcome will be s or f is

$$PSP = \sum_{k=1}^{N} p_k P(s|x_k) P(E_k) + \sum_{k=1}^{N} p_k P(f|x_k) P(X \backslash E_k).$$

We can predict outcomes with results better than chance if $PSP > p$.

We now show that the above definition for PSP yields the results previously obtained in both the Willoughby example (two possible stations of origin and one destination) and the spinner of Experiment 1.

5.2 Example 1: Willoughby (Two Possible Stations of Origin, One Destination)

Let us define the following notation:

$$x_1 = \text{train is at station } S_1 \text{ west of Willoughby,}$$

$$x_2 = \text{train is at station } S_2 \text{ east of Willoughby,}$$

$$s = \text{train goes toward Willoughby.}$$

We set $p_1 = p_2 = \frac{1}{2}$. We can now compute the following values. In the second and third lines, the notation is as in Section 2.

$$P(s|x_k) = 1, \quad P(f|x_k) = 0, \quad k = 1, 2,$$

$$E_1 = \{x \mid x \text{ is east of station } S_1\}, \quad P(E_1) = 1 - p,$$

$$E_2 = \{x \mid x \text{ is west of station } B\}, \quad P(E_2) = 1 - q.$$

We now find that

$$PSP = \frac{1}{2}(1 - p) + \frac{1}{2}(1 - q) = 1 - \frac{1}{2}(p + q) > \frac{1}{2}.$$

5.3 Example 2: Experiment 1 of Section 4.1

Defining x_1 and x_2 as in Example 1, we now define s to be the event that the train goes east. We still have $p_1 = p_2 = \frac{1}{2}$. We also have the following conditional probabilities:

$$P(s|x_1) = r, \quad P(f|x_1) = 1 - r,$$

$$P(s|x_2) = 1 - r, \quad P(f|x_2) = r.$$

With notation as in Experiment 1, we can now write

$$E_1 = \{x \mid x \text{ is east of station } S_1\}, \quad P(E_1) = 1 - p_1,$$

$$E_2 = \{x \mid x \text{ is east of station } S_2\}, \quad P(E_2) = 1 - p_3.$$

The relevant probabilities for the four possible scenarios are presented in Table 5.4. Since success and failure here are defined in terms of east and west, we have included a parenthesized s and f in the table for convenience. We can

TABLE 5.4.

Probabilities for the scenarios in the reinterpretation of Experiment 1

Station	Spinner	Direction	Pointer Correct?	Combined Probability
$S_1 \left(\frac{1}{2}\right)$	Red (r)	East (s)	$1 - p_1$	$\frac{1}{2}r\left(1 - p_1\right)$
$S_1 \left(\frac{1}{2}\right)$	Blue $(1 - r)$	West (f)	p_1	$\frac{1}{2}(1 - r)p_1$
$S_2 \left(\frac{1}{2}\right)$	Red (r)	West (f)	$1 - p_3$	$\frac{1}{2}r(1 - p_3)$
$S_2 \left(\frac{1}{2}\right)$	Blue $(1 - r)$	East (s)	p_3	$\frac{1}{2}(1 - r)(p_3)$

now compute *PSP*:

$$PSP = \frac{1}{2}r(1 - p_1) + \frac{1}{2}(1 - r)p_3 + \frac{1}{2}(1 - r)p_1 + \frac{1}{2}r(1 - p_3)$$
$$= \frac{1}{2}(r + p_1 - 2rp_1) + \frac{1}{2}(r + p_3 - 2rp_3)$$
$$= r + \left(\frac{1}{2} - r\right)(p_1 + p_3)$$
$$= r - \left(r - \frac{1}{2}\right)(p_1 + p_3) > r - \left(r - \frac{1}{2}\right) = \frac{1}{2}.$$

The definition of *PSP* gives the results previously obtained. Of course, the formula for the *PSP* given above admits an obvious generalization to multinomial trials.

Our results seem to imply that the way to predict the flip of a coin is to embed it in the middle of a suitable compound Bernoulli trial. It will be interesting to see where this idea leads.

References

[1] D. Blackwell. On the translation parameter problem for discrete variables. *Ann. Math. Statistics* **22** no. 3 (1951) 393–399.
[2] Leonard M. Wapner, *Unexpected Expectations: The Curiosities of a Mathematical Crystal Ball*, CRC Press, Boca Raton, FL, 2012.

6

COINS AND LOGIC

Tanya Khovanova

In memory of Raymond Smullyan.

Raymond Smullyan wrote numerous logic books with many delightful logic puzzles. Usually the action happens on islands where two kinds of people live: *knights* and *knaves*. Knights always tell the truth, knaves always lie. Smullyan coined the names for knights and knaves in his 1978 book *What Is the Name of This Book?* [11]. Here is a puzzle from that book.

Puzzle 1. *There are two people, A and B, each of whom is either a knight or a knave. A makes the following statement: "At least one of us is a knave." What are A and B?*

Solution. If A were a knave, then it would be true that at least one of A and B is a knave. That would imply that A spoke truthfully, which is a contradiction. So A must be a knight. Since A's statement must be true, it follows that B is a knave. □

What does this puzzle have to do with coins? Coin puzzles, in which a balance scale is used to ferret out a counterfeit coin among a collection of real ones, are a classic of recreational mathematics. We can say that a fake coin is like a liar and a genuine coin is like a truth-teller. Consider the following coin puzzle.

Puzzle 2. *You are given N coins that look identical, but one of them is fake and is lighter than the other coins. All real coins weigh the same. You have a balance scale that you can use twice to find the fake coin. What is the greatest number N of coins such that there exists a strategy that guarantees finding the fake coin in two weighings?*

This puzzle is a slight variation of the first and easiest coin puzzle, published by E. D. Schell in 1945 [10]. I defer the solution to the next section.

There are amusing parallels between coin puzzles and knight/knave puzzles, which we shall discuss in this chapter. Along the way, we shall consider a variety of interesting brainteasers. Since the solutions to the puzzles will often be necessary for later discussions, they will be presented right after the puzzle statement itself. Readers wanting to solve the puzzles on their own are encouraged to read with a piece of paper in hand, to cover up the solutions.

1 Some Coin Theory

Assume that our balance scale has two pans, and that each weighing involves placing the same number of coins on each pan. The outcome of one weighing is always one of these three types:

- "="—when the pans are balanced,
- "<"—when the left pan is lighter,
- ">"—when the right pan is lighter.

There are nine possible sequences for two outcomes:

$$== \quad =< \quad => \quad <= \quad <= \quad <> \quad >= \quad >< \quad >> \, .$$

Each sequence of outcomes identifies no more than one coin as being the fake. It follows that the number of coins we can sift in two weighings cannot be more than the number of sequences. This shows that for N as in Puzzle 2, we have $N \leq 9$.

In a moment, I shall give a strategy that works for nine coins, which will show that the solution to Puzzle 2 is $N = 9$. Let us first pause to define some useful terminology.

Let us distinguish two types of strategies: *adaptive* strategies, in which each weighing can depend on the results of all previous weighings, and *oblivious* (or nonadaptive) strategies, in which all the weighings must be specified in advance.

In any weighing, let us denote a coin's presence on the left pan by L. A coin's presence on the right pan is denoted by R, and a coin that remains outside the weighing is denoted by O. Every coin's path throughout a sequence of weighings can then be described by a string of Ls, Rs, and Os. Such a string is called the coin's *itinerary*.

In an oblivious strategy, the itinerary of every coin is known in advance. Now suppose two coins have the same itineraries. That means they are together in every weighing. It follows that if one of them is fake, we cannot identify which one of them it is.

Going back to Puzzle 2: Since there are nine possible itineraries, we cannot have an oblivious strategy that finds the fake coin in two weighings if there are more than nine coins. This provides an alternative way of understanding what we found when considering sequences of outcomes.

Itineraries in an adaptive strategy are trickier. Suppose in the first weighing we compare three coins versus three coins and the scale balances. We can find the fake coin in the second weighing if we compare two coins that were not on the scale in the first weighing. That means the three coins that were on the left pan during the first weighing have the same itineraries. The same is true for the three coins that were on the right pan during the first weighing. In this scenario, we found the fake coin, but the itineraries are not unique. However, there is something that *is* unique. The itinerary of the fake coin cannot be shared in this strategy with any other coin, for the same reason as before: The fake coin has to be distinguishable from other coins. Call the itinerary of the fake coin the *self-itinerary*.

Given an adaptive strategy, every coin has a unique self-itinerary that finds this particular coin if it is fake. Distinct coins have distinct self-itineraries, since if two coins have the same self-itineraries, then any adaptive strategy would proceed the same way whether the first or the second coin is fake. This implies that the strategy cannot distinguish which of them is fake.

There is a one-to-one correspondence between outcomes and self-itineraries. To see this, consider, for example, the specific outcome <=, and assume for the moment that it identifies the fake coin. Recalling that the fake coin is lighter than a real coin, the fake must have self-itinerary *LO*. That is, the fake appears first on the left pan, making it lighter, and then not on the scale at all, causing it to balance. This works both ways: given a self-itinerary, we can find the outcome string that points to this coin.

If we are looking for an oblivious strategy to find the fake among nine coins, each coin must be assigned a distinct itinerary. Let us number the coins $1 - 9$, and assign them the itineraries

$$LL, \ LR, \ LO, \ RL, \ RR, \ RO, \ OL, \ OR, \ OO,$$

respectively. Due to luck or magic or mathematics, these set of itineraries work: Each weighing has the same number of coins, three, on each pan.

Now, for the first weighing, place coins 1, 2, and 3 on the left pan and coins 4, 5, and 6 on the right pan. For the second weighing, place coins 1, 4, and 7 on the left pan and coins 2, 5, and 8 on the right pan. Note that coin 1 appears on the left pan both times. Coin 2 appears on the left pan in the first weighing and on the right pan in the second. Coin 3 appears on the left pan in the first weighing and is off the scale in the second. And so on.

Two possible scenarios that can result from this strategy are shown in Figures 6.1 and 6.2. Suppose the scale balances on the first weighing and then leans to the right on the second. (That is, we have the outcome sequence =<.) Then the fake coin must be off the scale in the first weighing and on the left

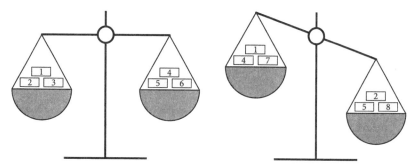

Figure 6.1. A possible sequence of outcomes for the weighings described in the solution to Puzzle 2. The fake coin is off the scale in the first weighing and on the left in the second. It therefore has self-itinerary *OL*, which corresponds to coin 7.

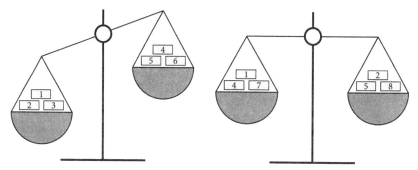

Figure 6.2. Another possible scenario for the sequence of weighings described in the solution to Puzzle 2. This time the fake has self-itinerary *RO*, which corresponds to coin 6.

pan in the second, giving the self-itinerary *OL*. This corresponds to coin 7. Alternatively, we might imagine that we get outcome sequence >=. In this case the fake coin is on the right in the first weighing and off the scale in the second. This gives the self-itinerary *RO*, which corresponds to coin 6.

We have proved that N cannot be more than 9, and we have now found an oblivious strategy that distinguishes among nine coins. Problem solved.

2 Normals and Chameleons

Some of Smullyan's puzzles featured normals, in addition to knights and knaves. Like normal people in life, they sometimes lie and sometimes tell the truth. Here is another puzzle from Raymond Smullyan's book with the wonderful self-referential title [11]:

Puzzle 3. *Consider a married couple, Mr. and Mrs. A. It is known that either both of them are normal, or one of them is a knight and the other a knave. They make the following statements: Mr. A: My wife is not normal. Mrs. A: My husband is not normal. What are Mr. and Mrs. A?*

Solution. Suppose Mr. A is a knave. Then his statement is false, implying that his wife is normal. This contradicts the given information. By similar reasoning, Mrs. A cannot be a knave. It follows that they are both normal. □

I was brought up on Raymond Smullyan's books. I was so happy when I met him for the first time at the Eighth Gathering for Gardner conference in 2008, where he played a kissing trick on me [4]. In 2012, I drove to Raymond's place in the Catskills. In Figure 6.3, we are about to go out to lunch and are discussing who is driving.

Back to normals. The following puzzle will be useful later.

Puzzle 4. *What one statement can a normal make to prove their identity?*

Solution. "I am a liar." A knight who said this would be lying, while a knave who said this would be telling the truth. □

What happens if a normal does not want anyone to know who they are? Then the interrogator is in trouble. The normal person can consistently behave as either a truth-teller or a liar.

What is a coin analog of a normal person? Let us define a *chameleon* coin to be one that can pretend to be either real or fake (lighter than a real coin).

Did you notice that coins and people have different names in these puzzles? It is conceivable to hear someone call genuine coins "truth-tellers" and fake coins "liars." But, really, it would be strange to call chameleon coins "normal coins." There is nothing normal about them.

What is a good question to ask about a chameleon? Suppose there are N coins and one of them is a chameleon, while the others are genuine. Can we find the chameleon? Actually, no. If the chameleon always pretends to be real, then it cannot be found. Likewise, the chameleon can hide among fake coins, never to be found. This is similar to a normal person who can pretend to be either a truth-teller or a liar. Luckily in most logic puzzles, there are several people, and the normal can be ratted out by someone else.

Suppose we have a mixture of real coins and fake coins in addition to one chameleon. We cannot find the chameleon. Can we find the fake coin in the presence of a chameleon? Again, no. If the chameleon always pretends to be fake, then we cannot prove that any particular coin is fake. What do we do? Here is the simplest possible question we can ask.

Question 1. *We have N total coins. All but two of them are genuine and weigh the same, but two of them are not genuine. One is a classic fake that is lighter, and*

Figure 6.3. Raymond Smullyan at his place in the Catskills, in 2012.

one is a chameleon. How can we find two coins that are guaranteed to include the classic fake in the smallest number of weighings?

If the chameleon always pretends to be fake, then the two coins that we find at the end will be the chameleon and the fake. If the chameleon sometimes pretends to be real, we might be able to find the fake coin, which is better than what is needed for this puzzle.

TABLE 6.1.
Empirical results for the chameleon question

Number of weighings	1	2	3	4	5	6	7	8	9	10
ITB	3	4	7	13	22	38	66	115	198	344
Two fake coins	3	4	7	13	22	38	65	113	194	341
Chameleon/fake	**2**	**4**	**6**	**11**	**20**	36	60	108	180	324

The technical details for this problem are formidable and can be found in the papers by Khovanova, Knop, and Polubasov (which also introduced the notion of a chameleon coin) [6], and Knop and Polubasov [8]. Here I will just present a summary.

There is a bound for how fast we can find our two coins: It cannot be faster than finding two fake coins among N coins. This is because the chameleon can pretend to be fake. The starting point for many coin-weighing problems is to count the number of outcomes. If there are w weighings, then there are 3^w possible outcomes. In the problem of two fake coins, each outcome points to two coins. Thus the total number N of coins that can be processed is bounded by $\binom{N}{2} \leq 3^w$. This bound is often called the information-theoretic bound (ITB).

Table 6.1 shows some empirical results. The first row shows the number of weighings, while the second row is the ITB on the number of coins we can process with that number of weighings. The third row shows the best known results for the number of coins we can process when there are two fake coins. The fourth row shows the comparable data when we have one fake and one chamelon. The numbers in bold are exact, derived from an exhaustive computer search. As you can see, the ITB for two fake coins is very close to the best known algorithm. You can also see that the chameleon makes life more difficult.

3 Alternators

Normal people are unpredictable. They can lie or tell the truth at their whim. We can simplify normals by making them more deterministic. *Alternators* switch between telling the truth and lying. Here is an alternator puzzle, whose provenance is not known.

Puzzle 5. *Folks living in Trueton always tell the truth. Those who live in Lieberg always lie. People living in Alterborough alternate strictly between telling the truth and lying. One night 911 receives a call. The caller says, "There's a fire, help!" The operator cannot identify the phone number, so she asks, "Where are you calling from?" The caller replies, "Lieberg." Assuming no one has overnight guests from another town, where should the firemen go?*

Solution. Only an alternator can say they are from Lieberg, since a truth-teller making that statement would be lying, while a liar making that statement would be telling the truth. The solution to the puzzle does not end here. Since the caller is an alternator whose second statement was false, it follows that the first statement was true. So, there really is a fire, and the firemen should go to Alterborough. ☐

Here is another puzzle related to alternators.

Puzzle 6. *How can we determine in two questions whether a person is a knight, a knave, or an alternator?*

Solution. A standard approach is to ask twice whether two plus two is four. ☐

This puzzle shows that, unlike a normal person, a lone alternator can always be found.

What is a coin analog of the alternator? Of course, it is a coin that switches its weight between being real and being lighter after each weighing. Such a coin will also be referred to as an *alternator*. This is the first time the names for coins and people match.

I do not know the history of people alternators. It is likely that in logic, the normals appeared before the alternators. It is not surprising that coin puzzles followed suit: The alternator coins appeared after the chameleons.

Anyway, I run a math club for middle school students at the Massachusetts Institute of Technology called PRIMES STEP. This is a junior program that branched out from another MIT youth program called PRIMES. Surprisingly, "PRIMES" is an acronym: Program for Research in Mathematics, Engineering, and Science for High School Students. A posteriori, it is less surprising that "STEP" is also an abbreviation: SolveTheorizeExploreProve.

In the 2015–2016 academic year, I gave the alternator puzzle as a research problem to my PRIMES STEP students who were in middle school: grades 6–8. Research is not just for people with PhDs! In Figure 6.4, the PRIMES STEP students give a talk about their research.

Unlike the case for the chameleon coin, one alternator coin among real coins can always be found. We can simply use a strategy that finds a fake coin among real coins and then repeat every weighing twice.

However, notice that alternator puzzles, whether with coins or with people, have a problem: The solver does not necessarily know the starting state of the alternator. We have three cases to consider when studying an alternator coin:

1. The starting state is known and lighter.
2. The starting state is known and real.
3. The starting state is unknown.

Figure 6.4. PRIMES STEP students give a talk.

We begin by counting the number of possible outcomes for the first—lighter—case. This number is no longer 3^n, since now not every outcome string is possible. The string cannot have two imbalances in a row, since after an imbalance, the alternator changes weight and pretends to be real. The next weighing has to balance, no matter whether the alternator is on the scale or not.

Finding the number of outcome strings that do not have two imbalances in a row is a combinatorial problem. The answer for n weighings is J_{n+2}, where J_n is the coolest sequence my students never heard of: the Jacobsthal numbers [9]:

$$0, \quad 1, \quad 1, \quad 3, \quad 5, \quad 11, \quad 21, \quad 43, \quad 85, \quad 171, \quad \ldots .$$

The sequence is defined by two initial terms and a recursion:

$$J_0 = 0, \quad J_1 = 1,$$
$$J_n = J_{n-1} + 2J_{n-2}, \quad \text{for } n > 1.$$

Let us see how this recursion arises when we count outcomes. If the first of the n outcomes is a balance, then the rest could be any string without two imbalances in a row. The number of such imbalances is given by J_{n-1}. If the first outcome is an imbalance, the next weighing must balance, and we can then have any outcome without two imbalances in a row after that. Since there are two types of imbalances, there are $2J_{n-2}$ possibilities in this scenario.

If the alternator starts in the real state, the first weighing cannot be an imbalance. So the count of outcomes of n weighings is J_{n+1}.

The cool part is that there is an adaptive strategy that processes this many coins. The theoretically possible number of outcomes is the precise bound.

What about when the starting state of the alternator is unknown?

The number of possible outcomes of n weighings in an unknown state is J_{n+2}: the same as for the lighter state. Does this mean that we can process J_{n+2} coins in n weighing when the alternator starts in the unknown state? This cannot be true. The unknown state cannot be easier than the real state. What are we missing here? When the alternator coin starts in the unknown state, if it is ever on the scale, we will not only find the coin, but after we find it, we will also know the initial starting state. Roughly speaking, each coin matches two different outcomes. I say "roughly," because it might be possible that the coin is never on the scale, and we find it by proving that all the other coins are real. This way we do not know the starting state of the alternator. The good news is that there can only be one such coin. That means we cannot process more than $(J_{n+2} + 1)/2$ coins in the unknown state.

However, the number of coins in the unknown state that we can process cannot be more than the number of coins in the real state, that is, J_{n+1}. It follows that we cannot process more unknown coins than the minimum of these two numbers: $(J_{n+2} + 1)/2$ or J_{n+1}. The cool property of the Jacobsthal numbers is

$$J_{n+1} = 2J_n + (-1)^n,$$

that is, the next Jacobsthal number is almost twice the previous number. In any case, the minimum is J_{n+1}, and this is how many unknown coins we can process in an adaptive strategy. You can find more details in our paper [1].

4 Oblivious Strategies For Finding Alternators

What about an oblivious strategy for the alternator? The STEP students did not find it. Is it possible to do? This is a difficult question, so I shall just make a few brief remarks here.

One problem is that in this scenario, outcomes do not uniquely define self-itineraries. For example, consider the outcome <==<. The alternator has to be on the scale three times: in the first and the last weighing and then once in between to change its state. Therefore, there are four possible itineraries corresponding to this sequence of outcomes:

LLOL   LROL   LOLL   LORL.

Which one should we choose?

We can try to choose itineraries in a systematic way. For example, suppose we choose *LLOL* as the itinerary corresponding to this sequence of outcomes.

We can decide that after the coin changes from light to real, we put it on the same scale again in the next weighing. Using this technique, one can develop oblivious strategies for alternators in known starting states by following the general approach we took in our solution to Puzzle 2.

However, finding oblivious strategies when the alternator begins in an unknown state is more complicated.

To see the difficulty, consider the outcome $\neq=\neq=\neq$, where $\neq$ indicates an imbalance. The coin pointed to by this outcome must start in the light state and be on the scale for every weighing. The same coin starting in the real state must have outcome: $=\neq=\neq=$. That means the eight outcomes of type $\neq=\neq=\neq$ (note that each $\neq$ represents two possibilities, depending on whether the scale leaned to the left or to the right) should correspond to the same coins as the four outcomes of type $=\neq=\neq=$. It follows that we have outcomes that cannot be matched to coins. Therefore, an oblivious strategy for the unknown starting state that is as good as an adaptive strategy does not exist.

What, then, is the best oblivious strategy? This is still an open problem. See the paper by Khovanova and Knop [5] for more details.

How do adaptive and oblivious strategies transfer to logic? For an oblivious strategy, you prepare your questions in advance. Let us try the following puzzle.

Puzzle 7. *You are on an island where knights, knaves, and alternators live. You meet two people. You want to figure out who they are in the smallest number of questions. The only type of question you can ask to one of them is: Is the other person a knight? You can replace "knight" in this question with either "knave" or "alternator." What oblivious strategy will work?*

Solution. Since there are two people, each of whom can be any of three types, there are nine cases in total. And since a binary question can only divide a sample space in half, we will need at least four questions. Here is one way to do it: We start by twice asking the first person whether the second is an alternator. Then we twice ask the second person whether the first one is a knight. If one of them changes their answer, we know that this person is an alternator, and we can then figure out the identity of the other person. If neither changes their answer, then neither is an alternator. The first two answers would then tell us the identity of the first person. Then we can deduce the identity of the second person from the last two answers. □

5 Heavier versus Lighter

Traditionally, fake coins are lighter. The reason is historical: Cheaper metals were lighter.

TABLE 6.2.
A strategy that works for Puzzle 8

First weighing	1, 2, 3, 4 versus 5, 6, 7, 8
Second weighing	1, 8, 9, 11 versus 3, 4, 5, 10
Third weighing	3, 6, 10, 12 versus 1, 4, 6, 11

We mathematicians like symmetry. We can imagine a fake coin that is heavier than the real one. Finding such a coin is not an interesting problem. We can use the same strategy as for a lighter fake coin, we just need to interpret the outcomes in the opposite way: $<$ means the fake heavier coin is on the right pan. As a result, most math problems about counterfeit coins assume that the fake coin is lighter.

However, there are puzzles where you need to find a fake coin that is heavier or lighter. The most famous is the twelve-coin problem [2] first introduced in 1946.

Puzzle 8. *There are twelve identical-looking coins. One of them is fake: It is either lighter or heavier than a real coin. Find the fake coin, and determine whether it is lighter or heavier by using a balance scale three times.*

Solution. Assume that the coins have been numbered 1–12. The strategy shown in Table 6.2 will get the job done. This strategy corresponds to itineraries:

$$LLR \quad LOO \quad LRL \quad LRR \quad RRO \quad ROR$$
$$ROL \quad RLO \quad OLL \quad ORO \quad OLR \quad OOL$$

for coins 1 through 12, with coins 1–6 in the top row, and coins 6–12 in the bottom row.

To see how this works, let us work through an example: The outcome *LLR* indicates that the fake coin was on the scale for all three weighings. Moreover, it was on the same pan in the first two weighings and then moved to the other pan for the third weighing. Only coin 1 answers to this description. And since this coin was always in the heavier pan, we conclude that the fake is heavier. □

The important thing to notice is that there are no two coins that are opposite each other each time they are on the scale. If there were two coins like that and one of them was fake, we would not be able to distinguish between them. The strategy will lead to the conclusion either that one of them is fake and lighter or that the other one is fake and heavier.

Also note that if we only need to find the fake and do not care whether it is heavier or lighter, then we could process a thirteenth coin as well. We assign this coin the itinerary *OOO*. If the scale balances all three times, then the thirteenth coin is the fake. We can also expand this problem to more weighings. If we need to both find and identify the fake coin, then we can process $(3^w - 3)/2$ coins in w weighings. If we just need to find the fake, then we can process one additional coin.

Returning to logic puzzles, what would be the analogs of lighter versus heavier fake coins? Both types of coins are "liars," after all.

It might look like logic is binary: truth versus lie, while, in contrast, coin weighing is ternary. If you think about it, then in real life, lies can take many forms. For example, a liar can say that two plus two is three, or that two plus two is seventeen.

So, let us introduce two different types of knaves. We define an *exaggerating liar* to be one who increases any number he mentions by 1, and a *diminishing liar* to be one who decreases any number by 1. Here is a sample puzzle:

Puzzle 9. *On an island there are three types of people: exaggerating and diminishing liars, and truth-tellers. You meet three islanders A, B, and C. They say the following:*

A: *There are three liars among us.*
B: *There is one liar among us.*

Who is who?

Solution. A's statement must be a lie. Since there are only three people, A can only be an exaggerating liar. Hence, there are actually two liars in the group. It follows that B must be a diminishing liar and C is a truth-teller. □

Consider another coin problem from the book *Mathematical Circles: Russian Experience* [3]:

Puzzle 10. *There are 101 coins, and only one of them differs from the other (genuine) ones by weight. We have to determine whether this counterfeit coin is heavier or lighter than a genuine coin. How can we do this using two weighings?*

Solution. First, let us try this for other numbers N of coins that are easier to think about than 101. If N is divisible by 3, then we divide the coins into three piles, each with the same number of coins. We compare the first pile to the second in the first weighing, and the first pile to the third in the second weighing. If both weighings are unbalanced, then the fake coin is in the first pile, and we know whether it is heavier or lighter. If one of the weighings balances, then the balancing piles do not contain the fake coin. Again, we know the answer.

If N is divisible by four, we divide the coins into four equal piles. We compare the first and the second pile against the third and the fourth. The weighing has to be unbalanced. We then compare the first pile against the second. If the second weighing balances, then the coin must be in the third or fourth pile, and we know whether it is heavier of lighter. If the second weighing is unbalanced, then the coin must be in the first or second piles, and again we know whether it is heavier or lighter.

For an arbitrary number of coins, we can try to merge the two solutions together. We divide all coins into three piles of size a, a, and b, where $a \leq b \leq 2a$. (For example, for 101 coins, we might use two piles of 30 coins, and then one pile of 41 coins.) In the first weighing, we compare the first two piles. If they balance, then the fake coin must be among the remaining b coins. Now pick any b coins from among those involved in the first weighing, and compare them to the remaining b coins. If instead the first weighing is unbalanced, then the remaining coins have to be real. For the second weighing we pick a coins from the remaining pile and compare them to one of the pans from the first weighing. □

Back to logic. To continue an analogy, I have to invent a puzzle in which we do not need to find liars but only need to determine whether the liars exaggerate or diminish.

Puzzle 11. *On this island of knights and knaves, the knaves are all of the same type: either exaggerating or diminishing. You meet five islanders. A says, "There are three truth-tellers among us." B says, "There are three liars among us." What type of liars live on this island?*

Solution. The total number of people by their account is six. So one of A or B is a truth-teller, and the other is an exaggerating liar. □

Going back to alternator coins: The alternator that switches between real and heavier is not interesting to study, as by symmetry, it is the same as the alternator that switches between real and lighter. But we can have an alternator that switches between lighter and heavier. We discuss this case in the next section.

6 Lighter-Heavier Coin

Let us consider a fake coin that changes its weight from heavier than real to lighter than real and back. It never weighs the same as the real coin. Call it an *oscillating fake coin*. A logic analog would be a liar who switches between exaggerating and diminishing. We call such a liar an *oscillating liar*.

Puzzle 12. *On this island, there are two types of people: knights and oscillating liars. You meet two islanders A and B. A says, "There is one truth-teller among us." B says, "There is one truth-teller among us."*
Who are A and B?

Solution. They cannot both be truth-tellers. Therefore, at least one of them is an oscillating liar. And as they both made the same statement, the other one is also a liar. □

In the next puzzle, we haven an osculating fake coin.

Puzzle 13. *There are twelve identical-looking coins. One of them is an oscillating fake. Find the fake coin, and identify whether it starts in the light or heavy state by using a balance scale three times.*

Solution. Why are there twelve coins? Does it remind you of anything? This puzzle can be solved with a wonderful argument that reduces its solution to the solution of Puzzle 8. Suppose we have a strategy, and each time we have an imbalance, we mentally reverse the direction of the imbalance. That means the coin does not change state. We just need to find the fake coin, which we do not know whether it is heavier or lighter. We can use exactly the same strategy as before. The argument works for any number of coins. □

7 A Lying Scale

We have been discussing how coins are like people, in that they can either lie or tell the truth. However, in coin-weighing problems, we have another player: the scale. The scale can also lie.

Since scales are not binary, we have to be specific when defining a lying scale. Here is the simplest definition: A lying scale produces the opposite output of the truth-telling scale when there is an imbalance. It produces a balance when the coins on both pans weigh the same, but when there is an imbalance, the scale tilts the wrong way. This is called a *reverse* scale. A scale that shows the correct results will be called a *true* scale.

Of course, we can quickly test a scale by placing one coin on one pan while leaving the other pan empty. To generate interesting puzzles, we need to assume that we are required to place the same number of coins on each pan.

Puzzle 14. *We have N > 2 coins, but one is fake and lighter than the other coins. We have one scale that is either true or reverse, but we do not know which. How many weighings do we need to decide whether the scale is true, if we are only allowed to put the same number of coins on the pans in any given weighing?*

Solution. There is a parallel between not knowing whether the fake coin is lighter or heavier and changing the scale from true to reverse. The algorithm is the same as in Puzzle 10, where we were determining whether one fake coin is heavier or lighter. That means two weighings are enough, assuming we have more than a total of two coins. □

We can also have a different definition of lying. For example, we can invent a scale that always produces a wrong result, any wrong result. Call it a *lying* scale.

Puzzle 15. *We have $N > 5$ coins, one of which is fake and lighter. We also have a scale that could be true or lying. Given that we only allow the same number of coins on both pans during a weighing, differentiate between a true and lying scale in three weighings.*

Solution. We compare coins 1 versus 2 in the first weighing, 3 versus 4 in the second weighing, and 5 versus 6 in the third weighing. If the scale is true, we will have at least two balances. If the scale is lying, we will have at least two imbalances. □

8 A Broken Scale

Let us consider a scale that behaves like a normal person: Sometimes the scale lies, and sometimes it tells the truth. But we should not call this scale a normal scale, as it is not normal at all. Call it a *broken* scale. A broken scale is a more realistic object than a coin that changes weight. Not surprisingly, the following puzzle appeared before chameleons were invented. I do not know the origins of this puzzle, but Knop wrote a very good paper on this subject in Russian [7].

Puzzle 16. *We have 3^{2n} identical-looking coins: One is fake and lighter. There are three balance scales, one of which is a broken scale. What is the smallest number of weighings that guarantee finding the fake coin?*

One idea for solving this puzzle comes from the theory of error-correcting codes. We can use an algorithm to find the fake coin on one scale and repeat each weighing on all three scales. If one of the scales disagrees with the other two, then that scale is the broken one. We can then speed up the rest of the process by continuing our search for the fake coin on one of the true scales. In the worst case, all the scales agree all the time, and we need $6n$ weighings to find the fake coin. In this case, we do not know which scale is broken. This algorithm finds the fake coin, but it is far from the optimal solution. Call this algorithm the *majority vote* strategy.

Let us try to find a more efficient approach. We do not know which scale is broken. That means we cannot be certain that every weighing is truthful. However, we can try to set up a system that catches a lie as quickly as possible. As soon as the broken scale is identified, we can use another scale to find the fake coin, as we already know how to do this. In all solutions I have seen, the worst case is when the broken scale gives a correct answer until the very end and then lies.

We want to find a bound on the number of weighings. We showed that we cannot guarantee that the broken scale never lies. Suppose the broken scale lies not more than once. As before, we can draw a parallel between this problem and another problem that is much easier: Find the fake coin, given one scale that is allowed to lie not more than once. Now we are all set to produce a bound. Suppose there is a strategy in this case that requires w weighings. Then the number of possible outcomes is 3^w. After we have found the fake coin, we can look back at all the weighings, and we will know which weighing lies. Any one of w weighings can lie, and there are two possibilities for what the lying weighing can show. It is also possible that there are no lies. That means $2w + 1$ different outcomes can point to the same coin. It follows that the number of coins we can process is not more than $3^w/(2w + 1)$.

The paper by Knop [7] finds a strategy for identifying the fake coin in Puzzle 16 in $3n + 1$ weighings. It might be possible to do better.

Let us consider the simplest example, when $n = 1$. This corresponds to nine coins. We know that on a truthful scale, we can process nine coins in two weighings. Therefore, the majority vote strategy allows us to find the fake coin in six weighings.

Our bound tells us that we need at least four weighings. Here is a strategy for four weighings: In the first weighing, we compare coins 123 versus 456 on the first scale. In the second weighing, we compare coins 147 versus 258 on the second scale. The first scale points to three coins a, b, and c as potentially being fake. The second scale also points to three coins: a, d, and e. The two lists of suspicious coins always overlap by one coin, namely, a. In any case, the fake coin has to be one of the five coins: a, b, c, d, e. For the third weighing, we use the third scale and compare coins b, c versus d, e. If the weighing balances, then any two of the three scales point to coin a as being fake. That means a is the fake coin. Suppose the third weighing is not balanced. Without loss of generality, suppose b and c are lighter than d and e. That means the third and the second scales contradict each other: one of them has to be lying. It follows that the first scale is a true scale. We can use it for the last weighing to find the fake coin among a, b, and c.

Now that we have lying scales and lying coins, is there a parallel in logic that combines both? We can say that scales are people and coins are facts that they know. A fake coin represents a wrong fact. Scales describe their opinions about facts that might be wrong. Ideas like this have already been introduced in logic, as discussed in the next section.

9 Sane and Insane People

Raymond Smullyan wrote a lot of books with logic puzzles. In his book *The Lady or the Tiger?* [12], he introduced *sane* and *insane* people.

Sane people believe that $2 + 2 = 4$, and insane people believe that $2 + 2 \neq 4$. In other words, sane people believe true statements, and insane ones believe false statements.

A chapter of Smullyan's book is devoted to an asylum, in which each member is either a doctor or a patient. All the doctors are supposed to be sane, and all the patients are insane. Inspector Craig is sent to inspect the asylum.

Puzzle 17. *What statement can a sane patient make to prove that they should be released?*

Solution. I am a patient. □

You might assume that insane people are similar to liars. Here is another puzzle.

Puzzle 18. *A member of the asylum says, "I believe I am a doctor." Can we conclude who they are?*

Solution. It might seem that we cannot. Indeed, both insane patients and sane doctors would say, "I am a doctor." But this statement is different. An insane patient believes he is a doctor. That means, he believes that he does not believe he is a doctor. That means, an insane patient would say, "I believe I am a patient." □

Now if we allow the members of the asylum to lie, we get the following. A sane liar and an insane truth-teller will both say that two plus two is not four. Compare this to the fact that a true scale produces an imbalance in the presence of a fake coin and a lying scale produces an imbalance when all the coins are real.

10 A Final Puzzle

Let me close with an especially amusing puzzle.

People are more flexible than scales in the sense that they are able to talk about a lot of things, while scales only compare weights. That means there is a greater variety of logic puzzles than coin-weighing puzzles. However, weighings are more structured and are easier to study.

In many logic puzzles, people talk about each other. But scales cannot talk about other scales. Or can they?

Puzzle 19. *There are three identical-looking balance scales that could be true or reversed. There are also $N > 2$ coins, and one of them is fake and lighter. What is the smallest number of weighings that you need to figure out the type of each scale? You can assume that the scales themselves all have the same weight, and that in* *this puzzle, we are not required to place the same number of coins on each pan in any weighing.*

Solution. This can actually be done in one weighing! Place one coin on one of the pans of scale one, place two coins on one of the pans of scale two, and then place scales one and two on opposite pans of scale three. This way each weighing is an imbalance, and you know how a true scale should interpret it. □

References

[1] B. Chen, E. Erives, L. Fan, M. Gerovitch, J. Hsu, T. Khovanova, N. Malur, A. Padaki, N. Polina, W. Sun, J. Tan, and A. The. Alternator coins. *Math Horizons* **25** no. 1 (2017) 22–25.

[2] F. J. Dyson. Note 1931—The problem of the pennies. *Math. Gaz.* **30** (1946) 231–234.

[3] D. Fomin, S. Genkin, and I. Inteberg. *Mathematical Circles: Russian Experience.* American Mathematical Society, Providence, RI, 1996.

[4] T. Khovanova. Raymond Smullyan's magic trick. Tanya Khovanova's Math Blog. Online at http://blog.tanyakhovanova.com/2010/05/raymond-smullyans-magic-trick. Last accessed June 2018.

[5] T. Khovanova and K. Knop. Coins that change their weights. arXiv:1611.09201 [math.CO], 2016.

[6] T. Khovanova, K. Knop, and O. Polubasov. Chameleon coins. arXiv:1512.07338 [math.HO], 2015.

[7] K. Knop. Weighings on "sly scales" [in Russian]. *Matematika v shkole* **2** (2009) 74–76.

[8] K. Knop and O. Polubasov. Two counterfeit coins revisited [in Russian]. Online at http://knop.website/math/ff.pdf. Last accessed June 2018.

[9] *The On-Line Encyclopedia of Integer Sequences.* Sequence A001045. Online at https://oeis.org. Last accesed June 2018.

[10] E. D. Schell. Problem E651—Weighed and found wanting. *Amer. Math. Monthly* **52** (1945) 42.

[11] R. Smullyan. *What Is the Name of This Book?* Prentice-Hall, Englewood Cliffs, NJ, 1978.

[12] R. Smullyan. *The Lady or the Tiger?* Knopf, New York, 1982.

PART II

✧✧✧✧✧✧✧

Games

7

◇◇◇◇◇◇◇◇◇◇◇◇◇◇◇◇◇◇◇◇◇◇◇◇◇◇◇◇◇◇◇

BINGO PARADOXES

Barry A. Balof, Arthur T. Benjamin, Jay Cordes,
and Joseph Kisenwether

Imagine playing a game of Bingo on a standard 5 × 5 card. When you complete your five-in-a-row, it is equally likely that your row is horizontal or vertical, with those probabilities each being greater than your row being diagonal. But now imagine that you walk into a large Bingo hall with 1000 cards in play, when suddenly someone shouts "Bingo!" indicating that they have 5 in a row (either horizontal or vertical or diagonal). Believe it or not, the horizontal and vertical bingos are no longer equally likely for the winning card. The winning cards are much more likely to include horizontal winners than vertical winners. Moreover, when the free space is used, the most likely situation for the winning card or cards is that they are all diagonal.

1 History of the Problem

This paradox was first observed when Joe Kisenwether was working for the gaming industry in Nevada. He had written software to play Bingo on a large scale, generating cards and choosing numbers randomly. It was brought to his attention that, in larger-scale games, horizontal wins were happening more often than vertical ones. Trusting in the randomness of the software, we looked to find a combinatorial explanation as to why this might have been the case.

Recall that traditional 5 × 5 Bingo cards have the following structure: There are five columns, labeled B, I, N, G, and O. Column B has five numbers chosen from 1 through 15. The numbers in columns I, N, G, and O come from the sets 16 through 30, 31 through 45, 46 through 60, and 61 through 75, respectively. The numbers can appear in any order within a column. Typically, column N has a free space in the third position. In a recent paper by Benjamin, Kisenwether, and Weiss [1], it was shown that when the free space was not used, and when all possible cards are used, then horizontal and diagonal winners would always happen at the same time, and the chance of a horizontal bingo was about three times more likely than that of a vertical bingo. A similar conclusion was reached by Dick Hess in his book *The Population Explosion and Other Mathematical Puzzles* [2] when the free space is utilized. We will consider both versions of the game (with and without the free space), explore the calculations in greater detail, and answer new questions.

2 Galactic Bingo

To get a better understanding of the properties of the winning card when a large number of cards are in play, we explore the ultimate expansion of the game. Instead of playing in a room with 1000 cards, we first consider the "Galactic" version of the game, where every possible card in the universe is in play. We will first look at the game when played without the free space. By counting the ways to pick numbers for each column, there are $(15 \times 14 \times 13 \times 12 \times 11)^5 = 6.077 \times 10^{27}$ possible Bingo cards, all of which are being used. Now, as soon as the caller has drawn at least one number from each of columns B, I, N, G, O, the game will end with horizontal (and diagonal) winners. However, as soon as five numbers have been chosen from any single column, the game will end with vertical winners. We note that in the Galactic version, it is impossible to have horizontal and vertical winners at the same time, since if the game ended with the number 75, it cannot simultaneously be the first time and the fifth time that a number from column O was called. Thus, in Galactic Bingo, the probability of a horizontal win is the probability that all five columns are called before any column is called five times.

We note first that the game will never last more than seventeen draws. In a worst-case scenario, one can draw four numbers from each of four columns without the possibility of a Bingo. We will focus first on the horizontal case. Suppose that we have ended the game on the $(k+1)$st draw. This means that in the first k balls drawn (with $4 \leq k \leq 16$), one column must have no numbers chosen, and the other columns each have between one and four numbers chosen. By noting how many of the first k balls come from each of the columns, we can associate a partition of the integer k. For example, if the winning ball came on the tenth draw and was from column G, and prior to that, the number of balls from columns B, I, G, and O were, respectively, 3, 1, 4, and 1, then we would associate the partition $\pi = (4, 3, 1, 1)$. To determine the probability of a horizontal win, we will examine all partitions of numbers k between 4 and 16 that have exactly 4 parts, with no part greater than 4. We say that $\pi = (\pi_1 \pi_2 \pi_3 \pi_4)$ is a *horizontal partition* if it satisfies $4 \geq \pi_1 \geq \pi_2 \geq \pi_3 \geq \pi_4 \geq 1$ and $\pi_1 + \pi_2 + \pi_3 + \pi_4 = k$.

Now we count the number of arrangements of the 75 Bingo balls that lead to each possible horizontal partition $\pi = (\pi_1, \pi_2, \pi_3, \pi_4)$. If π is a partition of the integer k, then the winning ball is chosen on the $(k+1)$st draw and can be chosen in 75 ways. Let us suppose that the winning ball was O75. For the partition $\pi = (4, 3, 1, 1)$ there are 12 ways to decide, among columns B, I, N, and G, which column has 4 numbers, which column has 3 numbers, and which columns have 1 number. Here, we say $m_\pi = 12$. Likewise $m_\pi = 12$ for $\pi = (4, 2, 2, 1)$ or $(3, 3, 2, 1)$, but $m_\pi = 4$ for $\pi = (3, 2, 2, 2)$. In general, m_π equals 1, 4, 6, 12, or 24, depending on whether our partition has 4 of a kind (like $(2, 2, 2, 2)$) or 3 of a kind (like $(2, 2, 2, 1)$)) or 2 pair (like $(3, 3, 2, 2)$) or one pair (like $(4, 2, 2, 1)$), or no pairs (like $(4, 3, 2, 1)$), respectively. Then we choose the requisite number of balls in each column from the 15 available. This can be done in

$$\binom{15}{\pi_1}\binom{15}{\pi_2}\binom{15}{\pi_3}\binom{15}{\pi_4}$$

ways. Once those k balls have been chosen, they can be arranged in $k!$ ways, and the balls that occur after O75 can be arranged in $(74 - k)!$ ways. Thus, since the balls can be arranged in 75! equally likely ways, the probability of achieving a horizontal bingo with partition type π is

$$P_\pi = \frac{75 m_\pi \binom{15}{\pi_1}\binom{15}{\pi_2}\binom{15}{\pi_3}\binom{15}{\pi_4} k! (74 - k)!}{75!}.$$

We can make a similar partition argument for a vertical bingo. Again, assume that the winning draw occurs in the $(k + 1)$st position. In this case, we want to designate one ball as the 'winning' ball, and to choose 4 other balls from that column. There may be as few as 0 or as many as 12 draws from the remaining columns. We cannot have any more than 5 balls from any one column, and we cannot have a ball in each of the four columns (if we did, then a ball from the winning column would have already given us a horizontal bingo). To that end, let $\pi = (\pi_1, \pi_2, \pi_3)$ be a partition of $k - 4$, with $0 \le k \le 16$ and $4 \ge \pi_1 \ge \pi_2 \ge \pi_3 \ge 0$, which we shall call a *vertical partition*. Reasoning as we did in the horizontal case, the probability of achieving a vertical Bingo with partition type π is

$$P_\pi = \frac{75 \binom{14}{4} m_\pi \binom{15}{\pi_1}\binom{15}{\pi_2}\binom{15}{\pi_3} k! (74 - k)!}{75!}.$$

To find the total probability of the Galactic game resulting in a horizontal or vertical Bingo, we need to add our probabilities over all partitions that satisfy the horizontal or vertical conditions. The problem of adding over all permutations is computationally long, but not insurmountable. There are exactly 35 horizontal and 35 vertical partitions. (The fact that these numbers are equal is not a coincidence, as shown by Benjamin, Kisenwether, and Weiss [1].) We use SAGE code to list the appropriate horizontal and vertical conditions and to compute P_π over all such partitions, The sums result in a probability of

75.178% of the galactic games resulting in a horizontal bingo, and 24.822% of the galactic games resulting in a vertical bingo. As a check, we note that the independently computed partition probabilities add to 100%.

2.1 Extensions of the Galactic Game

As an extension of the intuition behind the galactic game, it makes some probabilistic sense that we are more likely to draw one ball each from B, I, N, G, and O before drawing five of any one type of ball. Since the balls are being drawn without replacement, the likelihood of drawing, say, a second, third, fourth, and fifth B ball grows gradually smaller and smaller.

Our SAGE code and probability formulas allow for easy modifications of the game to more balls. Suppose that our game is still played on a 5×5 card, but now the numbers in the first column will range from 1 to t, the second column from $t+1$ to $2t$, and so on, with the last column having numbers from $4t+1$ to $5t$. The same partition analysis that we did on the Galactic game is valid, but now our formula for the horizontal and vertical probabilities look like

$$P_\pi(H) = \frac{5tm_\pi\binom{t}{\pi_1}\binom{t}{\pi_2}\binom{t}{\pi_3}\binom{t}{\pi_4}k!(5t-1-k)!}{(5t)!},$$

$$P_\pi(V) = \frac{5t\binom{t-1}{4}m_\pi\binom{t}{\pi_1}\binom{t}{\pi_2}\binom{t}{\pi_3}k!(5t-1-k)!}{(5t)!}.$$

We also looked at a case where there were "infinitely many" balls available per column. In this case, the game amounts to drawing balls labeled B, I, N, G and O until we get either one of each type or five of one type. On each draw, the probability of getting a ball from any given column is always $1/5$. (This extension was inspired by a problem of Velleman and Wagon [3].) In this case, our sum will still fall over the set of all horizontal partitions and vertical partitions, but it will only involve the arrangements of the types of balls drawn before the winning ball.

Again we are drawing the winning ball on the $(k+1)$st turn. Our probabilities in this case are calculated using multinomial coefficients as follows:

$$P_\pi(H) = \frac{m_\pi\binom{k}{\pi_1,\pi_2,\pi_3,\pi_4}}{5^k},$$

$$P_\pi(V) = \frac{m_\pi\binom{k}{4,\pi_1,\pi_2,\pi_3}}{5^k}.$$

We list the probabilities in Table 7.1, followed by a few remarks.

TABLE 7.1.
Galactic probabilities with t balls per column (no free space)

t	$P(H)$	$P(V)$
∞	0.668205043	0.331794957
25	0.717237894	0.282762106
15	0.751778611	0.248221389
12	0.774069504	0.225930496
5	0.938066921	0.061933079

Note that, as intuitively expected, the probabilities for a horizontal bingo go up as the number of balls in each column get fewer and fewer, with the extreme case being only five balls in each column. Note also that, even with "infinitely" many balls available in each column, the odds are still 2:1 in favor of a horizontal bingo happening before a vertical bingo in the Galactic game.

2.2 What About the Free Space?

Our computations and analysis thus far have been on Bingo cards with 25 numbers on them, 5 in each column. But what if each card has a free space in the center (as is usually the case with Bingo cards)? The game could now end in as few as four or as many as sixteen draws. To make our theoretical calculations for the Galactic game, we will still look at horizontal and vertical partitions, but we modify our definitions slightly. In these cases, we assume that the game is won on the kth draw. For a horizontal win to occur on someone's card, we need to draw one each of B, I, G, and O. Moreover, at least one of these must contribute only 1 ball (as the winning ball) to the k drawn balls. This will happen in at least four and at most thirteen draws from the columns B, I, G, and O.

Suppose that we draw n balls from the N column (with $0 \leq n \leq 3$) among our first k. The partition of B, I, G, and O contributions must form a partition $\pi = (\pi_1, \pi_2, \pi_3, \pi_4)$ of $k - n$, where $4 \geq \pi_1 \geq \pi_2 \geq \pi_3 \geq \pi_4 = 1$. Again, we use m_π to denote the type multiplier of partition π, and let w_π denote the number of parts of size 1 (that is, the number of columns that could have contributed the winning ball. Our horizontal probability is then given by

$$P_\pi = \frac{m_\pi w_\pi \binom{15}{n}\binom{15}{\pi_1}\binom{15}{\pi_2}\binom{15}{\pi_3}\binom{15}{\pi_4}(k-1)!(75-k)!}{75!}.$$

We sum these values over all horizontal partitions π (there are 20 of them), and then iterate that sum for $n = 0, 1, 2, 3$. Again, we turned to SAGE to work through the partitions and the sum, and these give a probability of getting a horizontal Bingo of 73.7342%

TABLE 7.2.
Galactic probabilities with t balls per column (with free space)

t	$P(H)$	$P(V)$
25	0.708786173	0.291213827
15	0.73734237	0.26265763
12	0.755786492	0.225930496
5	0.938066921	0.061933079

With the free space in play, we will get a vertical bingo as soon as we draw *either* 5 of B, I, G or O, *or* 4 Ns. Thus, we have two types of vertical partitions. For those partitions where $n \leq 3$, we need one part of size 5 (the winning column), and at most two other parts of size at most 4 (otherwise, we would have either a vertical bingo in another column or a horizontal bingo). We have 60 choices for the winning ball, which we assume occurs on the kth draw, with n balls from the N column, and let $\pi = (\pi_1, \pi_2)$, where $4 \geq \pi_1 \geq \pi_2 \geq \pi_3 = 0$ be a partition of $k - n$ (there are 15 such partitions). Then, the probability of a vertical bingo in a column other than N is given by

$$P_\pi = \frac{60 m_\pi \binom{15}{n}\binom{14}{4}\binom{15}{\pi_1}\binom{15}{\pi_2}(k-1)!(75-k)!}{(75)!}.$$

If our vertical bingo occurs in column N, and there are 15 possibilities for the winning ball, then we can draw up to 4 balls from up to three of the remaining columns to go with our 4 balls from the N column. Here, a vertical partition will be of the form $\pi = (\pi_1, \pi_2, \pi_3, \pi_4)$, where $\pi_1 + \pi_2 + \pi_3 + \pi_4 = k - 4$, and $4 \geq \pi_1 \geq \pi_2 \geq \pi_3 \geq \pi_4 = 0$. Our probability here of a vertical bingo with partition π (of which there are 35) is

$$P_\pi = \frac{15 m_\pi \binom{14}{3}\binom{15}{\pi_1}\binom{15}{\pi_2}\binom{15}{\pi_3}(k-1)!(75-k)!}{75!}.$$

Adding probabilities over both types of vertical partition gives a probability of 26.2658% of a vertical bingo being the winner in the galactic game.

Again, with the use of SAGE, we were able to generalize our results to more or fewer balls per column. We list the probabilities in Table 7.2, with remarks again to follow.

The row corresponding to the probabilities when $t = 5$ is a game played with only 24 balls and with the free space, owing to there being only 4 numbers available for the N column (we wanted the galactic game, in this case, to represent everyone having the same numbers in some permutation on their card). Note that the probabilities on the table overall, while still overwhelmingly in favor of a horizontal Bingo, are a bit more balanced between the two. This makes intuitive sense, as there are slightly higher odds of a vertical Bingo in the

N column (needing only 4 balls). Also of note is that the probabilities are exactly the same for the 5-ball case with or without the free space. This makes sense: Since everyone has the same set of numbers on their card, we could have called any number "first" and still had the same probabilities of a horizontal versus a vertical winner.

2.3 Is My Card a Winner?

Suppose that we are in the Galactic Bingo hall, and that a winner has been called. What is the probability that our card has a Bingo as well? And, perhaps more interestingly, is it more likely that our card has a Bingo if the called winner was horizontal or if it was vertical? Again, we can calculate over horizontal and vertical partitions.

Our probability will take the form of a weighted sum over the horizontal partitions. For each horizontal partition, we use our previously calculated probability that the partition occurs. We then multiply by the probability that our card has numbers already called in the appropriate row. As before, we fix the "winning ball," which has a 5/15 probability of being on our card. The chance of having already called numbers in the same row as the winner will be the number of balls called in each column divided by 15. In the notation of our earlier discussion, let $\pi = (\pi_1, \pi_2, \pi_3, \pi_4)$ be a horizontal partition with associated probability P_π. The probability of having a horizontal winner with partition π is then

$$W_\pi = \frac{5\pi_1\pi_2\pi_3\pi_4 P_\pi}{15^5}.$$

Using SAGE to help us sum over all horizontal partitions, we find that the chance of our card having a horizontal Bingo in the Galactic game is 0.009064%

Our vertical probability will, in practice, be much easier to calculate. If a vertical Bingo has been called, we will have that Bingo if our five numbers in the winning column match the five numbers called from that column, or a probability of $1/\binom{15}{5}$ or 1/3003. So, in this case,

$$W_\pi = \frac{P_\pi}{3003}.$$

Summing over all vertical partitions, we find that the chance of our card having a vertical Bingo in the Galactic game to be 0.008266%.

As before, our SAGE code is generalizable to determine these probabilities when there are more or fewer balls in each column. Table 7.3 gives the probabilities that an individual card is a winner in the Galactic case with 25, 15, 12, and 5 balls in each column. We use $P_W(H)$ and $P_W(V)$ to denote the probabilities that an individual card has a horizontal or a vertical winner, respectively. (Note

TABLE 7.3.
Probability that an individual card is a winner in the Galactic game (no free space)

t	$P(H)$	$P(V)$	$P_W(H)$	$P_W(V)$	Ratio (H/V)
25	0.717237894	0.282762106	6.67516E-06	5.32208E-06	1.254238637
15	0.751778611	0.248221389	9.06436E-05	8.26578E-05	1.00096612542
12	0.774069504	0.225930496	0.000286111	0.000285266	1.00002961935
5	0.938066921	0.061933079	0.028101877	0.061933079	0.453745837

TABLE 7.4.
Probability that an individual card is a winner in the Galactic game (with free space)

t	$P(H)$	$P(V)$	$P_W(H)$	$P_W(V)$	Ratio (H/V)
25	0.708786173	0.291213827	1.62554E-05	1.33975E-05	1.213319761
15	0.73734237	0.26265763	0.000148386	0.000136813	1.00084591993
12	0.755786492	0.225930496	0.000400001	0.000397885	1.00005317967
5*	0.938066921	0.061933079	0.024712828	0.061933079	0.399024701

that the infinite case does not make sense here, since with infinitely many cards in play, any individual card will win with probability 0.)

Note that in the classic game, again, any given card is more likely to have a horizontal than a vertical Bingo than in the Galactic game. As we add more possible balls to each column, that advantage is more pronounced. We have found that the horizontal and vertical odds are equal with 12 balls available in each column. In the case with only 5 balls in each column, note that when a vertical Bingo is called, all cards are guaranteed to have that Bingo, which helps explain the over 2-to-1 advantage of our card being successful with a vertical versus a horizontal bingo in that game.

A similar set of calculations can be done with the Galactic game involving the free space. Here, the calculations are a bit trickier, as we need to account for the possibilities of a horizontal Bingo whether using the free space or not. We present the sums in Table 7.4, in a similar manner to the game without the free space.

Again, with more balls per column, your card is more likely to be a winner with a horizontal Bingo. with horizontal and vertical being equal for 12 balls per column. In the minimal game, the likelihood that your card will have a vertical Bingo has an even greater advantage over the horizontal.

2.4 Generalizing the Game

The theoretical calculations given for the game discussed above are generalizable in a straightforward manner to games on cards other than a 5×5 card. In Table 7.5, we provide the probabilities of horizontal and vertical wins when the cards have shape $k \times k$, where the first column has k numbers from

TABLE 7.5.
Horizontal and vertical probabilities on other square cards ($3k$ balls per column)

Board size	$P(H)$	$P(V)$
2×2	0.5454545	0.4545455
3×3	0.5851505	0.4148495
4×4	0.6664540	0.3335460
5×5	0.7517786	0.2482214
6×6	0.8264577	0.1735424
7×7	0.8849543	0.1150458
8×8	0.9271252	0.07287515
9×9	0.9555891	0.04441141

the set $\{1, \ldots, 3k\}$, the second column has numbers from $\{3k+1, \ldots, 6k\}$, and so on.

Consider the smallest nontrivial Bingo game, where the card has shape 2×2, and the first column has the numbers 1 and 2 in some order, and the second column has numbers 3 and 4 in some order. Here, the galactic version of this game has just 4 cards. It is easy to see that there will always be a Bingo after two draws, and the probability of a vertical win is exactly $1/3$.

2.5 An Advantage for Vertical

Are there sizes of cards that will see an advantage for vertical in the Galactic game? And are there are shapes of cards (and numbers of balls) that have equal probabilities for horizontal and vertical? Trivially, if we play on a 1×5 card, then the galactic game will end with a vertical Bingo after 1 draw. For a 2×5 card, the probability of getting two numbers from the same column is intuitively higher than getting one from each column. These merit some combinatorial exploration.

Our probability formulas for horizontal and vertical wins in the Galactic game can be easily extended to nonsquare cards by changing the parameters on our horizontal and vertical partitions. Suppose that we are playing on an $r \times c$ Bingo card, with t balls available in each column. A *horizontal partition* will have exactly $c - 1$ parts, each of size at most $r - 1$. A *vertical partition* will have at most $c - 2$ parts, each of size at most $r - 1$. We can calculate the probability of any particular partition happening, again by specifying the winning ball, finding the multiplier m_π (a multinomial coefficient, based on the part sizes) to choose which columns will contribute balls, and then choosing and arranging the balls from the chosen columns:

$$P_\pi(H) = \frac{(tc)m_\pi \prod \binom{t}{\pi_i} k!(tc-k-1)!}{(tc)!},$$

TABLE 7.6.

Horizontal and Vertical Probabilities on non-square cards (15 balls per column)

Card Size	$P(H)$	$P(V)$
1×5	0	1
2×5	0.0439978	0.9560022
3×5	0.2589986	0.7410014
4×5	0.5351031	0.4774969
5×5	0.7517786	0.2482214

$$P_\pi(V) = \frac{(tc)m_\pi \binom{t-1}{r-1} \prod \binom{t}{\pi_i} k!(tc-k-1)!}{(tc)!}.$$

By way of example, we computed the probability of horizontal and vertical Bingos in the Galactic game of the $r \times 5$ case with 15 balls to choose from in each column. These are presented in Table 7.6.

Notice that horizontal has only a slight advantage in the 4×5 case. This suggests that we might be able to even the advantage by adding more balls per column. Note that this adjustment is not possible in the 5×5 case, as the horizontal probability is just over 2/3 in the limit. We can achieve a probability of a vertical Bingo of 0.5009 by playing a game on a 4×5 card with 28 balls in each column.

3 Finite Results

So how does real Bingo compare to Galactic Bingo? In Galactic Bingo, horizontal and vertical wins cannot happen at the same time, and diagonal wins always occur simultaneously with horizontal wins. Yet in a typical bingo game, with dozens, or hundreds, or even thousands of cards in play, it is possible for horizontal and vertical wins to happen at the same time, and for diagonal wins to occur with no horizontal wins.

When playing Solitaire Bingo, with just one card, we used dynamic programming to determine the probabilities of the seven different Bingo outcomes, namely, H, V, D, HD, HV, VD, and HVD. An outcome like HV indicates that when the card obtained a Bingo, it was simultaneously horizontal and vertical. Table 7.7 gives the exact results (rounded to four decimal places) for the single card game when the free space is not in use.

Table 7.8 gives the exact results for the single card game when the free space is used.

As expected, when a single card is in play, horizontal and vertical outcomes behave exactly the same, so $P(H) = P(V)$ and $P(HD) = P(VD)$, regardless of whether the free space is in use. Also, since there are five horizontal rows, five

TABLE 7.7.
Probabilities of Bingo combinations in the Solitaire game (no free space)

n	P(H)	P(V)	P(D)	P(HV)	P(VD)	P(HD)	P(HVD)	Total
1	0.3941	0.3941	0.1333	0.0497	0.0131	0.0131	0.0026	1

TABLE 7.8.
Probabilities of Bingo combinations in the Solitaire game (with free space)

n	P(H)	P(V)	P(D)	P(HV)	P(VD)	P(HD)	P(HVD)	Total
1	0.3587	0.3587	0.2062	0.0413	0.0162	0.0162	0.0028	1

vertical columns, and two diagonals, it makes sense that $P(H) = P(V) > P(D)$ in both cases and that $P(D)$ would be higher when the free space is used.

To understand finite Bingo with more than one card, we wrote a computer program that played the game 100,000 times with n cards, for various values of n. Each simulation used randomly generated cards, with the game being played until a Bingo is achieved on at least one card. When $n > 1$, it is possible for a Bingo to be achieved on more than one card. We say that a round has label H if all of the winning cards round are of type H. A round has label HV if any of the winning cards are of type HV or (more likely) there are winning cards of type H and winning cards of type V. The other five cases are interpreted similarly. Table 7.9 gives the results when the free space is not in use. The results are given as probabilities, rounded to three decimal places. The last row in the table is the Galactic case. In comparing the behavior of horizontal and vertical wins, the last two columns, labeled $R1$ and $R2$, are the HV ratio, given by

$$R1 = \frac{P(H) + P(HD)}{P(V) + P(VD)},$$

and the Galactic ratio, given by

$$R2 = \frac{P(H) + P(HD) + P(D)}{P(V)},$$

respectively. The first ratio compares wins that are essentially horizontal with those that are essentially vertical. The second quantity treats diagonal wins like horizontal wins, as is true in the galactic case, since the numerator considers cases where we have definitely obtained all five letters before any letter has appeared five times, and the denominator denotes the opposite situation.

Table 7.10 gives the results for traditional Bingo (where the free space is used).

TABLE 7.9.
Probabilities generated with multiple cards in play (no free space)

n	P(H)	P(V)	P(D)	P(HV)	P(VD)	P(HD)	P(HVD)	Total	R1	R2
1	0.39243	0.39488	0.13427	0.05010	0.01309	0.01295	0.00228	1.000	1	1.4
5	0.40971	0.38961	0.15221	0.02709	0.00838	0.01170	0.00130	1.000	1.1	1.5
10	0.41652	0.37834	0.15805	0.02419	0.00882	0.01281	0.00127	1.000	1.2	1.6
50	0.43019	0.35183	0.16309	0.02334	0.00884	0.02126	0.00145	1.000	1.25	1.7
100	0.43798	0.33600	0.16776	0.02239	0.00792	0.02646	0.00149	1.000	1.3	1.9
500	0.45842	0.29422	0.17013	0.01984	0.00748	0.04829	0.00162	1.000	1.7	2.3
1000	0.45864	0.28348	0.17736	0.01737	0.00621	0.06503	0.00166	1.000	1.9	2.5
5000	0.43272	0.26554	0.14483	0.00867	0.00328	0.14386	0.00110	1.000	2.1	2.7
10^4	0.396	0.260	0.129	0.005	0.002	0.207	0.000	1.00000	2.3	2.8
10^5	0.145	0.252	0.034	0.000	0.000	0.569	0.000	1.00000	2.8	3.0
10^{27}	0	0.248	0	0	0	0.752	0	1	3.0	3.0

TABLE 7.10.
Probabilities generated with multiple cards in play (with free space)

n	P(H)	P(V)	P(D)	P(HV)	P(VD)	P(HD)	P(HVD)	Total	R1	R2
1	0.35915	0.35775	0.20643	0.04109	0.01617	0.01674	0.00267	1.000	1	1.6
5	0.35434	0.34251	0.25619	0.01888	0.01026	0.01648	0.00134	1.000	1.1	1.8
10	0.34627	0.32661	0.28029	0.01687	0.00927	0.01952	0.00117	1.000	1.1	2.0
50	0.32405	0.29668	0.32225	0.01243	0.00885	0.03474	0.00100	1.000	1.2	2.3
100	0.31454	0.27782	0.34134	0.01177	0.00793	0.04567	0.00093	1.000	1.2	2.5
500	0.27748	0.25119	0.36121	0.00743	0.00527	0.09642	0.00100	1.000	1.5	2.9
1000	0.25365	0.25023	0.34968	0.00578	0.00423	0.13561	0.00082	1.000	1.5	3.0
5000	0.15653	0.26298	0.26514	0.00162	0.00079	0.31275	0.00019	1.000	1.8	2.8
10^4	0.103	0.265	0.196	0.000	0.000	0.435	0.000	1.000	2.0	2.8
10^5	0.002	0.266	0.008	0.000	0.000	0.723	0.000	1.000	2.7	2.8
10^{27}	0	0.263	0	0	0	0.737	0	1.000	2.8	2.8

Looking at the data in Tables 7.9 and 7.10, we see some interesting patterns. When the free space is not in use, we see that the R1 and R2 ratios increase pretty steadily as the number of cards increases, becoming essentially Galactic when somewhere between 100,000 and 1,000,000 cards. (This is an unrealistic size. According to the *Guinness Book of World Records*, the largest Bingo game had 70,000 people [4].) With 10,000 cards, there are still enough diagonal-only wins to delay the R1 ratio. But as *n* gets larger and larger, it becomes more and more likely that there will be multiple winners, which increases the probability of *HD*. When the free space is used, the game approaches Galactic proportions much

TABLE 7.11.
Use of the free space in simulated games

n	$P(H)$	$P(V)$	$P(D)$	$P(HV)$	$P(VD)$	$P(HD)$	$P(HVD)$	Total	R1	R2
1	0.11866	0.11813	0.20643	0.02306	0.01617	0.01674	0.00267	0.50186	1	2.9
5	0.13839	0.13657	0.25619	0.01120	0.01026	0.01648	0.00134	0.57043	1.1	3
10	0.14653	0.14161	0.28029	0.01062	0.00927	0.01952	0.00117	0.60901	1.1	3.15
50	0.16691	0.15021	0.32225	0.00863	0.00885	0.03474	0.00100	0.69259	1.3	3.5
100	0.17266	0.14986	0.34134	0.00877	0.00793	0.04567	0.00093	0.72716	1.4	3.7
500	0.17822	0.14464	0.36121	0.00615	0.00527	0.09642	0.00100	0.79291	1.8	4.4
1000	0.17276	0.14331	0.34968	0.00473	0.00423	0.13561	0.00082	0.81114	2.1	4.6
5000	0.11931	0.13480	0.26514	0.00144	0.00079	0.31275	0.00019	0.83442	3.2	5.2

TABLE 7.12.
Solo winners among n cards (100,000 simulations, no free space)

n	$P(H)$	$P(V)$	$P(D)$	$P(HV)$	$P(VD)$	$P(HD)$	$P(HVD)$	Total
1	0.3941	0.3941	0.1333	0.0497	0.0131	0.0131	0.0026	1
5	0.39937	0.36581	0.15090	0.01047	0.00329	0.00377	0.00013	0.93374
10	0.40227	0.34754	0.15611	0.00494	0.00190	0.00192	0.00003	0.91471
50	0.40566	0.30086	0.15938	0.00105	0.00040	0.00048	0	0.86783
100	0.40753	0.27422	0.16330	0.00048	0.00014	0.00028	0	0.84595
500	0.40618	0.20620	0.16240	0.00003	0.00003	0.00005	0	0.77489
1000	0.39282	0.17736	0.15904	0.00001	0.00001	0.00002	0	0.72926
5000	0.32371	0.09591	0.13084	0	0	0	0	0.55046
10^4	0.26899	0.03250	0.10756	0	0	0	0	0.40905
10^5	0.00037	0	0.00070	0	0	0	0	0.00107

more quickly. The most interesting feature about these numbers is that when n is between 100 and 1000, the most probable outcome is that all the winning cards are diagonal. We suspect that what is happening here is that for a card to win in this range, it has to be a lot luckier than the rest of the cards, which means that it probably uses its numbers pretty efficiently. As a result, there is a good chance that it uses the free space. And since the free space is not only used by one horizontal row and one vertical row, but also by both diagonals, it seems plausible that diagonal wins would become more common.

To explore that last point, we investigate how often the free space is used in a winning Bingo. When $n = 100$ cards are in play, our simulation showed that the free space was used 72.716% of the time. (Specifically, among the 31,454 H wins in Table 7.10, a free space was used 17,266 of those times; similarly, the free space was used 14,986 times among the V wins and 877 times among the HV wins; naturally, the free space was used in all of the wins of the type D, VD, HD, and HVD.) Table 7.11 gives the full set of data for how often the free space

TABLE 7.13.
Solo winners among n cards (100,000 simulations, with free space)

n	$P(H)$	$P(V)$	$P(D)$	$P(HV)$	$P(VD)$	$P(HD)$	$P(HVD)$	Total
1	0.3587	0.3587	0.2062	0.0413	0.0162	0.0162	0.0028	1
5	0.34655	0.32370	0.25176	0.00791	0.00410	0.00419	0.00014	0.93374
10	0.33562	0.30135	0.27354	0.00387	0.00216	0.00250	0.00001	0.91471
50	0.30770	0.25224	0.30588	0.00058	0.00041	0.00069	0	0.86783
100	0.29497	0.22346	0.31811	0.00029	0.00023	0.00029	0	0.84595
500	0.24948	0.16687	0.31551	0.00003	0	0.00001	0	0.77489
1000	0.22235	0.14437	0.29005	0	0.00001	0	0	0.72926
5000	0.12179	0.05960	0.17799	0	0	0	0	0.55046
10^4	0.07459	0.01680	0.11213	0	0	0	0	0.35938
10^5	0.00049	0	0.00082	0	0	0	0	0.00131

was utilized in traditional Bingo. It should be thought of as a refinement of the previous table.

How often is there just a single winner? The last column of Table 7.12 gives the overall probability of a single winner when n cards are in play, broken out over the seven types of Bingos. The first row is an exact calculation (as seen earlier), and the rest of the rows are based on simulation of 100,000 games. Table 7.12 gives the solo-winner statistics when the free space is not used.

Table 7.13 gives the solo-winner statistics when the free space is used.

Is there a bet that favors vertical? Well, as n gets really large, say, beyond 5000 cards, then if you had to bet whether *all* outcomes would be horizontal, or all vertical, or all diagonal, then the best bet is vertical, since horizontals and diagonals tend to happen together.

Acknowledgments

We thank Liam Lloyd for valuable programming assistance with this project.

References

[1] A. Benjamin, J. Kisenwether, and B. Weiss. The Bingo paradox. *Math Horizons* **25** no. 1 (2017) 18–21.

[2] D. Hess. *The Population Explosion and Other Mathematical Puzzles.* World Scientific, Singapore, 2016.

[3] D. Velleman and S. Wagon. *Unicycle or Bicycle: A Collection of Intriguing Mathematical Puzzles.* MAA Book series, forthcoming.

[4] Guinness World Records. Online at http://www.guinnessworldrecords.com/world -records/largest-game-of-bingo. Last accessed June 2018.

8

‹◇◇◇›

WIGGLY GAMES AND BURNSIDE'S LEMMA

David Molnar

In this chapter, I talk about abstract games, but mostly from a design perspective. I say little about gameplay and absolutely nothing about strategy. In particular, I focus on a family of tile-placement games in which the tiles are marked with arcs in such a way that as the game progresses, meandering paths are formed. I refer to these games as *wiggly games*.

The best-known of these is Tsuro, in which two to eight mean-spirited players scheme to build paths in such a way as to force their opponents' pawns off the board. Each player starts with a pawn on the perimeter of a six-by-six board and a hand of three tiles. After a tile is placed, any pawn situated adjacent to that new tile (whether or not that pawn belongs to the player who placed the tile) must move along the path created by the new tile until it reaches the end of that path. If that end of the path is at the edge of the play area, the player with that pawn is eliminated. In Figure 8.1, if the tile outlined in red is placed as shown on the left, the blue pawn must then travel along the quarter-circular arc, thus meeting its doom. If the tile is instead placed as shown on the right, the blue pawn will then travel along a different arc, to (temporary) safety.

Different tiles show different combinations of arcs. How many Tsuro tiles are there? Is every possible tile represented in the physical game? In any of the games discussed here, an enumeration of the tiles is well within the reach of a brute force approach. I will present a systematic enumeration of the Tsuro tiles, showing that all possible tiles are indeed found in the game. But the goal is not to count the tiles. The goal is to have a sense of *why* there are as many tiles as there are, which is not achieved by brute force, and to see how the enumeration problems for different wiggly games are connected. The techniques discussed here could also be used on larger problems (corresponding to physically impractical games) where brute force is not as feasible.

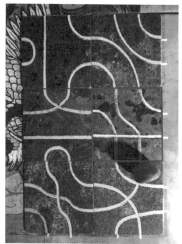

Figure 8.1. Life or death in Tsuro.

1 Wiggly Games

Our standard setup is this: The game contains a bunch of congruent tiles, with arcs drawn on them. Each arc has two ends, somewhere along the boundary of the tile. While the arrangements of arcs vary, the arcs always terminate at the same positions on the boundary, so that when an adjacent tile is laid, the arcs can join together to form a path. I refer to these positions on the boundary of the tile as the *sites*. It is assumed that each site will be the endpoint of exactly one arc. There may be additional restrictions—for example, arcs may or may not be allowed to cross. Also assume throughout that the tiles can be rotated before placement but not flipped over. It is common, but by no means necessary, that the game box contains all tiles possible under certain constraints.

The popular game Tsuro features square tiles, each marked with four arcs. Therefore, there are eight sites. The four arcs partition the eight sites into four pairs. Counting partitions of a fixed type is a standard enumeration problem, as discussed by Marcus [5] and many others. The multinomial coefficient

$$\binom{8}{2,2,2,2}$$

must be divided by 4! (since there are four "parts" of equal size), so the number of pairings is

$$\frac{8!}{4!2!2!2!2!} = \frac{8!}{8 \cdot 6 \cdot 4 \cdot 2} = 7 \cdot 5 \cdot 3 \cdot 1 = 105. \tag{1}$$

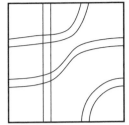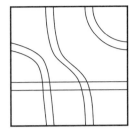

Figure 8.2. Distinct pairings can correspond to the same tile.

Counting the tiles, however, is not so simple. Consider Figure 8.2, which illustrates that different pairings can in fact correspond to different presentations of a single tile, in this case related by a 90° rotation. So there are fewer than 105 tiles.

The tile shown actually corresponds to four different pairings, but this must not be true of all the tiles, since 105/4 is not an integer. Some tiles correspond to only one pairing. What is needed is to create an *equivalence relation* on the pairings, so that each tile corresponds to an equivalence class. (The reader who is unfamiliar with equivalence relations may wish to consult Hammack [4].)

Here is the main point of the chapter: The problem of counting the tiles in such a game, when suitably framed, becomes a problem of counting *equivalence classes*, and there is a systematic method for doing so. This style of problem is often referred to as "counting modulo symmetry." To understand the problem better, I first solve the enumeration problem for a game in which the tiles have less symmetry, before returning to Tsuro.

2 Topology

In Markus Hagenauer's game Topology [3], the tiles are double-hexes, with ten edges joined by nonintersecting arcs. There is no movement in this game; players place three tiles per turn, scoring points by completing closed loops. The longer the loop is, the higher the score will be. In Figure 8.3, placing the highlighted tile will complete three loops. The outermost of these loops consists of five segments and encloses loops consisting of four and two segments. This will score eleven points.

The double-hex has only one nontrivial (direct) symmetry, the 180° rotation. Thus, we can partition the tiles into two sets: symmetric and asymmetric. To count the possible Topology tiles, we begin by assigning each tile a code. The restriction that the arcs cannot intersect presents a complication, but fortunately, it is a very well-known complication. There are ten sites. To encode a tile, walk around its boundary, using the rightmost edge as the starting point and proceeding counterclockwise. Label each site according to whether we have

Figure 8.3. The first scoring in a game of Topology.

Figure 8.4. A sample tile for Topology, marked so as to produce its code. Starting at the rightmost edge, a "(" denotes the beginning of an arc, and a ")" denotes the end.

already seen the arc ending at that site. Let us use the markers "(" for an arc we are seeing for the first time, and ")" for an arc we are revisiting. The highlighted tile from Figure 8.3 is shown in Figure 8.4 with these labels. It is assigned the code ((()))()().

No matter what tile we choose, we will eventually see both ends of each arc, so these codes will consist of exactly five left parentheses and five right parentheses. Moreover, at any point along our walk, there must have been at least as many left parentheses as right parentheses. Borrowing terminology from Marcus [5], let us refer to such a code as a "(-dominated" sequence. Conversely, given any (-dominated sequence of five left parentheses and five right parentheses, we can reconstruct a tile with that code. Such arrangements of parentheses are a common manifestation of the well-studied *Catalan numbers* [6]. Marcus [5] gives an elementary proof of the formula

$$C_n = \frac{1}{n+1}\binom{2n}{n},$$

where n is the number of pairs of parentheses. There are therefore $C_5 = 42$ codes for the Topology tiles.

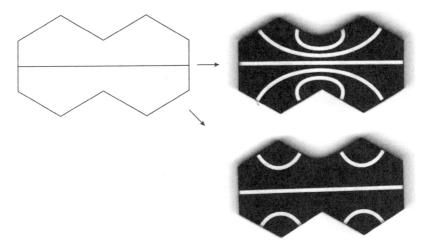

Figure 8.5. For any given centrally-symmetric arc, two symmetric tiles can be drawn.

Denote by A and S the number of asymmetric and symmetric tiles, respectively. The symmetric tiles have the same code in either orientation. The asymmetric tiles, when rotated $180°$, will pick up a different code. For example, the tile in Figure 8.4 is labeled $(((()))()()$; if it were rotated $180°$, then its code would be $(()())(()$. Since there are 42 codes, this gives us the relation $2A + S = 42$. However, we want the number of tiles, $A + S$. We can find this by determining the number of symmetric tiles.

Unfortunately, the parentheses codes are not good for detecting symmetry. The reader might verify that both $()(()())()$ and $(())((()))$ correspond to symmetric tiles. Consider, though, the effect of a $180°$ rotation on a symmetric tile: Arcs are rotated onto arcs. Since there are five arcs, and five is odd, some arc must be rotated onto itself. However, this centrally-symmetric arc must be unique, since the arcs do not intersect.

Any of the centrally-symmetric arcs divides the tile into two equal halves. There are only two possible arrangements of arcs on each half-tile: those corresponding to $(())$ and $()()$, as illustrated by Figure 8.5. Symmetry dictates that whichever arrangement is used on one half of the tile is repeated on the other. Therefore, there are $5 \cdot 2 = 10$ symmetric tiles. As $2A + S = 42$, it follows that[1]

$$A + S = \frac{1}{2}(42 + 10) = 26. \tag{2}$$

[1] Note that while there are 52 physical tiles in the available set, there are not in fact (as stated in the rules) two of each, as certain tiles that make it more difficult to close loops have been underrepresented.

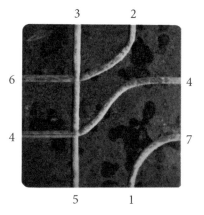

3   2

6   4

4   7

5   1

Figure 8.6. Assignment of a code for a Tsuro tile. Each digit gives the distance to the other end of the given arc, moving counterclockwise.

3 Tsuro

Since the arcs on a Tsuro tile may cross, the encoding used for Topology is not sufficient to reconstruct the tile: Even a string such as (()) has two interpretations. A more detailed code is necessary to make this distinction. Any presentation of a tile will be encoded by a string of eight digits. Each site can be assigned a single digit, indicating how many sites away from the other end of the arc that site is, assuming a counterclockwise orientation. This is demonstrated in Figure 8.6. The first digit of a code will correspond to the right site on the bottom edge, and the following digits are read off counterclockwise. The tile shown thus has the code 17423645.

Any such code allows us to recover the pairing of the eight sites discussed in Section 1, and conversely, any pairing of the sites can be used to construct the code. It then follows from equation (1) that there are 105 codes. Some codes, however, correspond to the same tile: We can rotate a tile before placing it on the board. This changes the code, but it does not change what we have in our hand. Accordingly, let us consider the corresponding codes to be *equivalent*, (denoted $\sim$). For example, $17423645 \sim 45174236$, as shown in Figure 8.7. (A complete definition of equivalence follows.)

Observe that the code 45174236 is obtained from the code 17423645 by moving the pair of digits 45 from the end to the beginning. Likewise, moving the first two digits from the beginning to the end corresponds to a 90° rotation clockwise, and shifting four digits corresponds to a 180° rotation.

Codes corresponding to mirror images should, however, not be considered equivalent, for the simple reason that it is not acceptable to play a tile upside-down. Consider now the complete set of 105 codes. Define two codes to be

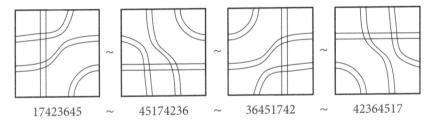

$$17423645 \quad \sim \quad 45174236 \quad \sim \quad 36451742 \quad \sim \quad 42364517$$

Figure 8.7. The equivalence class of a particular code.

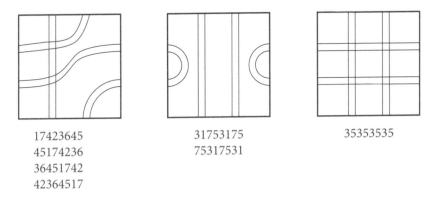

17423645
45174236
36451742
42364517

31753175
75317531

35353535

Figure 8.8. Three Tsuro tiles and their equivalence classes of codes.

equivalent if one can be obtained from the other by moving an *even* number of digits from one end to the other, without altering the order of the digits moved. Some codes are equivalent to three others, (as in Figure 8.7) but others are equivalent to one or no other.

4 Burnside's Lemma

In equation (2), enumerating the Topology tiles, we had $A + S = \frac{1}{2}(42 + 10) = 26$. Observe that each code for an asymmetric tile is part of the 42 but not part of the 10, and thus it only counts as $\frac{1}{2}$ in the final result. To find the analogous result for Tsuro tiles, note that since there are four symmetries, each tile is associated with one, two, or four codes.

Now, let X_k denote the number of codes fixed by a $(90k)°$ rotation, $k = 0, 1, 2, 3$. (Here we conflate the operation on the codes of moving two digits with the geometric transformation on the tile, as demonstrated in Figure 8.2.) The leftmost tile in Figure 8.8 is part of X_0 only; it has no symmetry. The middle tile is counted by X_0 and X_2, and the tile on the right is counted by X_0, X_1, X_2, and

X_3. So if there are a tiles of the first type, b of the second, and c of the third, it follows that

$$X_0 = 4a + 2b + c, \qquad X_1 = c, \qquad X_2 = 2b + c, \qquad X_3 = c.$$

The number of equivalence classes of codes is then

$$a + b + c = \frac{1}{4}(X_0 + X_1 + X_2 + X_3). \tag{3}$$

Equations (2) and (3) are both special cases of Burnside's Lemma, also known as Pólya's enumeration theorem, reportedly due to Frobenius. A full statement and proof can be found in a standard abstract algebra text, such as the one by Fraleigh [2], but I want to discuss it without the added overhead of group actions. We have a set X of "codes." An equivalence relation is defined on the codes, wherein two codes are equivalent if one can be transformed into the other in a prescribed way. If these transformations are indexed by some (finite) set of integers J, and X_k denotes the number of codes in X that are fixed by the kth transformation, then the number of equivalence classes modulo symmetry is given by

$$\left| X/{\sim} \right| = \frac{1}{|J|} \sum_{k \in J} X_k. \tag{4}$$

The proof of formula (4) is a straightforward double-counting argument, considering the collection of ordered pairs (x, k), where x is a code fixed by transformation k. Again, Fraleigh [2] is a good resource. We are now ready to complete the enumeration.

5 Tsuro Resolved

Returning to equation (3), we have that X_0 (the number of codes fixed by doing nothing) is 105. Since a 90° rotation and a 270° rotation are inverses, any tile fixed by one is fixed by the other. It follows that $X_1 = X_3$.

For a code to be fixed by a 90° rotation, it must consist of the same pair of digits repeated four times. It also must be the case that the code contains no 2s or 6s. As shown in Figure 8.9, a code with 90° symmetry containing a 2 would correspond to a tile in which two different arcs meet at the same site. There are no such tiles. Therefore, the only codes with this symmetry are 17171717, 35353535, 44444444, 53535353, and 71717171. That is, $X_1 = X_3 = 5$.

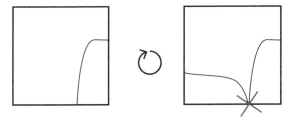

Figure 8.9. There are no codes with a 2 or a 6 having 90° symmetry.

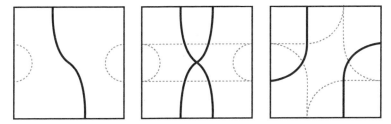

Figure 8.10. Three incomplete tiles.

The analysis required for codes fixed under a 180° rotation is slightly more elaborate. Such a code must consist of some block of four digits repeated twice. Let us condition on the number of 4s, which is necessarily even:

- There is exactly one code with eight 4s, namely, 44444444.
- There are no codes with six 4s: once six 4s are inserted into a code, there are two sites remaining, which are a distance four apart, and thus must also be 4s.
- There are also no codes with two 4s. Assume, as in the left of Figure 8.10, that a 4-4 arc has been drawn. Only a 4-4 arc is its own image under a 180° rotation, so when any subsequent arc is drawn, its image must then be drawn as well. This leaves two sites open, which again are a distance four apart.
- If a code contains four 4s, there are two 4-4 arcs, as shown in the center in Figure 8.10. The remaining four sites can be paired off in three ways, but one of these ways creates two more 4-4 arcs. The two 4-4 arcs can be chosen in $\binom{4}{2} = 6$ ways, so there are $6 \cdot 2 = 12$ codes with four 4s.
- If a code contains no 4s, then no arc is its own image under a 180° rotation. There are six possible first digits for the code, not including 4. Once this first digit is chosen, two arcs are fixed, as on the right in Figure 8.10. As in the previous case, this leaves four remaining sites, which can be paired off in two ways. So there are also 12 codes with no 4s.

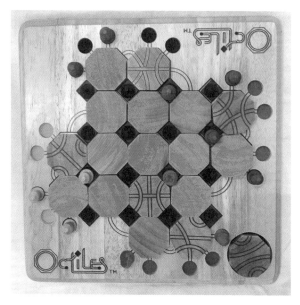

Figure 8.11. An Octiles game after the first five turns. Red has moved twice.

This gives us $X_2 = 1 + 12 + 12 = 25$. Therefore, the number of equivalence classes of Tsuro codes, according to equation (3), is

$$\frac{1}{4}(105 + 5 + 25 + 5) = 35.$$

6 Octiles

While there are many other wiggly games of interest, I finish with a quick look at a game that, as far as the enumeration is concerned, is related to Tsuro in an interesting way. The game Octiles features, as one might imagine, octagonal tiles. These tiles are each adorned with four arcs. Octiles is a race game in which two to four players maneuver across an evolving—and wiggly—landscape, as illustrated in Figure 8.11. Movement is similar to that of Tsuro, but only one pawn moves per turn, belonging to the player who has just played a tile. Complete rules can be found at the website boardspace [1].

The tiles can be coded in exactly the same way as for Tsuro: There are eight sites along the boundary of the tile, and each can be assigned a digit from 1 to 7. (Use the bottom edge for the first digit, and proceed counterclockwise.) However, the equivalence relation has changed. Figure 8.12 demonstrates that

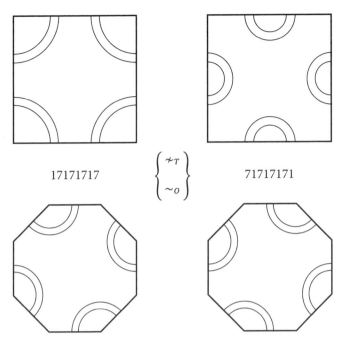

17171717
$$\left\{ \begin{matrix} \nearrow_T \\ \sim_O \end{matrix} \right\}$$
71717171

Figure 8.12. Certain codes are equivalent for Octiles but not for Tsuro.

the codes 17171717 and 71717171 are equivalent in Octiles, while they were not in Tsuro. The equivalence relation for Octiles, $\sim_O$, is thus coarser than the equivalence relation for Tsuro, $\sim_T$. So there will be fewer equivalence classes.[2]

Two codes are equivalent under $\sim_O$ if one can be turned into the other by moving *any* number of digits from one end of the code to the other. It is natural to associate the moving of a single digit with a 45° rotation. Now let Y_k denote the number of codes invariant under a $(45k)°$ rotation, $k = 0, 1, 2, ..., 7$. Formula (4) tells us that the number of equivalence classes is

$$\frac{1}{8}(Y_0 + Y_1 + Y_2 + Y_3 + Y_4 + Y_5 + Y_6 + Y_7). \tag{5}$$

All codes are fixed by doing nothing, so $Y_0 = 105$. However, if an odd number of digits are shifted and the code is unchanged, then all the digits must be the same. This is only possible if all the digits are 4. So $Y_1 = Y_3 = Y_5 = Y_7 = 1$. Since Octiles and Tsuro share the same codes, we have already done the rest of the work: Y_2 and Y_6 are the number of codes invariant under a 90° rotation. But this

---

[2] You wouldn't know it by holding the box, though; Octiles is much heavier.

was counted by X_1, which is 5. Lastly, Y_4 is the number of codes invariant under a 180° rotation, which is X_2, or 25. Plugging into equation (5), we see that the number of tiles is

$$\frac{1}{8}(105 + 1 + 5 + 1 + 25 + 1 + 5 + 1) = 18.$$

Acknowledgments

I thank the designers of the games discussed herein. Some credit should also be given to boardgamegeek.com users mrraow, for the wonderful phrase "wiggly games"; dsr15, for posing the original question about Tsuro; and Gendo_Ikari82, for reminding me that I had answered it.

References

[1] D. Dyer. *The Rules for Octiles*. Online at http://www.boardspace.net/octiles/english/rules.html. Last accessed June 2018.

[2] J. Fraleigh. *A First Course in Abstract Algebra*, seventh edition. Pearson, London, 2003.

[3] G. Breitenbach, and M. Hagenauer. Topology. Online at https://nestorgames.com/rulebooks/TOPOLOGY_EN.pdf. Last accessed June 2018.

[4] R. Hammack. *Book of Proof*, second edition. Online at https://www.people.vcu.edu/~rhammack/BookOfProof/. Last accessed June 2018.

[5] D. Marcus. *Combinatorics: A Problem-Oriented Approach*. Mathematical Association of America, Washington, DC, 1998.

[6] Sequence A00108. In *The On-Line Encyclopedia of Integer Sequences*. Online at https://oeis.org. Last accessed June 2018.

9

LOSING AT CHECKERS IS HARD

Jeffrey Bosboom, Spencer Congero, Erik D. Demaine,
Martin L. Demaine, and Jayson Lynch

Winning isn't for everyone. Sometimes, you want to let your little sibling/ nibling/child win, so that they can feel the satisfaction, encouraging them to play more games. How do you play to lose? This idea is formalized by the notion of the *misère* form of a game, where the rules are identical except that the goal is to lose (have the other player win) according to the normal rules [1,2]. Although your little relative may *want* to win, it's probably unreasonable to assume that they play optimally, so in the misère game, we not only aim to lose but also to prevent the opponent from losing.

Checkers (in American English; "draughts" in British English) is a classic board game that many people today grow up with. Although there are several variations, the main ruleset played today is known as American checkers and English draughts, and the game has had a World Championship since the 1840s [12]. Computationally, the game is *weakly solved*, meaning that its outcome from the initial board configuration (only) has been computed [10]: If both players play optimally to win, the outcome will be a draw.

But what if both players play optimally to lose? (Again, your little relative may not aim to lose, but you want to lose no matter what they do, so this "worst-case" mathematical model more accurately represents the goal.) The misère form of checkers is known as *suicide checkers, anti-checkers, giveaway checkers*, and *losing draughts* [12]. This game is less well studied. No one knows whether the outcome is again a draw or one of the players successfully losing. There are a few computer players, such as Suicidal Cake [5], that occasionally compete against each other.

1 Results

1.1 Complexity of Winning

In this chapter, we consider a few checkers variants, in particular, suicidal checkers, from a *computational complexity* perspective. Computational complexity allows us to analyze the best computer algorithm to optimally play a game, and in particular, how the time and memory required by that algorithm must *grow* with the game size. For this analysis to make sense, we need to generalize the game beyond any constant size, such as twelve pieces on an 8×8 board. The standard generalization for checkers is to consider an $n \times n$ board, and an arbitrary position of pieces on that board (imagining that we reached this board configuration by several moves from a hypothetical initial position). The computational problem is to decide who wins from this configuration assuming optimal play.

In this natural generalization, Robson [9] proved that normal checkers is *EXPTIME-complete*, meaning in particular that any correct algorithm must use time that grows exponentially fast as n increases—roughly needing c^n time for some constant $c > 1$. EXPTIME-completeness is a precise notion we will not formalize here, but it also implies that checkers is one of the hardest problems that require exponential time, in the sense that all problems solvable in exponential time (which is most problems we typically care about) can be converted into an equivalent game of checkers. So maybe all that time spent playing checkers as a kid wasn't a waste... (There are many such complexity results about many different games; see, e.g., Hearn and Demaine [6] and Demaine [3].)

One of the checkers variants we study in this chapter adds one rule: Every move must capture at least one piece. Thus, a player loses if they cannot make a jumping move. This *always-jumping* checkers is not very interesting from the usual starting configuration (where no jumps are possible), but it is interesting from an arbitrary board configuration. In Section 5, we show that always-jumping checkers is *PSPACE-complete*—a complexity class somewhere between "easy" problems (P) and EXPTIME-complete problems. Formally, always-jumping can be solved using polynomial space (memory)—something we don't know how to do for normal checkers, where many moves can be made between captures—and always-jumping checkers is among the hardest such problems. This result suggests that always-jumping checkers is somewhat easier than normal checkers, but not by much.

1.2 Complexity of Losing

But what is the computational complexity of suicide checkers? This question remains unsolved.

In this chapter, we study a closely related problem: What is the computational complexity of deciding whether you can lose the game "in one move" (that is, make a move that forces your opponent to win in a single move)? We call this the *lose-in-1* variation of a game. In Section 3, we prove that lose-in-1 checkers is *NP-complete*—a complexity class somewhere in between "easy" problems (P) and PSPACE-complete problems. In particular, assuming a famous conjecture (the Exponential Time Hypothesis), NP-complete problems (and thus also PSPACE-complete problems) require nearly exponential time–roughly 2^{n^ε} for some $\varepsilon > 0$.

Hardness of losing checkers in one move is particularly surprising, because *winning* checkers in one move (*mate-in-1*) is known to be easy. An early paper about the complexity of normal checkers gave an efficient algorithm for deciding mate-in-1 checkers, running in time linear in the number of remaining pieces [4]. Namely, for each of your pieces, draw a graph whose vertices represent reachable positions by jumping, and whose edges represent (eventually) jumpable pieces. Then, assuming all opponent pieces appear as edges in this graph, winning the game is equivalent to finding an *Eulerian path* in the graph (a path that visits each edge exactly once) that starts at the piece's initial position. Finding Eulerian paths goes back to Euler in 1736, though the problem was not fully solved until 1873 by Hierholzer [7], who also gave a linear-time algorithm.

Proving that lose-in-1 checkers is an NP problem instead of, say, PSPACE or EXPTIME, involves extending this mate-in-1 result.

1.3 Complexity of Cooperating

The final checkers variants we consider in this chapter are *cooperative* versions, where the two players cooperate to achieve a common goal, effectively acting as a single player. The same proofs described above also show NP-completeness of *cooperative win-in-2* checkers—where the players together try to eliminate all pieces of one color in two moves—and *cooperative always-jumping* checkers— where the players together try to eliminate all pieces of one color using only jumping moves. In Section 5-5, we describe these results in more detail.

1.4 Checkers Font

In Section 6, we give a checkers *puzzle font*: a series of cooperative always-jumping checkers puzzles whose solutions trace out all twenty-six letters (and ten numerals) of the alphabet. See Figures 9.7 and 9.8. We hope that this font will encourage readers to engage with the checkers variants studied in this chapter. An animated font is also available as a free web app.[1]

2 Checkers Rules and Variants

First let us review the rules of American checkers/English draughts. Two players, Black and Red, each begin with 12 pieces placed in an arrangement

---

[1] Available at http://erikdemaine.org/fonts/checkers/. Last accessed January 2019.

on the dark squares of an 8×8 checkerboard. Black moves first, and play takes place on the dark squares only. Initially, pieces are allowed to move diagonally forward (i.e., toward the opponent's side) to other diagonally adjacent dark squares, but not backward. If a player's piece, an opponent's piece, and an empty square lie in a line of diagonally connected dark squares, the player must "jump" over the opponent's piece and land in the empty square, thereby capturing the opponent's piece and permanently removing it from the board. Furthermore, jumps must be concatenated if available, and a player's turn does not end until all possible connected jumps have been exhausted. If a piece eventually reaches the last row on the opponent's side of the board, it becomes a "king," which enables movement to any diagonally adjacent dark square, regardless of direction. The goal of each player is to capture all of the opponent's pieces.

As in Robson and Fraenkel et al. [4,9], we generalize the game to an arbitrary configuration of an $n \times n$ board. We still assume that the next player to move is Black.

Next we describe the two main variants of checkers considered in this chapter.

First, in *lose-in-1* checkers, one player tries to make a single move such that, on the following turn, the opponent is forced to capture all of the first player's remaining pieces with a sequence of jumps.

Second, in *always-jumping* checkers, we add one rule: A piece must jump an opponent's piece on every move. In other words, pieces are now forbidden from moving to another square without a jump; the first player with no available jumps (or no pieces) loses. As in standard checkers, jumps must be concatenated in a single turn as long as there is another jump directly available.

In Section 6, we define and analyze cooperative versions of these two checkers variants, where players work collaboratively to make one of them to win (without particular regard to *who* wins). These variants arise if the players hate draws, or if they want to make the game end quickly.

To improve readability in the following proofs and figures, we rotate the checkerboard 45°, so that pieces move horizontally and vertically, rather than diagonally.

3 Lose-in-1 Checkers Is NP-Complete

In this section, we prove that the following computational decision problem is NP-complete:

Problem 1 (Lose-in-1 checkers). *Given a checkers game configuration, does the current player have a move such that the other player must win on their next move?*

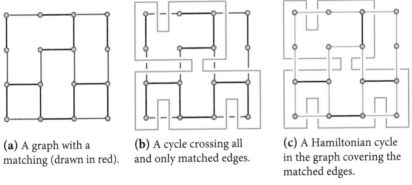

(**a**) A graph with a matching (drawn in red).

(**b**) A cycle crossing all and only matched edges.

(**c**) A Hamiltonian cycle in the graph covering the matched edges.

Figure 9.1. An example of Planar Forced-Edge Hamiltonicity (a) satisfying the promise (b), and its solution (c).

In general, proving NP-completeness consists of two steps: (1) proving "NP-hardness" (Theorem 2 below) and (2) proving "membership in NP" (Theorem 3 below) [3]. The first step establishes a *lower bound* on complexity—technically, that the problem is as hard as all problems in the complexity class in NP. The second step establishes an *upper bound* on complexity—that the problem is no harder than NP.

Without getting into the definition of NP, we can prove NP-hardness of a problem by showing that it is at least as hard as a known NP-hard problem. We prove such a relation by giving a *reduction* that shows how to (efficiently) convert the known NP-hard problem into the target problem. In our case, we will reduce from the following problem; refer to Figure 9.1 for an example.

Problem 2 (Planar Forced-Edge Hamiltonicity).   *Given a max-degree-3 planar undirected graph[2] and a perfect matching in that graph,[3] is there a cycle that visits every edge in the matching?*

The graph and matching must also satisfy the promise that we can draw a cycle that crosses every matched edge (possibly multiple times) without crossing any nonmatched edges. (An equivalent promise is that the nonmatched cycles do not enclose any vertices of the graph, i.e, form faces of the graph.)

---

[2] A graph is *max-degree-3* if every vertex has at most three incident edges. A graph is *planar* if it can be drawn in the plane, with vertices as points and edges as curves (or straight lines), such that the edges intersect only at common endpoints.

[3] A *perfect matching* in a graph is a set of edges ("matched edges") such that every vertex is the endpoint of exactly one matched edge.

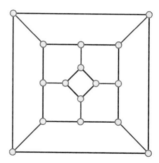

Figure 9.2. A graph and matching not satisfying the promise of Planar Forced-Edge Hamiltonicity: The middle square of unmatched edges forms a separator that prevents any cycle from visiting both the diagonal and orthogonal matched edges.

This particular problem has not yet been shown to be NP-hard, but it follows from an existing proof (another reduction from another known NP-hard problem):

Theorem 1. *Planar Forced-Edge Hamiltonicity is NP-complete.*

Proof. This result follows from Plesník's proof [8] of NP-hardness of Hamiltonicity in directed graphs where every vertex either (1) has two incoming edges and one outgoing edge or (2) has one incoming edge and two outgoing edges. Any edge from a type-1 vertex to a type-2 vertex is forced to be in the Hamiltonian cycle. By inspecting Plesník's reduction, no two vertices of the same type are adjacent (the two types form a bipartition). Thus, the forced edges form a perfect matching in the graph. By further inspection, we can verify that the reduction always produces graphs that satisfy the promise. (The promise is not true in general; see Figure 9.2 for a counterexample.) □

Now we give a reduction from Planar Forced-Edge Hamiltonicity to lose-in-1 checkers:

Theorem 2. *Lose-in-1 checkers is NP-hard.*

Proof. Our goal is to convert an instance of Planar Forced-Edge Hamiltonicity into an equivalent instance of lose-in-1 checkers. For example, the instance in Figure 9.1 will convert into the checker board shown in Figure 9.3.

The overall structure of the reduction is to create two parity classes on the checkerboard, where each class contains one piece of one player and all other pieces of the other player. If Black can find a Hamiltonian cycle of concatenated

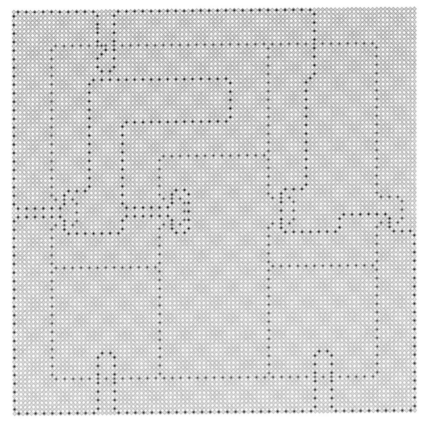

Figure 9.3. The checkerboard produced by the reduction for the example graph and cycle shown in Figure 9.1. The motive pieces are near the upper-left horizontal edge.

jumps on its turn, then Red will be forced to win by capturing all remaining black pieces in the second parity class; otherwise, some black pieces will remain, because Red cannot traverse the entire cycle.

To produce a checkers instance from a Planar Forced-Edge Hamiltonicity instance, we begin by embedding the 3-regular planar undirected graph in the planar grid. Theorem 1 of Storer [11] establishes that a graph with n vertices can always be embedded in an $n \times n$ grid. We scale the grid embedding by 27 and place red pieces at every second point along the edges. We modify the vertical matched-edge segments by moving the middle three pieces in each segment six columns to the right, then adding three red pieces in the same parity class in the rows immediately above the upper moved piece and below the lower moved piece. We call the resulting structures *tabs* (by analogy with the tabs and pockets of jigsaw puzzles). In the figures, this grid embedding consists of the red pieces on cyan squares.

The Forced-Edge Hamiltonicity instance includes the promise that we can draw a cycle that crosses all matched edges and no unmatched edges. We construct such a cycle using the following algorithm:

1. Take the subgraph of the dual graph[4] given by the duals of the matched edges. By the promise, this graph is connected.
2. Until this graph is a tree, choose an edge in a cycle, and replace one endpoint of the chosen edge with a new vertex.
3. Double the tree to form a cycle.

Each matched edge has *cycle-crossing points* at which the cycle may cross. For the horizontal edges, the cycle-crossing points are the outer two pieces of the three center pieces in the edge; for the tabs, they are the middle piece of both horizontal sections of each tab. We lay out this cycle in the checkerboard, starting from an edge that is crossed twice consecutively (there is at least one such edge, corresponding to a leaf in the doubled tree). At each such immediate recrossing, after entering at a cycle-crossing point, the cycle steps into the face for a limited distance, then leaves via the other cycle-crossing point (see Figure 9.4). At other crossings, the cycle walks the boundary of the face, starting with the adjacent edge nearest the cycle-crossing point, inset by 7 spaces from horizontal edges and 8 from vertical edges (being off by one is necessary for the parity to work out). This inset ensures that the cycle can enter and leave at immediate recrossings without interfering with the boundary walk. The walk crosses subsequent edges, using the first cycle-crossing point encountered (and possibly resumes in the same direction after reentering via the other cycle-crossing point on that edge). The cycle walks the outer face in the same way as any other face. We place black pieces at every second point on the cycle, in the opposite parity class from the red empty pieces (so they are adjacent, not coincident, at edges). In the figures, these are the black pieces on orange squares.

We now place the *motive* pieces, one per player, which will be the pieces moving during their turns. Only the motive pieces are positioned to make captures, so because captures must be made if available, only the motive pieces will move during the next two plies. We place Black's motive piece between two red pieces in the cyan parity class along a forced edge, where one of the adjacent red pieces is the point of a cycle crossing. The cycle crossing is moved one unit to now cross at the black motive piece; we can do this without interfering with the rest of the instance, because the crossings were placed at least three spaces apart in the previous step. (See Figure 9.5.) Red's motive piece is then placed between two black pieces in the orange parity class. Both motive

---

[4] The dual graph has a (dual) vertex for each face, and a (dual) edge between two (dual) vertices corresponding to faces that share an edge.

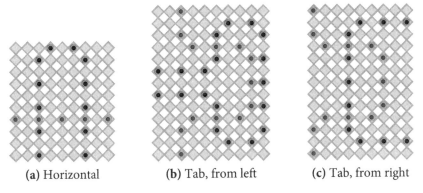

(a) Horizontal   (b) Tab, from left   (c) Tab, from right

Figure 9.4. Behavior of the cycle when immediately recrossing edges.

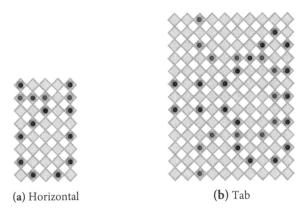

(a) Horizontal   (b) Tab

Figure 9.5. Placing the black motive piece.

pieces are kings, allowing them to move in any direction while traversing the graph or cycle.

If the input Forced-Edge Hamiltonicity instance contains a Hamiltonian cycle including all the forced edges, then Black can capture with the motive piece all red pieces along all forced edges in the cyan parity class, with the motive piece returning to its initial position. Red is then forced to capture the entire orange cycle, which contains all black pieces (including the black motive piece), and win the game. Thus the lose-in-1 instance has an answer of YES.

If the input Forced-Edge Hamiltonicity instance does *not* contain a Hamiltonian cycle including all the forced edges, Black will not be able to capture any red pieces on at least one of the forced edges in the cyan parity class. Then Red's capture sequence will end, because the red piece at a cycle crossing of that forced edge will block its motive piece from continuing to capture, so Red will not win the game with this move. Thus, the lose-in-1 instance has an answer of NO. □

The second step is to show membership in NP. The goal here is to show that, when the answer to the problem is YES, then there is a succinct certificate (essentially, a solution) that easily proves that the answer is YES.

Theorem 3. *Lose-in-1 checkers is in NP.*

Proof. The certificate is Black's move (capture sequence)—whose length is less than the number of pieces—that forces Red to win. We give an efficient (polynomial-time) algorithm to verify such a certificate. First we execute the capture sequence to recover the board state at the beginning of Red's turn. Now we need to decide whether Red has the option to *not* win in one move, in which case, the certificate is invalid. If Red cannot capture, then this is the case: Red either cannot move at all (and thus loses) or any move does not immediately win. Thus assume that Red can capture (otherwise return INVALID). For each red piece r that can capture, we will check whether moving r can lead to Red "getting stuck," leaving some black pieces uncaptured.

Construct the graph for solving mate-in-one from Fraenkel et al. [4] as described earlier in section 1.2, with vertices representing reachable positions for r by jumping and edges representing (eventually) jumpable black pieces. For the certificate to be valid, there must be at least one winning move using r, so the graph must include all black pieces as edges, and it must have an Eulerian path starting at r (otherwise return INVALID). Thus, the graph has zero or two vertices of odd degree, and if there are two, one of them must be r.

If all vertices of the graph have even degree, then check whether the graph minus the starting vertex for r has any cycles. If it does have cycles, then Red can avoid that cycle and cover the rest of the graph, thereby getting stuck at the start vertex. Conversely, if Red can get stuck, then there must be a cycle in the remainder, because the graph is Eulerian. Therefore we can detect validity in this case.

If the graph has two vertices r, s of odd degree, then connect them both to a new vertex r', and apply the algorithm for the previous case starting from r'. Cycles in this graph including r' are equivalent to paths in the old graph starting at r and ending at s. Therefore we can again detect validity. □

4 Always-Jumping Checkers Is PSPACE-Complete

In this section, we prove that the following computational decision problem is PSPACE-complete:

Problem 3 (Always-jumping checkers).   *Given a checkers game configuration in which all moves must capture a piece, does the first player have a winning strategy?*

Theorem 4.   *Always-jumping checkers is PSPACE-complete.*

Again, the proof consists of two parts—PSPACE-hardness and membership in PSPACE. In this case, however, membership in PSPACE is straightforward, because always-jumping checkers falls into the class of bounded 2-player games [6]: Each move strictly decreases the number of pieces, so the number of moves is bounded by the initial number of pieces. To prove PSPACE-hardness, we reduce from a known PSPACE-hard problem called Bounded 2-Player Constraint Logic (B2CL) [6]. For this proof, we assume familiarity with Hearn and Demaine [6] or the corresponding lectures of Demaine [3].

Proof. To reduce from B2CL, we construct VARIABLE, FANOUT, CHOICE, AND, and OR gadgets in always-jumping checkers that simulate the corresponding gadgets from B2CL. Refer to Figure 9.6 and the following definitions.

Wire.   A wire is simply a line of alternating red pieces and empty spaces. A black piece can traverse the entire length of a wire during a single move. As every piece is a king, a wire can turn at any empty square along its length.

Parity shift.   During a single turn, a piece is confined to one of four parity classes (one of the four squares in a 2×2 block of squares). In most of our gadgets, the parity of the incoming and outgoing pieces differs. We handle this with a SHIFT gadget, as shown in Figure 9.6a (which can also be seen as removing one of the output wires from a CHOICE gadget). Repeated applications of the SHIFT gadget (and its 90° rotation) can move a piece to any parity class of the board, so a combination of wires, turns, and shifts can move a piece to an arbitrary dark square.

VARIABLE.   When the game begins, players take turns choosing to either activate or deactivate the VARIABLE gadgets, shown in Figure 9.6b. If Black moves first on a given variable, he must continue capturing red pieces along the wire, corresponding to setting the variable to true. If Red moves first, she can capture the black piece, corresponding to setting the variable to false. Once set, a variable never changes value.

FANOUT, CHOICE, AND, OR.   These gadgets are all based on a similar construction. Triples of red pieces prevent the black pieces from capturing, and pairs of black pieces prevent the red pieces from capturing. When a black piece arrives

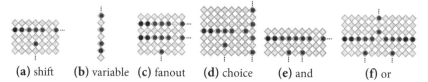

(a) shift   **(b)** variable   **(c)** fanout   **(d)** choice   **(e)** and   **(f)** or

Figure 9.6. Gadgets in our reduction from bounded 2-player constraint logic [6] to always-jumping checkers. For example, Black can get a piece to the top output of the OR gadget only if he can get a piece to one of the inputs, and Black can get a piece to the top output of the AND gadget only if he can get a piece to both of the inputs. Gadgets (b)–(e) also form a reduction from bounded nondeterministic constraint logic [6] to cooperative always-jumping checkers.

via an input wire, it captures the middle piece of some triple(s), allowing other black pieces to leave the gadget along the output wires. In the FANOUT gadget, the arriving black piece breaks both triples, allowing a black piece to exit along each output wire. CHOICE also has two output wires, but only one black piece can leave, so Black must choose which wire to activate. In the AND gadget, both black pieces must arrive to break their respective triples, while in the OR gadget, the arrival of a black piece along either wire enables a black piece to leave. Even if both OR inputs are activated, the output wire is consumed when the first black piece leaves.

Board layout. Our reduction translates the given B2CL graph directly using the gadgets above, adding turns and shifts as necessary for layout. Black is the first player to move in the resulting checkers game.

In B2CL, the first player wins if he can flip a particular edge (specified as part of the instance), which corresponds to activating an AND output wire in the checkers game. That AND output wire leads to a long series of SHIFT gadgets, giving Black a large number F_B of free moves if Black can activate the AND output.

In B2CL, Red needs to be able to pass, so we also provide Red with a large number F_R of free moves using a collection of F_R isolated gadgets, consisting of a pair of red pieces adjacent to a single black piece: (Recall that the player to move loses if he or she cannot capture.) We set F_R to twice the number of gadgets not counting the long chain of SHIFT gadgets, which is an upper bound on the number of black moves if the formula is not satisfied. We then set $F_B = 2 F_R$, so that Red runs out of moves if Black can activate the desired series of SHIFT gadgets. Therefore, Black wins the checkers game exactly when the first player wins the B2CL instance, so always-jumping checkers is PSPACE-complete.   □

5 Cooperative Versions Are NP-Complete

In cooperative lose-in-1 (win-in-2) checkers, players collaborate to force one of them to win in two moves. The computational decision problem for this game asks whether the current player has a move such that the other player can win on their next move, which is almost identical to the competitive lose-in-1 decision problem. Indeed, the same reduction as in Section 3 shows cooperative lose-in-1 (win-in-2) checkers is NP-complete.

In cooperative always-jumping checkers, players collaborate to remove all pieces of one color, and so the game becomes essentially a single-player puzzle. Because at least one piece is still consumed on every turn, the game is bounded, so it is in NP. Furthermore, we can prove NP-hardness by reducing from Bounded Nondeterministic Constraint Logic (Bounded NCL), which is NP-complete [6]. It happens that the same FANOUT, CHOICE, AND, and OR gadgets from Section 4 can form this reduction, which proves that cooperative always-jumping checkers is NP-complete.

6 Checkers Font

In Figure 9.7, we show a collection of cooperative always-jumping checkers puzzles whose solutions trace letters of the alphabet and digits 0–9, as shown in Figure 9.8. Each puzzle is a plausible end game of cooperative always-jumping checkers. In each puzzle, Black moves first and, in any solution that eliminates all pieces of one color, the paths traced by the movement of the pieces combine to create the letter or digit. In some puzzles, Black wins (e.g., 'b' and 'w'), while in other games, Red wins (e.g., 'r' and 'y').

7 Open Problems

The main remaining open problem is to analyze the computational complexity of suicide (misère) checkers. A natural conjecture is that this game is EXPTIME-complete, like normal checkers.

Another natural open problem is to analyze the computational complexity of cooperative checkers, where the players together try to eliminate all pieces of one color. We have shown that this problem is NP-complete when limited to just two moves or when we add the always-jumping rule.

A final open problem is the computational complexity of *mate-in-2* checkers: Making a move such that, no matter what move the opponent makes, you can win in a second move. Given that mate-in-1 checkers is easy, while lose-in-1 is NP-complete, we might naturally conjecture that mate-in-2 checkers is NP-complete, but our proofs do not immediately apply.

Figure 9.7. A puzzle font for checkers. Each letter and digit is represented by an 8 × 8 cooperative always-jumping checkers puzzle, where the goal is to eliminate all pieces of one color. Black always moves first, but does not always win. The movement of the pieces in a puzzle solution traces the letter or digit, as revealed in Figure 9.8. See http://erikdemaine.org/fonts/checkers/ for an interactive version.

Figure 9.8. Solved font: Paths traced by the solutions to the 8 × 8 cooperative always-jumping checkers puzzles shown in Figure 9.7.

Acknowledgments

This work was done in part during an open problem session about hardness proofs, which originally grew out of an MIT class (6.890, Fall 2014). We thank the other participants of the session, in particular Mikhail Rudoy, who proved Theorem 3 and allowed us to include his proof in this chapter.

References

[1] E. R. Berlekamp, J. H. Conway, and R. K. Guy. *Winning Ways for Your Mathematical Plays,* second edition. A. K. Peters, Wellesley, MA, 2001–2004.

[2] J. H. Conway. *On Numbers and Games,* second edition. A. K. Peters, Boca Raton, FL. 1976.

[3] E. D. Demaine. Algorithmic lower bounds: Fun with hardness proofs. Online at http://courses/casil.mit.edu/6.890/. Last accessed January 2019.

[4] A. S. Fraenkel, M. R. Garey, D. S. Johnson, T. Schaefer, and Y. Yesha. The complexity of checkers on an $N \times N$ board—preliminary report. In *Proceedings of the 19th Annual Symposium on Foundations of Computer Science,* IEEE Computer Society, Washington, DC, 1978, 55–64.

[5] M. Fierz. Suicide checkers. Online at http://www.fierz.ch/suicidecheckers.php. Last accessed June 2018.

[6] R. A. Hearn and E. D. Demaine. *Games, Puzzles, and Computation.* A. K. Peters, Boca Raton, FL, 2009.

[7] C. Hierholzer. Über die Möglichkeit, einen Linienzug ohne Wiederholung und ohne Unterbrechung zu umfahren. *Math. Annalen* **6** no. 1 (1873) 30–32. (Sadly, Hierholzer did not live long enough to actually finish this paper; it was reconstructed from memory by Jacob Lüroth.)

[8] J. Plesník. The NP-completeness of the Hamiltonian cycle problem in planar digraphs with degree bound two. *Infor. Processing Lett.* **8** no. 4 (1979) 199–201.

[9] J. M. Robson. N by N Checkers is EXPTIME complete. *SIAM J. Comp.* **13** no. 2 (1984) 252–267.

[10] J. Schaeffer, N. Burch, Y. Björnsson, A. Kishimoto, M. Müller, R. Lake, P. Lu, and S. Sutphen. Checkers is solved. *Science* **317** no. 5844 (2007) 1518–1522.

[11] J. A. Storer. On minimal-node-cast-planar embeddings. *Networks* **14** no. 2 (1984) 181–212.

[12] Wikipedia. Draughts. Online at https://en.wikipedia.org/wiki/Draughts. Last accessed June 2018.

10

◇◇

CHUTES AND LADDERS WITHOUT CHUTES OR LADDERS

Darren B. Glass, Stephen K. Lucas, and Jonathan S. Needleman

The board game Chutes and Ladders dates back to ancient India, but was popularized in the United States by the Milton Bradley company in the middle of the twentieth century. The basic setup of the game is simple: There are squares numbered from one to one hundred, and players use a spinner to obtain an integer generated uniformly at random between one and six. They then move forward this number of spaces. The winner is the first player to make it through the complete board, but there are two caveats. First, the player must land exactly on the last square, and if they overshoot, then they lose their turn. Second, landing on certain squares causes the player to move either up the board via a "ladder" or down the board via a "chute".

Some mathematicians have attempted to model this game and to answer different questions about it. As far as we can tell, Daykin, Jeacocke, and Neal [5] were the first authors to model the game using Markov processes, a technique that several others have used in the past. More recently, Cheteyan, Hengeveld, and Jones [3] used a Markov model to explore how varying the size of the spinner would affect the expected length of a game. In particular, they noticed that longer spinners have higher expected values that allow one to move to the end of the board faster, but at the same time, such spinners make it harder to land exactly on the final square and finish the game. Their work found where the trade off occurred between these two competing ideas. For the traditional board, they find that the optimal spinner size is 15.

Connors and Glass [4] followed up on this paper and showed (among other things) that the expected number of turns it takes to win a game on a board with p squares is the same with a spinner of size p or size $p - 1$. Their analysis starts with assuming a board with no chutes or ladders, adds a ladder, and then generalizes to an arbitrary number of chutes or ladders. After a presentation on this work at the MOVES 2015 conference, we started discussing some related questions. Those discussions led to this chapter.

We begin by assuming there are no chutes or ladders on a board with p squares. One moves forward a number of spaces chosen uniformly at random from $\{1, 2, \ldots, s\}$ until one lands exactly on the final square p. We are interested primarily in answering two questions: What is the probability that any particular square is visited when moving through the board? And, what is the expected number of moves it will take for a player to reach the final square?

One approach would be to use elementary combinatorics or generating functions to find the probability distribution that k dice can sum to each number up to p for $k = 1, 2, 3, \ldots$, and sum. This approach is cumbersome for large values of p and s, so we instead develop analytic solutions using recurrence relations. Althoen et al. [1] claim that if $s = 6$ and $p = 100$, then a game with no chutes or ladders is expected to take "almost exactly 33" moves, but our work shows that the actual answer is much closer to $100/3$.

In the final section, we look further at what would happen if one has a single ladder or single chute on the board. In particular, we determine the optimal placement for a given ladder of length L if one's goal is to minimize the expected length of the game. To do this, we first generalize our earlier results to compute the probability of landing on any given square on a board with exactly one ladder or chute.

1 Probability of Visiting a Square

1.1 Recurrence Relations

Let a_i be the probability that square i is visited in a game. We wish to calculate, or at least approximate, the value of a_i for all i.

We begin by looking at the squares at the beginning of the board, where we can give exact values for a_i. To calculate a_i for $1 \le i \le s$, note that there are $\binom{i-1}{j-1}$ ways of landing on square i in exactly j moves, based on the squares that one lands on leading to this square. Moreover, each of these paths occurs with probability exactly $(1/s)^j$. So we obtain

$$a_i = \sum_{j=1}^{i} \left(\frac{1}{s}\right)^j \binom{i-1}{j-1} = \frac{1}{s}\left(1 + \frac{1}{s}\right)^{i-1} = \frac{(s+1)^{i-1}}{s^i}.$$

For values of i that are not too close to either end of the board, you can reach square i from squares $i-1, i-2, \ldots, i-s$ with a single roll of the die, each with probability $1/s$. As a result, we have the following recurrence relation for $s < i \le p - s$:

$$a_i = \frac{1}{s}(a_{i-1} + a_{i-2} + \cdots + a_{i-s}).$$

Squares numbered less than or equal to s have different relations, where a square can be visited by a direct move from starting off the board, or a move from a previous square. Thus

$$a_1 = \frac{1}{s}, \quad a_2 = \frac{1}{s}(1+a_1), \quad a_3 = \frac{1}{s}(1+a_1+a_2), \quad \cdots$$

$$a_s = \frac{1}{s}(1+a_1+a_2+\cdots+a_{s-1}).$$

An easier approach is to use the same recurrence relation for all squares and encompass all the initial special cases using appropriate initial conditions, so

$$a_i = \frac{1}{s}\sum_{j=1}^{s} a_{i-j} \text{ for } 1 \le i \le p-s,$$

with

$$a_0 = 1, \text{ and } a_{-1} = a_{-2} = \cdots = a_{-s+1} = 0.$$

From both computational and theoretical viewpoints, it is more convenient to start with a first element, so with a shift $b_i = a_{i+s}$, we have

$$b_i = \frac{1}{s}\sum_{j=1}^{s} b_{i-j} \text{ for } s+1 \le i \le p, \text{ with } b_1 = b_2 = \cdots = b_{s-1} = 0, \text{ and } b_s = 1.$$

The recurrence relation becomes more complicated when close to the end of the board. Since we have to finish at square p, if we are less than s squares from square p, we won't move if the roll is too large. As a result, the contribution to the probability of landing on a later square is $1/t$, where t is the number of reachable squares, not $1/s$. So our final recurrence relation becomes

$$c_i = \sum_{j=1}^{s} \frac{c_{i-j}}{\min(s, p+s-i+j)},$$

with

$$c_1 = c_2 = \cdots = c_{s-1} = 0, \text{ and } c_s = 1,$$

for $i = s+1, s+2, \ldots, s+p$. The minimum in the denominator is made more complicated by the shift so that the first square is at position $s+1$.

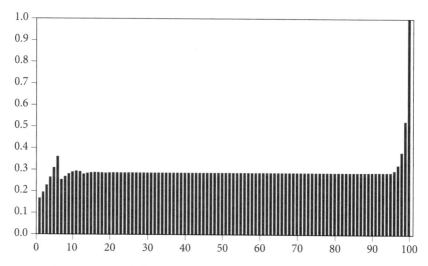

Figure 10.1. Probability of visiting each square with $s = 6$ and $p = 100$.

1.2 Numerical Results

Before we attempt to prove any analytic results, it is useful to look at the numerical results that one can obtain by using the recurrence relations given in the previous section. In particular, Figure 10.1 shows the probability of visiting a given square using a six-sided die on a one hundred square board. The probability of square one is $1/6$, and it increases to $16807/46656 \approx 0.36023$ at square six. After this, squares take more than one roll to reach, and the probability reduces, then approaches a constant. The probability suddenly increases at square 96, since there are only five possible locations from square 95, and eventually reaches the expected probability one at square 100.

Solving our recurrence relation using standard double-precision arithmetic, the constant appears to be $2/7$. Using exact integer arithmetic, we find that square 50 has probability of being visited

$$\frac{23093751944542498452496533649790941 0567}{80828127746476406064313960045653629337 6},$$

and as square number increases, so does the number of digits in the fractions. However, converted to a 20 decimal-place approximation, square 50s probability is

$$0.28571430006368345737,$$

the same as 2/7 to seven decimal places. The further we go, the closer the fraction is to 2/7, and at square 95, they agree to thirteen decimal places. Experimentally, we can verify that this asymptotic behavior in the probability of visiting a square occurs for all possible die sizes s. It appears that in general, the probability approaches $2/(s+1)$, which we will prove in the next section.

1.3 Proofs

As discussed earlier, the sequence $\{b_i\}$ satisfies a certain recurrence relation. The main result of this section will be to give the limit of a sequence satisfying this recurrence for any choice of initial conditions. We will then use it to solve explicitly for the probability of landing on any given square during a game.

Theorem 1. *Let a sequence $\{b_i\}$ be defined by a set of initial conditions $b_1, \ldots, b_s$ and the recurrence relation $b_i = \frac{1}{s} \sum_{j=1}^{s} b_{i-j}$ for $i > s$. Then this sequence converges to the value $\frac{2}{s(s+1)} \sum_{k=1}^{s} k b_k$.*

Proof. We note that the sequence is defined by a linear homogenous recurrence relation of order s with constant coefficients. It follows from standard results in the field that the solutions are of the form $b_i = \sum_{j=1}^{s} C_j t_j^i$, where the t_js are the roots of the characteristic polynomial

$$P(t) = t^s - \frac{1}{s}t^{s-1} - \frac{1}{s}t^{s-2} - \cdots - \frac{1}{s}t - \frac{1}{s},$$

and the C_js are constants that depend on the initial conditions.

In general, the roots of $P(t) = 0$ cannot be found in closed form and often come in complex pairs. However, it is immediately obvious that $P(1) = 0$. Dividing $P(t)$ by $\frac{t-1}{s}$ gives $Q(t) = st^{s-1} + (s-1)t^{s-2} + \cdots + 2t + 1$. The roots of $Q(t)$ will precisely give us the roots of $P(t)$ other than $t = 1$, so we turn our attention to $Q(t)$.

Recall the Gauss-Lucas Theorem (see, for example Marden [6, p. 22]), which states that the convex hull of the zeros of a polynomial contains the zeros of the derivative of the polynomial. Note that one antiderivative of $Q(t)$ is

$$t^s + t^{s-1} + \cdots + t^2 + t + 1 = \frac{t^{s+1} - 1}{t - 1},$$

whose roots are the $(s+1)^{th}$ roots of unity other than 1 itself. Thus, the Gauss-Lucas Theorem tells us that the roots of $Q(t)$ must have modulus less than 1.

As a result, we can find that a closed form for the b_i is given by

$$b_i = C_1 + \sum_{j=2}^{s} C_j t_j^i,$$

where each $|t_j| < 1$, and in particular, as i increases, we have that $b_i \to C_1$. Given the initial conditions, one could attempt to solve for the value of C_1 by solving the system of linear equations above. However, this would require solving an increasingly nasty system of linear equations depending on roots that cannot be found in closed form, and the solution for each value of s would be independent. Luckily, we can find this constant without explicitly solving these systems.

For our b recurrence relation, let $L_{\vec{v}} = \lim_{i \to \infty} b_i$, where the list of initial conditions is $\vec{v} = \langle b_1, b_2, \ldots, b_s \rangle \in \mathbb{R}^s$. We have already proven such a limit exists. Since our recurrence relation is linear, for $\vec{v}_1, \vec{v}_2 \in \mathbb{R}^s$ and $c_1, c_2 \in \mathbb{R}$, we have that

$$L_{c_1 \vec{v}_1 + c_2 \vec{v}_2} = c_1 L_{\vec{v}_1} + c_2 L_{\vec{v}_2}.$$

Finally, let $\vec{e}_k \in \mathbb{R}^s$ be the vector that has a 1 in position k and 0 in all other positions, and note that the initial conditions for our problem are $\vec{e}_s$.

Now consider $L_{\vec{e}_{k+1} - \vec{e}_k}$. The initial conditions are

$$b_k = -1, b_{k+1} = 1, \text{ and } b_j = 0, \text{ for } 1 \le j \le s, j \ne k, k+1.$$

Substituting into the recurrence relation, cancellation of the plus and minus 1s shows that additionally,

$$b_i = 0 \text{ if } s+1 \le i \le s+k.$$

Combining these equations, we have

$$b_{k+1} = 1, \text{ and } b_i = 0, \text{ for } k+2 \le i \le s+k.$$

Since shifting the initial conditions to the right has no effect on the limiting value, and this combination of initial values is $\vec{e}_1$, we have $L_{\vec{e}_{k+1} - \vec{e}_k} = L_{\vec{e}_1}$. Expanding gives

$$L_{\vec{e}_1} = L_{\vec{e}_{k+1} - \vec{e}_k} = L_{\vec{e}_{k+1}} - L_{\vec{e}_k},$$

so

$$L_{\vec{e}_k} = (L_{\vec{e}_k} - L_{\vec{e}_{k-1}}) + (L_{\vec{e}_{k-1}} - L_{\vec{e}_{k-2}}) + \cdots + (L_{\vec{e}_2} - L_{\vec{e}_1}) + L_{\vec{e}_1}$$
$$= k L_{\vec{e}_1}.$$

TABLE 10.1.
Rates of convergence to a constant with various die sizes s

s	2	3	4	5	6
$|C_2|$	0.5	0.57735	0.642	0.6920	0.7303

In addition, if all the initial conditions are constant, our recurrence relation gives all constant terms, so $L_{\langle 1,1,\dots,1\rangle} = 1$, and

$$1 = L_{\langle 1,1,\dots,1\rangle} = \sum_{k=1}^{s} L_{\vec{e}_k} = \sum_{k=1}^{s} k L_{\vec{e}_1} = \frac{s(s+1)}{2} L_{\vec{e}_1}.$$

So $L_{\vec{e}_1} = \dfrac{2}{s(s+1)}$, and

$$L_{\vec{v}} = \sum_{k=1}^{s} b_k L_{\vec{e}_k} = \sum_{k=1}^{s} b_k k L_{\vec{e}_1} = \frac{2}{s(s+1)} \sum_{k=1}^{s} k b_k.$$

□

Note that the rate of convergence is the modulus of the second largest root of $Q(t) = 0$. Table 10.1 shows the modulus of the second largest root of $Q(t) = 0$ for various s, which shows the actual rate of convergence to a constant. The larger the die is, the slower the convergence will be, so this asymptotic result will not be applicable when the die size is close to the size of the board. A (pessimistic) upper bound on the largest modulus can be found using the classic Enestrom-Kakeya theorem (see Anderson et al. [2] for an exposition of this result) as $(s-1)/s$.

Corollary 1. *The probability of visiting a square $i < p - s$ in a game with no chutes or ladders approaches $2/(s+1)$ when using an s-sided die, as long as p is sufficiently large.*

Proof. Since we are staying away from the end of the board, we can use the b recurrence relation for our probability of visiting a square. We are then looking for the limiting value $L_{\vec{e}_s}$, which by Theorem 1 is $2/(s+1)$. □

This result confirms the numerical observations of the previous section.

1.4 End-Game

Now that we understand what happens to the value of a_i for i that are before the end of the game, we wish to look at the values of a_i for $i > p - s$. Looking back to the numerical results portrayed in Figure 10.1, the probability of landing at squares near the end of the board increases from the asymptotic $2/(s+1)$ to 1. To see this analytically, recall from Section 1.1 that if d_i is the probability of landing on square i, then for $i > p - s$, we have

$$d_i = \sum_{j=1}^{i-p+s-1} \frac{d_{i-j}}{p-i+j} + \sum_{j=i-p+s}^{s} \frac{d_{i-j}}{s}.$$

If we assume the asymptotic result $d_i = 2/(s+1)$ for values of $i \leq p - s$ in the second sum, then we get

$$d_i = \frac{p-i+1}{s} \frac{2}{s+1} + \sum_{j=1}^{i-p+s-1} \frac{d_{i-j}}{p-i+j}.$$

After a change of variables where we let $a = i + s - p - 1$, and $k = a + 1 - j$, this becomes

$$d_{p-s+a+1} = \frac{2(s-a)}{s(s+1)} + \sum_{k=1}^{a} \frac{d_{p-s+k}}{s-k},$$

for $a = 1, 2, \ldots, s - 1$, and $d_{p-s+1} = 2/(s+1)$. To solve, note that

$$d_{p-s+b+1} - d_{p-s+b} = \frac{2(s-b)}{s(s+1)} + \sum_{k=1}^{b} \frac{d_{p-s+k}}{s-k} - \frac{2(s-b+1)}{s(s+1)} - \sum_{k=1}^{b-1} \frac{d_{p-s+k}}{s-k}$$

$$= -\frac{2}{s(s+1)} + \frac{d_{p-s+b}}{s-b},$$

or

$$d_{p-s+b+1} = \frac{s-b+1}{s-b} \cdot d_{p-s+b} - \frac{2}{s(s+1)}.$$

Then

$$d_{p-s+2} = \frac{s}{s-1} \cdot \frac{2}{s+1} - \frac{2}{s(s+1)} = \frac{2s^2 - 2s + 2 \cdot 1}{s(s+1)(s-1)},$$

$$d_{p-s+3} = \frac{s-1}{s-2} \cdot \frac{2s^2 - 2(s-1)}{s(s+1)(s-1)} - \frac{2}{s(s+1)} = \frac{2s^2 - 4s + 2(1+2)}{s(s+1)(s-2)},$$

$$d_{p-s+4} = \frac{2s^2 - 6s + 2(1+2+3)}{s(s+1)(s-3)},$$

and in general for all $0 \le a \le s - 1$, we have

$$d_{p-s+1+a} = \frac{2s^2 - 2as + a(a+1)}{s(s+1)(s-a)},$$

which can be proven by induction or substitution back in the recurrence relation. This implies, for example, that $d_p \approx 1$ and $d_{p-1} \approx \frac{1}{2} + \frac{1}{s(s+1)}$.

For example, with $s = 6$, $p = 100$, and $d_{95} = 2/7$, we get $d_{96} = 31/105$, $d_{97} = 9/28$, $d_{98} = 8/21$, $d_{99} = 11/21$, and $d_{100} = 1$. These values match the double-precision numeric results and are extremely close to the exact integer arithmetic solutions apart from d_{100}, which (as we would hope) is exactly 1 in both cases.

2 Average Number of Moves

As well as the probability of visiting a square, we are interested in the average number of moves to complete a board of length p with an s-sided die. One choice, which is the approach of Cheteyan et al. [3], is to set up a Markov chain for moving along a board and use the appropriate theory to find the average number of moves. Our case is simpler, since there are no chutes or ladders, and with a six-sided die and one hundred squares, we get the following average number of moves.

$$\frac{77793808048991155069512637767746406705805011 7 \cdots 99952210986407}{233381424146973203195262584004221615132438739 \cdots 7639351484416}.$$

Converted to decimal, this is

$$33.333333333333370756088277230775175163654 19\ldots,$$

which is remarkably close to $100/3$.

2.1 Numerical Results

As discussed, one can use computational software to find the expected length of a game for specific values of s and p. Doing so leads to the (decimal) results and their approximations in Table 10.2. As with the probability of visiting a square, we see that the actual results are very close to nice rational approximations. We explore these values further in the next section.

Our observations that these exact numerical values were very close to simple rational numbers led to this chapter.

TABLE 10.2.
Average length of a game with various s and p, decimal approximation and simple rational approximation

s	p	Decimal approximation	Rational Approximation
2	100	67.5555555555555555555555555555485434586	608/9
3	100	51.833333333333333333333333324532	311/6
4	100	42.79999999999999999999623930575	214/5
5	100	37.1111111111111110226200277976	334/9
6	100	33.333333333333707560882772308	100/3
7	100	30.749999999977319122910955801	123/4
8	100	28.962962962908224215038451593	782/27
9	100	27.733333333984130262640875652	416/15
10	100	26.909090912457471350869519499	296/11
2	101	68.222222222222222222222222222222257282707	614/9
3	101	52.333333333333333333333324940	157/3
4	101	43.199999999999999999345366233	216/5
5	101	37.444444444444443129652103489	337/9
6	101	33.619047619047644392317347805	706/21
7	101	30.999999999998446348548630232	31
8	101	29.185185185168059857160935432	788/27
9	101	27.933333335573226983436743734	419/15
10	101	27.090909095264353061784167149	298/11
6	102	33.904761904761899679558924689	712/21
6	103	34.190476190476174006825546093	718/21
6	104	34.476190476190469198655199941	724/21

2.2 Recurrence Relations

Let E_{ps} be the average number of moves on a board with p using an s-sided die. As an alternative to the Markov chain approach, we can build a recurrence relation for E_{ps}. If you have a one-square board, you have a $1/s$ chance of reaching the end (in one move) and an $(s-1)/s$ chance of wasting the move and staying put. Since rolls are independent, this says that

$$E_{1s} = \frac{1}{s} \cdot 1 + \frac{s-1}{s}(1 + E_{1s}), \quad sE_{1s} = 1 + (s-1) + (s-1)E_{1s}, \quad E_{1s} = s,$$

as verified by simulation and Markov chain. Then with two squares, you could roll a 1 to square one, which means one roll plus average number of rolls for a one-square board. Or you could roll a 2, which takes you to square two, the end, or not move at all, so

$$E_{2s} = \frac{1}{s}(1 + E_{1s}) + \frac{1}{s} \cdot 1 + \frac{s-2}{s}(1 + E_{s2}),$$

$$sE_{2s} = 1 + s + 1 + (s-2) + (s-2)E_{2s},$$

$$E_{2s} = s,$$

which is valid for $s \geq 2$. If $s \geq 3$, then similarly,

$$E_{3s} = \frac{1}{s}(1 + E_{2s}) + \frac{1}{s}(1 + E_{1s}) + \frac{1}{s} \cdot 1 + \frac{s-3}{s}(1 + E_{3s}),$$

$$sE_{3s} = 1 + s + 1 + s + 1 + s - 3 + (s-3)E_{3s},$$

$$E_{3s} = s.$$

By induction, we have that $E_{ps} = s$. An alternative approach, taken in Connors and Glass [4], is to argue that if one can win beginning on the first roll, then at any point in the game, there is a unique roll that will win the game, so the probability of winning on a given roll is precisely $\frac{1}{s}$. Straightforward results on Bernoulli trials then show that the expected number of rolls it will take to win the game is s.

If $p > s$, then more than one roll is needed to reach the end, so this approach does not directly work. However, we can compute that

$$E_{ps} = \frac{1}{s}(1 + E_{p-1,s}) + \frac{1}{s}(1 + E_{p-2,s}) + \frac{1}{s}(1 + E_{p-3,s}) + \cdots + \frac{1}{s}(1 + E_{p-s,s}),$$

or

$$E_{ps} = 1 + \frac{1}{s}(E_{p-1,s} + E_{p-2,s} + \cdots + E_{p-s,s}),$$

which is a linear nonhomogeneous recurrence relation of order s with constant coefficients and initial conditions

$$E_{1s} = E_{2s} = \cdots = E_{ss} = s.$$

With $s = 6, 10$, the recurrence relation gives the results in Figure 10.2. Note that after an initial flat part due to the initial conditions, the solutions appear to look like straight lines very quickly.

For small values of s, one can explicitly solve these recurrence relations. For example, we can compute explicitly that

$$E_{p2} = \frac{8 + 6p}{9} - \frac{8}{9}\left(-\frac{1}{2}\right)^p,$$

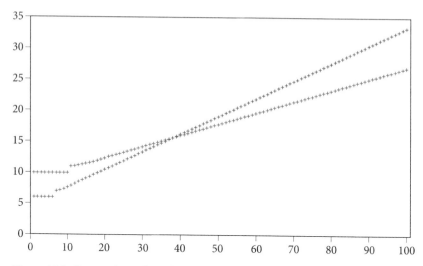

Figure 10.2. Expected number of moves (y-axis) on a board with $s=6$ and $s=10$, $p=1,2,\ldots,100$ (x-axis).

and

$$E_{p3} = \frac{11+3p}{6} - \frac{20+23i\sqrt{2}}{24}\left(\frac{-1}{1+i\sqrt{2}}\right)^{p-1} - \frac{20-23i\sqrt{2}}{24}\left(\frac{-1}{1-i\sqrt{2}}\right)^{p-1}.$$

Similarly,

$$E_{p4} \approx \frac{14+2p}{5}, \quad E_{p5} \approx \frac{34+3p}{9}, \quad E_{p6} \approx \frac{100+6p}{21},$$

where the exponential decay becomes less pronounced as s increases but is still significant. In fact, since these results for E_{ps} are only valid for $p > s$, the decaying terms are practically nonexistent in Figure 10.2. As expected, these results match the Markov chain results, but unfortunately were found case by case using Maple.

We will fix s and consider what happens to the value of E_{ps} as p varies. To do this, let us define the auxiliary variable

$$e_p = E_{ps} - \frac{2p}{s+1}.$$

Substituting into our recurrence relation, we get

$$e_p = E_{ps} - \frac{2p}{s+1}$$

$$= 1 + \frac{1}{s}(E_{p-1,s} + E_{p-2,s} + \cdots + E_{p-s,s}) - \frac{2p}{s+1}$$

$$= 1 + \frac{1}{s}\left(e_{p-1} + \frac{2(p-1)}{s+1} + \ldots + e_{p-s} + \frac{2(p-s)}{s+1}\right) - \frac{2p}{s+1}$$

$$= \frac{1}{s}(e_{p-1} + e_{p-2} + \cdots + e_{p-s}) + 1 + \frac{2}{s(s+1)}((p-1) + \ldots + (p-s) - ps)$$

$$= \frac{1}{s}(e_{p-1} + e_{p-2} + \cdots + e_{p-s}),$$

with

$$e_1 = s - \frac{2 \cdot 1}{s+1}, \quad e_2 = s - \frac{2 \cdot 2}{s+1}, \quad \ldots, \quad e_s = s - \frac{2 \cdot s}{s+1}.$$

In particular, this is the same recurrence relation as in Theorem 1, and therefore for large i we have that e_i approaches $L_{\vec{v}}$ with $\vec{v} = \langle e_1, e_2, \ldots, e_s \rangle$. In particular, we have

$$\lim_{i \to \infty} e_i = L_{\langle e_1, e_2, \ldots, e_s \rangle} = \frac{2}{s(s+1)} \sum_{k=1}^{s} k e_k$$

$$= \frac{2}{s(s+1)} \sum_{k=1}^{s} k\left(s - \frac{2k}{s+1}\right)$$

$$= \sum_{k=1}^{s} \frac{2}{s+1} \cdot k - \frac{4}{s(s+1)^2} \cdot k^2$$

$$= \frac{2}{s+1} \cdot \frac{s(s+1)}{2} - \frac{4}{s(s+1)^2} \cdot \frac{s(s+1)(2s+1)}{6}$$

$$= s - \frac{2(2s+1)}{3(s+1)} = \frac{3s^2 - s - 2}{3(s+1)} = \frac{(3s+2)(s-1)}{3(s+1)}.$$

As a result, we have the asymptotic result

$$E_{ps} \to \frac{(3s+2)(s-1)}{3(s+1)} + \frac{2p}{s+1} = \frac{(3s+2)(s-1) + 6p}{3(s+1)} \quad \text{as} \quad p \to \infty,$$

which agrees with all the specific values derived earlier.

One of way interpreting this result is to note that with an s-sided die, the expected value of a spin is $(s+1)/2$, so we would expect $2/(s+1)$ spins to advance one square, or $2p/(s+1)$ to advance p squares. The additional term is the number of additional spins to reach the end of the board due to the requirement that one must roll until one lands on the end square exactly. It makes sense that this additional term depends only on s, since the end-game only affects spins on the last $s-1$ squares, independent of board length. In fact, we have

$$\frac{(3s+2)(s-1)}{3(s+1)} = s - 1 - \frac{s-1}{3(s+1)}.$$

2.3 Best Spinner

For the classic Chutes and Ladders board, Cheteyan et al. [3] showed that the minimum average number of spins of 25.81 was obtained by using choosing the value $s = 15$. We can do the same here and identify the best-size spinner for each board size p to minimize the average number of spins. One approach uses explicit calculated results, like those in Figure 10.2, which show that a spinner of size 6 is better than a spinner of size 10 for boards of sizes up to 38.

Alternatively, if one uses the approximation $\widetilde{E}_{ps} = \frac{(3s+2)(s-1)+6p}{3(s+1)}$ from the previous section, then it is a straightforward calculus exercise to find the value of s that minimizes $\widetilde{E}_{ps}$ for a fixed value of p. Doing so gives the following result:

$$\tilde{s} = \sqrt{2/3 + 2p} - 1, \qquad \widetilde{E}_{min} = \frac{12p+4}{\sqrt{18p+6}} - \frac{7}{3}.$$

Figures 10.3 and 10.4 compare these asymptotic choices to the actual computed minimum for various board sizes, where the red line is the asymptotic, and blue crosses are the actual results. Rounding the asymptotic approximation spinner size does an excellent job.

3 One Chute or Ladder

For the remainder of the chapter, we consider how the addition of a single chute or ladder will affect the results of the previous sections.

3.1 Probability of Visiting a Square

Putting a ladder from square m to square n, $n > m$, changes our recurrence relations for the probability of visiting a square by explicitly setting $a_m = 0$ and

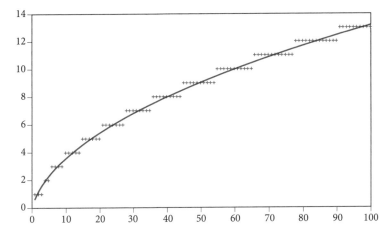

Figure 10.3. Comparing approximated $\tilde{s}$ (solid line) with optimal s (+ symbols)(y-axis) for $p = 1, 2, \ldots, 100$ (x-axis).

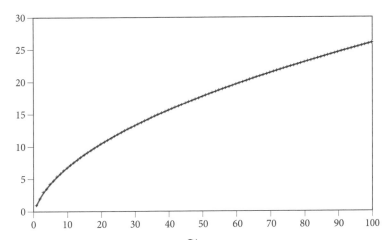

Figure 10.4. Comparing approximated $\widetilde{E_{ps}}$ (solid line) with optimal E_{ps} (+ symbols) (y-axis), $p = 1, 2, \ldots, 100$ (x-axis).

$$a_n = \frac{1}{s}(a_{n-1} + a_{n-1} + \cdots + a_{n-s}) + \frac{1}{s}(a_{m-1} + a_{m-2} + \cdots + a_{m-s}),$$

where one can get to square n by either landing at square m and going up the ladder, or moving from a previous square as usual. For example, Figure 10.5 shows the probability of visiting a given square using a six-sided die on a one hundred square board with a single ladder from square 30 to square 70. We can see that after the base of the ladder, there is a drop and an eventual asymptotic

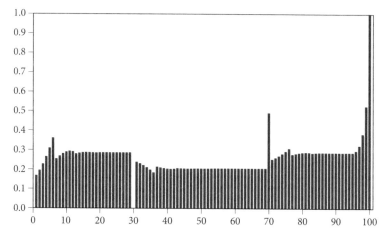

Figure 10.5. Probability of visiting each square with $s = 6$, $p = 100$, and a ladder from square 30 to square 70.

probability below that of 2/7 before the ladder. Then there is the expected spike at the top of the ladder, and new asymptotic behavior. Numerically, the asymptotic value after the base of the ladder, looks to be 10/49, and after the top of the ladder it returns to 2/7. In fact, the following theorem holds for this case.

Theorem 2. *Given a board with a single ladder, the probability of visiting a square using an s-sided die approaches $(2s - 2)/(s + 1)^2$ sufficiently after the base of the ladder (if the ladder is long enough) and approaches $2/(s + 1)$ after the top of the ladder (if the board is long enough).*

Proof. Because the recurrence relation defining the probabilities is linear, we know that given two different sets of initial conditions $\vec{u}$ and $\vec{v}$, we have $L_{\vec{u} + \vec{v}} = L_{\vec{u}} + L_{\vec{v}}$. Following the notation of section 1.3, we define b_i to be the probability of landing on square i on a board with no chutes or ladders, and we let c_i be the probability of landing on square i on a board with a single ladder from square m to square n. In particular, we see that $c_i = b_i$ for $i < m$, and $c_m = 0$. For $m < i < n$, we have that c_i is defined by the linear recurrence $c_i = \frac{1}{s} \sum_{j=1}^{s} c_{i-j}$, and in particular, it follows from Theorem 1 that if n is sufficiently large, then the c_i will converge to

$$L_{\langle b_{m-s+1}, \ldots, b_{m-1}, 0 \rangle} = L_{\langle b_{m-s+1}, \ldots, b_{m-1}, b_m \rangle - \langle 0, \ldots, 0, b_m \rangle}$$

$$= L_{\langle b_{m-s+1}, \ldots, b_{m-1}, b_m \rangle} - L_{\langle 0, \ldots, 0, b_m \rangle}$$

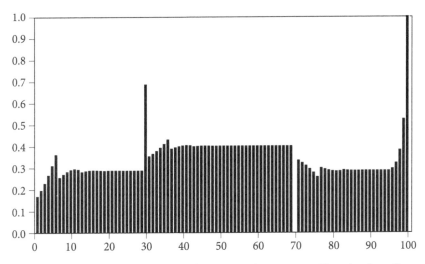

Figure 10.6. Probability of visiting each square with $s = 6$, $p = 100$, and a chute from square 70 to square 30.

$$= \frac{2}{s+1} - \frac{2}{s(s+1)} sb_m.$$

If m is sufficiently large, then we can assume that b_m is close to $2/(s+1)$, which implies that the sequence converges to $(2s-2)/(s+1)^2$, as desired.

To compute what happens after the top of the ladder, we note that $c_n = \frac{1}{s}\sum_{j=1}^{s} c_{i-j} + b_m$. In particular, if we look sufficiently past the top of the ladder, then the sequence will converge to

$$L_{\langle c_{n-s+1},...,c_{n-1},c_n\rangle} = L_{\langle c_{n-s},...,c_{n-1},0\rangle} + L_{\langle 0,0,...,b_m\rangle}$$

$$= \left(\frac{2}{s+1} - \frac{2}{s(s+1)} sb_m\right) + \frac{2}{s(s+1)} sb_m$$

$$= \frac{2}{s+1}. \qquad \square$$

We can get a similar result for boards with a single chute. For example, Figure 10.6 shows the probabilities of landing on every square on a board with a single chute from square 70 to square 30. Since the recurrence relation for the probability at square 30 depends on probabilities from squares 24 to 29 as well as 64 to 69, it is no longer explicit. As a result, the recurrence relations need to be set up as a system of equations that is solved using Gaussian elimination.

Theorem 3. *Given a board with a single chute both of whose ends are sufficiently far from the ends of the board, the probability of visiting a square using an s-sided die approaches $2/(s-1)$ sufficiently after the base of the chute (if the chute is long enough), and it approaches $2/(s+1)$ after the top of the chute.*

Proof. The proof is similar to the one for Theorem 2; in particular, assume that before the bottom of the chute, the probability of landing on a square can be approximated by $b = 2/(s+1)$ and between the two ends of the chute, the probability converges to some value c. Then the probability of landing at the bottom of the chute is $b + c$, and we deduce that

$$c = L_{\langle b,\dots,b,b+c\rangle} = L_{\langle b,\dots,b,b\rangle} + L_{\langle 0,\dots,0,c\rangle} = \frac{2}{s+1} + \frac{2}{s+1} \cdot c,$$

from which we compute that $c = 2/(s-1)$, as desired. After the top of the chute,

$$L_{\langle c,\dots,c,0\rangle} = L_{\langle c,\dots,c,c\rangle} - L_{\langle 0,\dots,0,c\rangle} = \frac{2}{s-1} - \frac{2}{s+1} \cdot \frac{2}{s-1} = \frac{2}{s+1}.$$

□

It is interesting to note that the probability of landing on the square immediately after the top of the chute can be approximated as

$$c_t = \frac{1}{s} \sum_{i=1}^{s} c_{t-i} = \frac{1}{s}(c + c + \cdots + c + 0) = \frac{(s-1)c}{s} = \frac{2}{s},$$

which explains why the probability for square 71 in Figure 10.6 is, in double-precision arithmetic, 0.333333026841736, which is very close to 1/3.

We note that the proofs of these theorems will generalize in a sense to give the following result.

Theorem 4. *For any board with any number of chutes or ladders, the probability of landing on any square sufficiently beyond the top of the final chute/ladder and before square $p - s$, if any such square exists, will converge to $2/(s+1)$.*

While this result will help explain the end-game probabilities in many situations, figuring out what precisely happens in between is a more complicated question that we save for future work.

3.2 Optimal Placement of a Single Ladder

In this section, we consider the question of how one should place a single ladder of a given length to minimize the expected length of a game. The approach we take is to study how adding a single ladder at various locations changes the expected length of the game compared to a board with no chutes or ladders at all.

There are two reasonable places for the ladder to be positioned. The first is for the base of the ladder to be placed on the square with highest probability of being landed on. The second is for the top of the ladder to be on the final square. As long as the ladder is short enough relative to the size of the board that the asymptotic results in previous sections can be used, we will be able to show that these are the only two cases that need to be considered.

Observe that when a ladder is landed on, the length of the board is reduced by L. So if the ladder has its base on square k and $k + L < p$, then the ladder is expected to approximately reduce the game length by

$$E_{maxp} \approx a_k \frac{2L}{s+1},$$

where a_k is the probability of landing on square k. By maximizing a_k, we minimize the game length for situations where the top of the ladder is not close to end of the board.

Recall that for values of k with $s < k \leq p - s$, we have that the sequence a_k is recursively defined by taking the average of the previous s terms. In particular, we see that $a_k \leq \max\{a_1, \ldots, a_s\}$. Moreover, it follows from earlier results that for $i \leq s$, we have that $a_i = \frac{(s+1)^{i-1}}{s^i}$, which is maximized when $i = s$.

Next, consider when the end of the ladder is at the end of the board. Assuming that the ladder is short relative to the length of the board, we have seen that there is approximately a $2/(s+1)$ chance of landing on the base of the ladder. If we use this value for the probability, then the ladder is expected to reduce the game length by

$$E_{end} = \frac{2}{s+1} E_{Ls} = \frac{2}{s+1} \left(\frac{(3s+2)(s-1)}{3(s+1)} + \frac{2L}{s+1} \right).$$

It remains to determine which is greater between E_{end} and E_{maxp}. In particular, if

$$\frac{2}{s+1} \left(\frac{(3s+2)(s-1)}{3(s+1)} + \frac{2L}{s+1} \right) > 2L \frac{(s+1)^{s-2}}{s^s},$$

then the ladder should be placed at the end of the board. Solving for L in this inequality, we find

$$L \leq \frac{(3s+2)(s-1)}{3\left(1+\frac{1}{s}\right)^s - 6}.$$

For example, when $s = 6$, we have $L \leq 63.9\ldots$. This tells us that if we are playing on a sufficiently large board and a spinner of size six, then the optimal placement of a ladder of length $L \leq 63$ has its top at the end of the board, while if the ladder is of length $L \geq 64$, then we should place the base of the ladder on square six. Note that this transition value is an increasing function in the value of s, and therefore, it is almost always optimal to place a single ladder at the end of the board.

If we ask the analogous question for the placement of a single chute, we can compute that if E_{ps} is the expected length of a game on a board with no chutes or ladders and $C_{ps}^{x,L}$ is the expected length of a game with a single chute from square x to square $x - L$, then we have

$$C_{ps}^{x,L} = E_{ps} + a_x(C_{p-L,s}^{L+1,1} - 1),$$

where as above, a_x is the probability of landing on square x when playing on a game with no chutes or ladders. In particular, to minimize the length of the game, we should place the top of the chute at the square we land on with smallest probability. To maximize the length of the game, we should place the top at the square we are most likely to land on. Results of the previous sections imply that the square we are most likely to land on is square $p - 1$. Finding the square we are least likely to land on, but which would still have the bottom of the chute on the board, is a more complicated question that will be saved for future work.

References

[1] S. C. Althoen, L. King, and K. Schilling. How long is a game of Snakes and Ladders? *Math. Gazette* **77** (1993) 71–76.

[2] N. Anderson, E. B. Saff, and R. S. Varga. On the Enestrom-Kakeya theorem and its sharpness. *Linear Algebra Appl.* **28** (1979) 5–16.

[3] L. A. Cheteyan, S. Hengeveld, and M. A. Jones. Chutes and ladders for the impatient. *College Math. J.* **42** (2001) 2–8.

[4] D. Connors and D. Glass. Chutes and Ladders with large spinners. *College Math. J.* **45** (2014) 289–295.

[5] D. E. Daykin, J. E. Jeacocke, and D. G. Neal. Markov chains and Snakes and Ladders. *Math. Gazette* **15** (1967) 313–317.

[6] M. Marden. *Geometry of Polynomials*. Math. Surveys 3, American Mathematical Society, Providence, RI, 1966.

11

BUGS, BRAIDS, AND BATCHING

Michael P. Allocca, Steven T. Dougherty, and Jennifer F. Vasquez

Sorting problems are ubiquitous in mathematics and many natural sciences: From computer science, to coding theory, to DNA modeling, efficient sorting solutions are of paramount importance. Recent developments in molecular biology have connected measures of so-called evolutionary distance between organisms with linear chromosomes to classic sorting problems involving permutations. Analogously, we can connect ideas of evolutionary distance between organisms with circular chromosomes to sorting problems through circular permutations.

Games provide significant insight into the nature of sorting problems and help develop intuitive solution strategies. In this chapter, we explore three games using variations of "Japanese ladders," which are a clever way to view permutations. The first game invites the player to engage in "batching" in the sense of traditional permutations. The second game enhances the first by introducing directional signage, which provides an alternative model for braids. The third game is a circular version of the first, where one might imagine a ladder placed on a soup can. This analogously models circular permutations, in which sorting questions connect to questions of evolutionary distances among organisms with circular chromosomes ("bugs"). Optimal solutions to all of these games correspond to most-efficient sorting strategies. Playing these games most efficiently will uncover rich mathematics.

1 Batching

Let us start with a simple game. Draw three lines, which we shall call *poles*, numerically labeling them on top in order. This is shown on the left in Figure 11.1. Now label the bottom of each pole randomly—say, 2, 3, and 1, as shown in the middle of Figure 11.1

The object of the game is to draw rungs across each line to create ladders with the following rules:

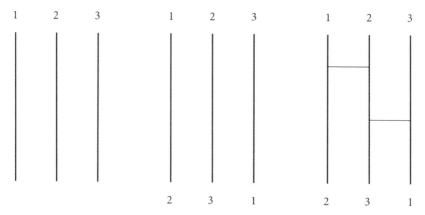

Figure 11.1. A simple line-drawing game. On the left, we draw three lines, called poles, and number them sequentially. In the middle, the bottoms of the poles are labeled in random order. On the right, we have one possible solution to the game.

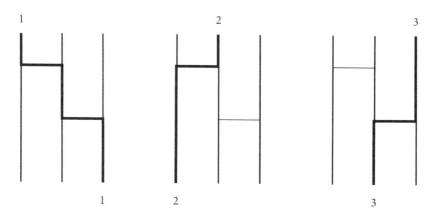

Figure 11.2. Tracing paths from the top numbers to their counterparts on the bottom.

- Each number on top "falls" downward.
- If the number encounters a rung, it must cross it.
- All numbers on top must "land" in the place labeled by the same number on the bottom.

According to these rules, a player may then choose to connect the first two poles with a rung, followed by another rung connecting the next two, as shown on the right in Figure 11.1. This ladder is a proper solution, since each number follows a path that adheres to the rules listed above, as shown in Figure 11.2.

We examine an example with four poles in Figure 11.3.

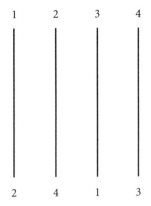

Figure 11.3. Playing the game with four poles.

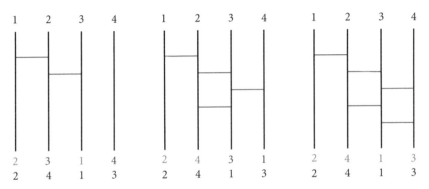

Figure 11.4. One technique for finding a solution to the four-pole puzzle. Green numbers indicate that the corresponding top number arrives at the correct destination, while red numbers indicate that the corresponding top number arrives at an incorrect destination.

A natural strategy that many people attempt is to first try to connect one of the numbers that is farthest from its final pole, and then repeat this technique to connect the remaining paths. This is illustrated in Figure 11.4.

The mathematics behind this game is quite simple. The above example corresponds to the permutation

$$\begin{pmatrix} 1\ 2\ 3\ 4 \\ 3\ 1\ 4\ 2 \end{pmatrix},$$

or $(1, 3, 4, 2)$ in cycle notation. The rungs in this example correspond to transpositions in the following manner (reading from top to bottom):

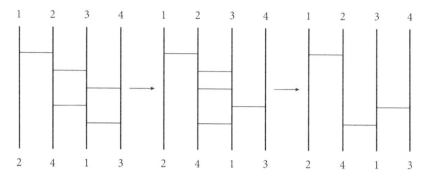

Figure 11.5. Finding a more efficient solution to the four-pole puzzle than the one shown in Figure 11.4.

$$\underbrace{(3,4)(2,3)(3,4)(2,3)(1,2)}_{\text{equivalent to } (1,3,4,2)} .$$

In fact, this connection lends itself to a more formal definition.

Definition 1. *Let A be a multiset with elements from $\{1, 2, \ldots, n\}$. A Japanese ladder (Amidakuji) is a representation of a permutation by the product*

$$\prod_{i \in A}(i, i+1).$$

Hence, every ladder has a corresponding permutation. Further, every permutation has a corresponding ladder that may be constructed by writing the permutation as a product of adjacent transpositions and associating each of these with rungs. This connection is not unique, however. This may lead us to question whether the solution to the previous ladder was optimal and ultimately discover a more efficient solution, as demonstrated in Figure 11.5.

In terms of permutations, the first ladder modification uses the fact that

$$(2,3)(3,4)(2,3) = (3,4)(2,3)(3,4),$$

or more generally, that

$$(i,i+1)(i+1,i+2)(i,i+1) = (i+1,i+2)(i,i+1)(i+1,i+2)$$

for any $1 \leq i \leq n-2$. The second modification exploits the fact that $(2,3)^2 = e$, the identity, or more generally, $(i,i+1)^2 = e$. This is a strategy that many beginners use to "clean up" nonoptimal ladder solutions.

The classic problem of sorting permutations leads to an algorithm that generates an optimal solution to any Japanese ladder. A thorough analysis is given in Dougherty and Vasquez [2]. For completeness, we include a synopsis of the technique below.

Assume $a_1, a_2, \ldots, a_n$ are the numbers written on the bottom of the rungs. An *inversion* is any incidence of a_i, a_j with $i > j$, but $a_i < a_j$. Let k be the number of inversions in the ordered n-tuple $(a_1, a_2, \ldots, a_n)$. Then k is the minimum number of inversions needed to solve the Japanese ladder. Consider the n-tuple $(a_1, a_2, \ldots, a_n)$. Beginning from the left, find the first inversion and transpose the elements.

Given a permutation $\pi \in \mathcal{S}_n$, write

$$\pi = \begin{pmatrix} 1 & 2 & \cdots & n \\ \pi(1) & \pi(2) & \cdots & \pi(n) \end{pmatrix}.$$

Define

$$K = |\{(i,j) \mid i < j, \pi(i) > \pi(j)\}|.$$

The number K is the number of inversions in π. To find the inverse of π, reverse the rows of the permutation to obtain

$$\begin{pmatrix} \pi(1) & \pi(2) & \cdots & \pi(n) \\ 1 & 2 & \cdots & n \end{pmatrix}.$$

Now, rearrange the top row, so that it is in order, keeping $\pi(i)$ in the same column as i. Finally, define

$$K' = |\{(i,j) \mid \pi(j) < \pi(i), j > i\}|.$$

We have $K = K'$, which gives an explicit algorithm for constructing an optimal ladder for π.

This algorithm can be seen as "combing" the permutation. Given $\pi \in \mathcal{S}_n$, consider

$$\begin{pmatrix} \pi(1) & \pi(2) & \cdots & \pi(n) \\ 1 & 2 & \cdots & n \end{pmatrix},$$

and using transpositions of the form $(k, k+1)$, move the columns so that the top row becomes $1, 2, \ldots, n$.

For a more detailed explanation of Japanese ladders and an example of this algorithm in action, see Dougherty and Vasquez [2].

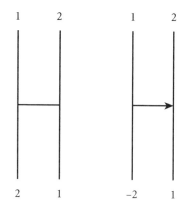

Figure 11.6. On the left is a normal Japanese ladder. On the right is an enhanced ladder. Notice that the 1 crosses in the direction of the arrow and therefore remains a 1. The 2 crosses in the opposite direction of the arrow, and therefore becomes a −2.

2 Braids

We now consider an enhancement of the Japanese ladders game. Suppose we change the rungs to arrows and also add the following rules:

- If a number crosses a rung in the opposite direction of the arrow, then the number is negated at the base of its ending pole.
- If a number crosses a rung in the same direction of the arrow, then the number is unchanged.

We call this new type of figure an *enhanced ladder*, as illustrated in Figure 11.6.

This enhancement increases the difficulty of the ladder game considerably. A player may intuitively consider her strategies for traditional Japanese ladders and use them as a starting point for enhanced ladders. Consider the example in Figure 11.7.

One may opt to begin by placing the rungs according to the optimal placement algorithm for the normal Japanese ladders game. From there, the player may experiment with arrow directions to achieve the desired order of positive and negative signs in the bottom row. This approach is shown in Figure 11.8.

This appears to be an optimal strategy. Indeed, the enhanced ladder in Figure 11.8 cannot be solved in less than three rungs. However, this strategy fails for many other enhanced ladders. For example, consider Figure 11.9.

Some examples of this game may also be found in Dougherty and Vasquez [2]. However, to develop sophisticated strategies for this game, we must examine the underlying mathematics. Analogous to the relationship between Japanese

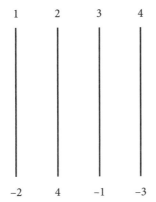

Figure 11.7. A starting position for a four-pole run of the enhanced game.

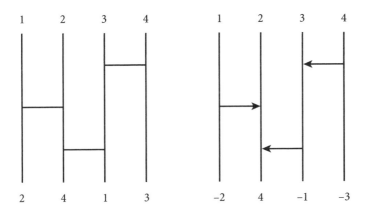

Figure 11.8. A solution to the game presented in Figure 11.7. On the left is a solution to the Japanese ladder game with the underlying permutation. On the right is a solution to the enhanced game.

ladders and permutations, one may suspect there is a relationship between enhanced ladders and signed permutations.

Definition 2. *A signed permutation is a map $\alpha : \{1, 2, \ldots, n\} \rightarrow \{\pm 1, \pm 2, \ldots, \pm n\}$ such that if $\alpha(i) = k$, then $\alpha(i') \neq \pm k$ whenever $i \neq i'$.*

Signed permutations may be understood in the alternative context of Coxeter groups of type B [3].

There are $n!2^n$ signed permutations of $\{1, 2, \ldots, n\}$. The key difference in comparison to Japanese ladders and permutations is that not all signed permutations are realizable by enhanced ladders. For example, there are $2!2^2 = 8$

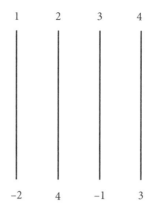

Figure 11.9. How many rungs are there in the optimal solution?

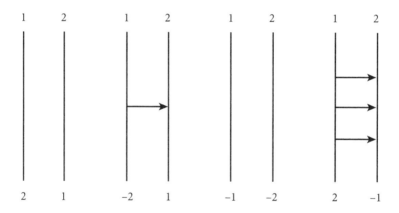

Figure 11.10. The elements of $\mathcal{E}_2$.

signed permutations of $\{1, 2\}$, yet only four of them are realizable by enhanced ladders, shown in Figure 11.10.

In the context of our examination of the game, we will make a distinction between the group of signed permutations (denoted $\mathcal{E}'_n$) and the group of signed permutations that are realizable by enhanced ladders (denoted $\mathcal{E}_n$). It is important to note that the product of signs in the final state (that is, the bottom row) of an enhanced ladder must equal $(-1)^k$, where k is the number of rungs in the ladder.

The enhanced ladder game is also naturally connected to topics in contemporary braid theory.

1 2 i $i+1$ n

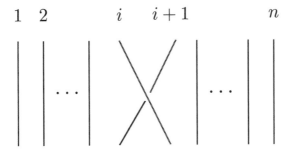

Figure 11.11. A representation of the braid generator σ_i.

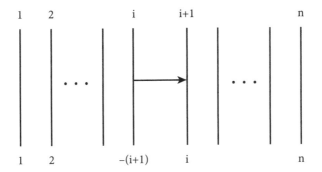

Figure 11.12. A ladder representation of the braid generator σ_i.

Definition 3. *The braid group, $\mathcal{B}_n$, is generated by the elements $\sigma_1, \sigma_2, \ldots, \sigma_{n-1}$ such that the following relations hold:*

- $\sigma_i \sigma_j = \sigma_j \sigma_i, \ |i - j| \geq 2;$
- $\sigma_i \sigma_{i+1} \sigma_i = \sigma_{i+1} \sigma_i \sigma_{i+1}, \ 1 \leq i \leq n - 2.$

You can visualize these generators as "braid twists," as in Figure 11.11.

The product of braids may be visualized by stacking these figures in order vertically. There is a natural connection to enhanced ladders if you denote the adjacent transposition $(i \rightarrow i+1, i+1 \rightarrow -i)$ as σ_i and compare Figure 11.11 to the ladder associated with σ_i in Figure 11.12.

In fact, the map $\varphi : \mathcal{B}_n \rightarrow S_n$ defined by $\sigma_i \rightarrow (i, i+1)$, where S_n is the symmetric group of order $n!$, is a homomorphism. The kernel of φ, denoted $\mathcal{P}_n$ (also known as the pure braid group), can be visualized as all twists that leave strands in their original place, as illustrated in Figure 11.13.

The following short exact sequence is an integral part of contemporary braid theory:

$$1 \longrightarrow \mathcal{P}_n \longrightarrow \mathcal{B}_n \longrightarrow S_n \longrightarrow 1.$$

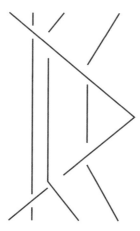

Figure 11.13. A pure braid.

However, given the connection between the braid group and the group of signed permutations that are realizable by enhanced ladders, one might consider the following to be a more natural version of this short exact sequence:

$$1 \longrightarrow \mathcal{P}_n \longrightarrow \mathcal{B}_n \longrightarrow \mathcal{E}_n \longrightarrow 1.$$

With this modification, the order of $\mathcal{P}_n$ decreases, and we will label the new kernel as $\mathcal{K}_n$. This establishes the following new short exact sequence:

$$1 \longrightarrow \mathcal{K}_n \longrightarrow \mathcal{B}_n \longrightarrow \mathcal{E}_n \longrightarrow 1.$$

For a more detailed analysis of the braid theory behind this game and further exposition of this short exact sequence, see Allocca et al. [1].

3 Bugs

We now consider how the dynamics of the Japanese ladders game would change if we were to place them on a cylindrical surface. Informally, one may imagine drawing a Japanese ladder on a standard sheet of paper with the poles drawn vertically, then connecting the area adjacent to the leftmost and rightmost poles by taping the left and right sides of the paper over a soup can.

This creates a much more challenging game, in which efficient solution strategies are highly nontrivial. As one might suspect, this game is deeply connected to the mathematics of circular permutations, which are ordered arrangements of a set in a circular pattern. The search for optimal solutions is analogous to efficient sorting algorithms, which is a contemporary open problem.

As an example, consider this circular permutation on four objects:

We will adopt the following one-line notation, reading elements clock-wise: $\sigma = [1342]$. It is worth noting that as a group of permutations, S'_n, these are still isomorphic to S_n. There is a small difference in generating sets, considering

$$\{(1, 2), (2, 3), \ldots, (n-1, n)\}$$

versus

$$\{(1, 2), (2, 3), \ldots, (n-1, n), (n, 1)\}.$$

We will call $(n, 1)$ the "cyclically" adjacent transposition.

It is also important to note that there are $2n$ equivalent expressions of a single circular permutation of n objects, representing the $2n$ permutations of the dihedral group. For example, $[1342]$ would represent the same circular permutation as

[2134] [4213] [3421] [1243] [2431] [4312] [3124],

as illustrated in Figure 11.14.

This highlights the need to consider equivalence classes in both the circular ladder game and in sorting circular permutations. We formalize this as follows. Let $L(n)$ be the set of all circular permutations of n objects written in second-line notation, that is, $L(n)$ is generated by $\mathcal{L}$, where

$$\mathcal{L} = \{\sigma_1, \sigma_2, \ldots, \sigma_{n-1}, \sigma_n\}.$$

If $\sigma, \tau \in L(n)$, we define $\sigma \sim \tau$ iff $\sigma = \gamma\tau$, where γ is any element of the dihedral group D_{2n}. We let $L'(n) = L(n)/\sim$. Notice that $|L(n)| = n!$, and

$$|L'(n)| = \frac{n!}{2n} = \frac{(n-1)!}{2}$$

when $n \geq 3$ and $|L'(n)| = 1$ for $1 \leq n \leq 2$.

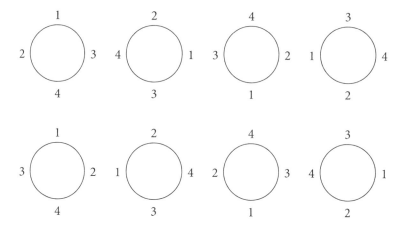

Figure 11.14. The same circular permutation.

In the context of the circular ladder game, an algorithm for finding the most efficient solution is unknown. More significantly, given n poles, it is also unknown what the minimum number of rungs will be. We consider the analogous open problem in sorting circular permutations.

Problem 1. *Find an upper bound on the minimum number of adjacent transpositions in L required to factor an arbitrary circular permutation in $L'(n)$.*

These ideas connect naturally to some strategies used by molecular biologists studying evolution. An organism's genetic information is stored in its chromosomes, which consist of dual strands of nucleotides called DNA. Genes are the segments of DNA containing information on constructing other cells. The set of all chromosomes of an organism is called its genome. Mutations tend to occur either at the level of nucleotides, or, more significantly, as larger-scale genome rearrangements, including deletions, duplications, translocations, and inversions/reversals. By focusing solely on inversions, or adjacent transpositions, we may connect these ideas naturally to permutations. When a mutation occurs, genes are the DNA segments that are most likely to be rearranged intact. From a mathematical perspective, it is reasonable to view the "evolutionary distance" between two organisms as the extent of "disorder" of gene arrangements in their respective genomes. It is natural to model this distance using permutations, enumerating each gene as necessary. Further, since mutations are rare occurrences, some models only consider the smallest number of rearrangements when comparing genomes. This leads to a combinatorial optimization problem: Given two genomes and a set of possible evolutionary events, find the shortest sequence of events transforming one into the other. This is precisely the question posed by finding the most efficient strategy for completing a Japanese ladder, answered by inversion numbers and the underlying algorithms in classical computer science.

While it is useful when considering some evolutionary distance models involving organisms with linear chromosomes, some chromosomes (primarily found in bacteria) are circular. This is where efficient solutions to the circular ladder game can have deep implications for application.

We now adopt a graph-theoretic approach to sorting circular permutations efficiently. Let $G(n) = (V, E)$ be a simple graph, where $V = L'(n)$ as defined earlier, and two vertices $v_1, v_2 \in V$ are adjacent if v_1 can be formed from v_2 using one of the transpositions in

$$\mathcal{L} = \{\sigma_1, \sigma_2, \ldots, \sigma_{n-1}, \sigma_n\}.$$

That is, two vertices are adjacent if and only if one is formed from the other by making a (cyclically) adjacent swap. Recall the following definitions central to graph theory.

Definition 4. *Let $d(v_1, v_2)$ be the minimum path distance between v_1 and v_2. The eccentricity of a vertex v, denoted by $\epsilon(v)$, is the largest distance between v and any other vertex, that is,*

$$\epsilon(v) = \max_{v_i \in V} d(v, v_i).$$

The radius, r, is the minimum eccentricity of any vertex, that is,

$$r = \min_{v \in V} \epsilon(v).$$

The diameter, d, is the maximum eccentricity of any vertex, that is,

$$d = \max_{v \in V} \epsilon(v).$$

Theorem 1. *Every vertex in the graph $G(n)$ has the same eccentricity. Further, the radius and the diameter of graph $G(n)$ are equal.*

Problem 1 can now be restated as follows.

Problem 2. *Determine the diameter of the graph $G(n)$.*

This is trivial for $n = 2$ and $n = 3$; each of these graphs has only one vertex and has diameter 0. For $n = 4$, we have three vertices, which represent the

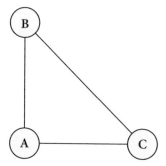

Figure 11.15. Graph for $n = 4$.

classes: $A = [1, 2, 3, 4]$, $B = [1, 2, 4, 3]$, and $C = [1, 4, 2, 3]$. The graph is illustrated in Figure 11.15.

Hence the diameter for the graph when $n = 4$ is 1. Note that this is significantly lower than the bound of $\left\lfloor \frac{n^2}{4} \right\rfloor = 4$ that we would have predicted using only cyclically adjacent transpositions. Also note that the degree of every vertex here is 2. That is, every vertex has two edges incident to it, hence $G(4)$ is a regular graph. One may naturally wonder whether $G(n)$ is always regular, and whether there is a relationship between n and the degree of each vertex. The graphs are somewhat anomalous for $n \leq 4$, since there are so few equivalence classes in $L'(n)$. The relationship becomes more clear, however, for $n > 4$.

Theorem 2. *If $n > 4$, then $G(n)$ is a regular graph, and the degree of each vertex is n.*

Proof. There are n transpositions of the form $(1, 2), (2, 3), \ldots, (n, n-1), (n, 1)$ that can be applied to any circular permutation without duplication as long as $n > 4$. $\square$

This immediately establishes the following.

Corollary 1. *For $n > 4$ the graph $G(n)$ has $\dfrac{(n-1)!(n)}{4}$ edges.*

Proof. There are $\dfrac{(n-1)!}{2}$ vertices with degree n, and so there are $\dfrac{\frac{(n-1)!}{2}(n)}{2}$ edges. $\square$

Note that every element of $L'(n)$ has two representatives that begin with 1. We will use one of these as representatives of each class. Notice that the three representatives of the vertices of $L'(4)$ are

$$[1, 2, 3, 4], \quad [1, 2, 4, 3], \text{and} \quad [1, 4, 2, 3].$$

These can be found by taking the identity class when $n = 3$ (namely, $[1, 2, 3]$) and placing "4" in each of the three positions. In general,

$$|L'(n+1)| = n|L'(n)|.$$

The elements of $L'(n+1)$ are formed by taking each element of $L'(n)$ and placing the element $n + 1$ in each of the n possible positions, giving that there are n elements in $L'(n+1)$ for each element of $L(n)$. Notice that these n vertices form a cycle in $G(n+1)$. This is seen in the example for the graph for $n = 4$ (Figure 11.15). Let v_1 and v_2 be two adjacent vertices in $G(n)$. Then let $v_{1,1}, v_{1,2}, \ldots, v_{1,n}$ be the n vertices formed in the above manner from v_1 in $G(n+1)$ and $v_{2,1}, v_{2,2}, \ldots, v_{2,n}$ be the n vertices formed in the above manner from v_2 in $G(n+1)$. Then without loss of generality, we can assume that $v_{1,i}$ and $v_{2,i}$ are adjacent in $G(n+1)$ for all i except when $n + 1$ is placed in the middle of the original swap.

For example, consider the vertices $[1, 4, 2, 3]$ and $[1, 2, 4, 3]$ in $G(4)$. These are adjacent via σ_2. Then we have the following adjacencies in $G(5)$:

$$[1, 5, 4, 2, 3] \leftrightarrow [1, 5, 2, 4, 3]$$
$$[1, 4, 5, 2, 3] \nleftrightarrow [1, 2, 5, 4, 3]$$
$$[1, 4, 2, 5, 3] \leftrightarrow [1, 2, 4, 5, 3]$$
$$[1, 4, 2, 3, 5] \leftrightarrow [1, 2, 4, 3, 5]$$

These are not the only adjacencies that occur, since the degree of the new vertices is n.

For $n = 5$ there are 12 vertices; we label them as follows:

$$A = [1, 2, 3, 4, 5] \quad B = [1, 2, 3, 5, 4] \quad C = [1, 2, 5, 3, 4] \quad D = [1, 5, 2, 3, 4]$$
$$E = [1, 2, 4, 3, 5] \quad F = [1, 2, 4, 5, 3] \quad G = [1, 2, 5, 4, 3] \quad H = [1, 5, 2, 4, 3]$$
$$I = [1, 4, 2, 3, 5] \quad J = [1, 4, 2, 5, 3] \quad K = [1, 4, 5, 2, 3] \quad L = [1, 5, 4, 2, 3].$$

The graph $G(5)$ is seen in Figure 11.16.

To find the diameter of this graph, by 1, it suffices to find the eccentricity of vertex A. The vertex A is 1 away from five vertices; 2 away from five vertices; and distance 3 away from one vertex, namely, vertex J. Hence the diameter of

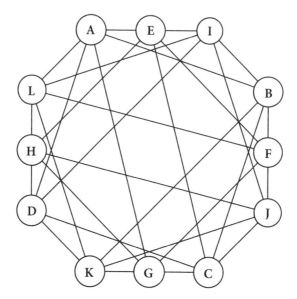

Figure 11.16. Graph for $n = 5$.

this graph is 3. Note that 3 is significantly lower than the previously predicted bound of $\left\lfloor \frac{n^2}{4} \right\rfloor = 6$.

Note the following interesting properties of the graph $G(n)$.

Theorem 3.   *For $n > 4$, there are at least $\frac{n(n-3)}{2}$ 4-cycles.*

Proof. The path formed by

$$(k, k+1)(s, s+1)(k, k+1)(s, s+1),$$

where $k, k+1 \neq s, s+1$ gives a 4-cycle. There are $\frac{n(n-3)}{2}$ such possible cycles. □

This immediately establishes the following.

Corollary 2.   *For $n > 4$, the minimum cycle in $G(n)$ is of length 4.*

We also identify a lower bound on the number of $(n-1)$-cycles.

Theorem 4.   *For $n > 4$, at each point in $G(n)$, there are at least n $(n-1)$-cycles at each vertex.*

Proof. These cycles are formed by

$$(1,2)(2,3)(3,4)\dots(n-1,n),$$
$$(2,3)(3,4)\dots(n-1,n)(n,1),$$
$$(3,4)\dots(n,1)(1,2),$$
$$\vdots$$
$$(n,1)(1,2)\dots(n-2,n-1).$$

□

We now use the idea of building $G(n+1)$ from $G(n)$ to establish our main result.

Theorem 5. *Given a circular permutation $\pi \in L'(n)$, the maximum number of adjacent transpositions needed to factor π is bounded by $\lfloor \frac{(n-1)(n-2)}{4} \rfloor$.*

Proof. Let $\pi = [a_1, a_2, \dots, a_n] \in L'(n)$. There are $2n$ elements in its equivalence class consisting of n rotations and n flips. Specifically,

$$R = \{[a_1, a_2, \dots, a_n], [a_2, a_3, \dots, a_1], \dots, [a_n, a_1, \dots, a_{n-1}]\}$$
$$F = \{[a_n, a_{n-1}, \dots, a_1], [a_{n-1}, a_{n-2}, \dots, a_n], \dots, [a_1, a_n, \dots, a_2]\}.$$

We aim to show that $\epsilon(\pi)$ is at most $\lfloor \frac{(n-1)(n-2)}{4} \rfloor$.

Now let $\Gamma(n)$ be the graph where the vertices are the elements of S_n, and two vertices corresponding to α_1, α_2 are adjacent if there is an adjacent transposition $\sigma_i \in L$ with $\alpha_1 = \sigma_i\alpha_2$. Then $G(n)$ is the quotient graph of $\Gamma(n)$ with the relation $\sim$ defined previously.

Notice that in R, there is an element of the form

$$\pi_1 = [1, b_1, \dots, b_{n-1}].$$

Moreover, there is a corresponding $\pi_2 \in F$ of the form

$$[1, b_{n-1}, \dots, b_1].$$

Now consider the following circular permutations in $L(n-1)$:

$$\tau_1 = [b_1, \dots, b_{n-1}] \quad \text{and} \quad \tau_2 = [b_{n-1}, \dots, b_1].$$

Let

$$I_1 = |\{i, j \mid i < j, \tau_1(i) > \tau_1(j)\}|$$

and

$$I_2 = |\{i, j \mid i < j, \tau_2(i) > \tau_2(j)\}|.$$

Note that I_1 is the classical inversion number of the permutation $(b_1, \ldots, b_{n-1})$, and thus gives an upper bound on the maximum distance between τ_1 and any other vertex in $\Gamma(n-1)$. Similarly, I_2 gives an upper bound on the maximum distance between τ_2 and any other vertex in $\Gamma(n-1)$. Furthermore,

$$I_1 + I_2 = \sum_{i=1}^{n-2} i = \frac{(n-1)(n-2)}{2}.$$

Also, I_1 and I_2 are integers, so

$$\min\{I_1, I_2\} \le \frac{1}{2} \frac{(n-1)(n-2)}{2} = \frac{(n-1)(n-2)}{4}.$$

These correspond to a path in $G(n)$ between π_1 and π_2. Hence, the maximum number of adjacent swaps needed to factor π is bounded by $\lfloor \frac{(n-1)(n-2)}{4} \rfloor$. □

Example 1. *Consider the permutation* $[3, 4, 1, 2, 5]$. *Cycle this, and let* $\pi_1 = [1, 2, 5, 3, 4]$. *Its classic inversion number is 2, the same as* $[2, 5, 3, 4]$. *Now let* $\pi_2 = [1, 4, 3, 5, 2]$. *This has classic inversion number 4, the same as* $[4, 3, 5, 2]$. *Moreover,*

$$4 + 2 = 6 = \frac{(n-1)(n-2)}{2}$$

for $n = 5$.

It is worth noting that the bound in 5 is $\lfloor \frac{(4-1)(4-2)}{4} \rfloor = 1$ for $n = 4$, and $\lfloor \frac{(5-1)(5-2)}{4} \rfloor = 3$ for $n = 5$, as we hoped in the previous examples.

4 Conclusion

As we have demonstrated here, the strategies used to play the games have many interesting and deep connections that can lead to solutions to classical mathematical problems. In particular, using a graph-theoretic approach, we have established a bound on the number of adjacent transpositions needed to sort a circular permutation, $\lfloor \frac{(n-1)(n-2)}{4} \rfloor$. This is lower than the currently known bound.

References

[1] M. Allocca, S. Dougherty, and J. F. Vasquez. Signed Permutations and the Braid Group. *Rocky Mountain J. Math.* **47** no. 2 (2017) 391–402.

[2] S. Dougherty and J. F. Vasquez. Amidakuji and games. *J. Recreational Math.* **37** no. 1 (2008) 46–56.

[3] J. E. Humphreys. *Reflection groups and Coxeter groups.* Cambridge Studies in Advanced Mathematics, vol. 29, Cambridge University Press, Cambridge, 1990.

PART III

Algebra and Number Theory

12

◇◇

THE MAGIC OF CHARLES SANDERS PEIRCE

Persi Diaconis and Ron Graham

1 Introduction

Charles Sanders Peirce (1839–1914) was an impossibilist: impossible to understand and impossible to ignore. One of the founders of the American school
of philosophy known as pragmatism (with Oliver Wendell Holmes, William
James, and John Dewey), Peirce is revered as the father of semiotics. He
made serious contributions to mathematical logic. He spent years as a working
physicist and geologist. He was an early contributor to statistics, using kernel
smoothers and robust methods 100 years before they became standard fare.
A spectacular appreciation of Peirce's contributions is in Louis Menand's 2001
Pulitzer Prize–winning book, *The Metaphysical Club: A Story of Ideas in America*. Indeed, if you get nothing more from our article than a pointer to this item,
you will thank us!

At the same time, Peirce is very difficult to parse. He wrote in a convoluted,
self-referential style, went off on wild tangents, and took very strong positions.
The breadth of his interests makes Peirce hard to summarize; we find the
following effort of Menand [38, p. 199] useful.

> What does it mean to say we "know" something in a world in which things
> happen higgledy-piggledy? Virtually all of Charles Peirce's work—an enormous
> body of writing on logic, semiotics, mathematics, astronomy, metrology, physics,
> psychology, and philosophy, large portions of it unpublished or unfinished—
> was devoted to this question. His answer had many parts, and fitting them all
> together—in a form consistent with his belief in the existence of a personal God—
> became the burden of his life. But one part of his answer was that in a universe in
> which events are uncertain and perception is fallible, knowing cannot be a matter
> of an individual mind "mirroring" reality. Each mind reflects differently—even
> the same mind reflects differently at different moments—and in any case reality
> doesn't stand still long enough to be accurately mirrored. Peirce's conclusion was
> that knowledge must therefore be social. It was his most important contribution
> to American thought, and when he recalled, late in life, how he came to formulate

it, he described it—fittingly—as the product of a group. This was the conversation society he formed with William James, Oliver Wendell Holmes, Jr., and a few others in Cambridge in 1872, the group known as the Metaphysical Club.

Should a modern reader take Peirce seriously? For example, in Menand's summary above, Peirce comes across as staunchly anti-Bayesian in his approach to uncertainty. At least one of your authors is a Bayesian [11]. Is it worth figuring out Peirce's views?

It turns out that Peirce had a lifelong fascination with card tricks. At the end of his life, he wrote a 77-page paper detailing some of his magical inventions. We thought, "This we can evaluate!" Are the tricks amazing and performable? No. Are they original and interesting? Yes.

Peirce's magical manuscript, "Some amazing mazes" [35], is no more readable than his semiotics. Its centerpiece is a deadly 15-minute effect with cards repeatedly dealt into piles and picked up in odd ways. The conclusion is hard to appreciate. He had an awareness of this, writing "please deal the cards carefully, for few would want to see it again." Despite all this, Peirce's methods contain four completely original ideas. Each of these can be developed into a charming, performable trick. The ideas have some mathematical depth; along the way Peirce gives card trick proofs of Fermat's little theorem and the existence of primitive roots.

It is our task to bring Peirce's ideas into focus. In what follows, the reader will find Peirce's

- cyclic exploitation principle,
- dyslexia effect,
- packet pickup, and
- primitive arrangement principle.

We try to make both the mathematical ideas and magical applications transparent. Along the way, new tricks and math problems surface.

Peirce's trick is explained, more or less as he did it, in the following section. We then break it into pieces in Sections 3–6 and 8. In each, we have tried to take Peirce's idea and make a good trick. We also develop the mathematical underpinnings. If this juxtaposition between magic and mathematics seems strange to you, take a look at Section 7 before proceeding. This presents one of our best tricks, "Together again." The reader mostly interested in magic should look at "Concentration," our development of Peirce's dyslexic principle in Section 4. The "Tarot trick," an application of Peirce's cyclic principle in Section 3, is also solid entertainment. If you are interested in magic, please read Section 5, our attempt to make a one-row version of the dyslexic principle. It still needs help to make it a good trick.

The mathematics involved goes from low to high, from the simplest facts of number theory and permutations through probabilistic combinatorics and the

Riemann hypothesis. We have tried to entwine it with the magic and make it accessible but, if it's not your thing, just skip over it.

The final section comes back to the larger picture of evaluating Peirce's work through the lens we have chosen. One of Peirce's gurus was the American polymath Chauncey Wright. They corresponded about card tricks. In an appendix, we unpack some of their tricks, discussed 50 years before "Some amazing mazes."

There is an enormous Peirce literature, in addition to his thirteen volumes of collected work [19, 34]. Surely the best popular introduction is Menand [38]. This may be be supplemented by Kaag [39]. For Peirce's card trick (and a friendly look at his weirder side), see Martin Gardner's *Fractal Music, Hypercards and More … Mathematical Recreations from SCIENTIFIC AMERICAN Magazine* [16, chap. 4]. Gardner's *The Whys of a Philosophical Scrivener* [17, chap. 2]—"Truth: Why I Am Not a Pragmatist"—is a useful review of the ups and downs of pragmatism. His *Logic Machines and Diagrams* [18] gives an extensive review of Peirce's many contributions to logic. Eisemann [20,21] gives a number theoretic analysis of Peirce's card trick; Alex Elmsley's "Peirce arrow" and "Through darkest Peirce" in Minch [22, 23] take the magic further. For Peirce and statistics, see Meier and Zabell [30], Stigler [43], and Diaconis and Saloff-Coste [10].

The Essential Peirce [32,33] is a careful selection of Peirce's key writings. The Wikipedia page for Peirce points to numerous further paths of investigation: the website of The Charles S. Peirce Society and their quarterly, *Transactions*; the website of Commens: Digital Companion to C. S. Peirce; and the website of the Peirce Edition Project all being rewarding destinations.

2 Peirce's "Some Amazing Mazes"

In this section, we first describe what the trick looks like and then give an explanation of why it works.

2.1 What the Trick Looks Like

The routine that Peirce performed "at the end of some evening's card play" looked like this: The performer removes the ace through king of hearts and the ace through queen of spades from a normal deck of cards, arranging each packet in order A, 2, 3, …, where the ace is the top card when the packet is placed face down. The twelve hearts, ace through queen, are put in a face-down pile on the table, off to one side. The king of hearts is placed off to the side as well in a separate pile. The twelve spades are held face down, as if about to be dealt in a card game. They are dealt face up into two piles, say left, right, left, right, …, so the piles look like those in Figure 12.1.

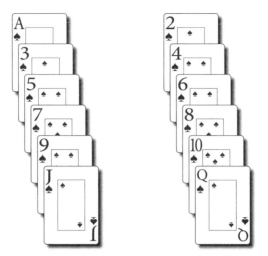

Figure 12.1. The spades dealt into two piles.

The performer stops before placing the last card (the black queen) and discards it (face up) to one side, replacing it with the top card of the heart pile (the ace). Now the two face-up piles of six cards are gathered together, the left pile is placed on top of the right, and then the whole packet is turned over.

This whole procedure is repeated 12 times (!!!). Each time, the final dealt card will be black. It is placed onto the growing discard pile, face up, and replaced by the current top card of the heart pile. At first, it is not surprising that the last card is black, since all or most of the dealt pile is, as well. But as more and more red cards are switched in, it becomes more surprising. On the twelfth deal, all the blacks will have been replaced by reds. This is the ending of the first part of the trick. Perhaps it is wise for the performer to point this out, saying something like, "You win if the last card is red—nope, not this time. Well, let's just add another red winner for you and try again." After this, the performer places the red king with the rest of the hearts on the top of the heart packet and turns the whole packet face down.

The second part of the trick begins now. The black packet that accumulated as cards were discarded is taken up, turned face down, and dealt (left to right) in a long row on the table. The red packet is turned face down and "mixed" by dealing it into a number of hands suggested by a spectator (e.g., seven hands). These cards are then picked up to form a single pile via "Peirce's pickup" (explained in Section 6). This may be repeated a time or two, and the red packet may even be cut. However, at the end, the performer openly cuts the red packet to bring the king of hearts to the bottom and then places this packet face down.

The denouement of the second part of the trick has begun. There is a row of black cards on the table and a pile of red cards. All cards are face down. The performer announces she will make but one simple adjustment and move a certain number of black cards cyclicly, from the left side of the black row to the right. (Say, three cards are moved; the actual number depends on the spectator's choices via a rule explained later.) The trick concludes with what we will call "Peirce's dyslexic effect." The performer claims that the black card will reveal the location of any red card. "What red card do you want to find?" Suppose the spectator says "6." The performer counts over six from the left end of the black row and turns up the card in that position: let's say it's a 9. After announcing "the cards say your card is nine down in the red packet" and counting down without disturbing the order, the ninth card is turned up, and it is indeed the requested 6. This may be repeated several times, each time the row of black cards "knows" where any requested red card is located. During the repetitions, if desired, the packet of reds can be dealt into a number of piles, finishing with a Peirce pickup.

This is the conclusion of the second part (and indeed, of the complete routine). Let us try to explain how strange an ending it is. In a standard, modern card trick, when a row of cards is revealed, it often has a strikingly simple order. For example, a classical trick has the deck cut into four piles, and the top card of each pile is turned up to show an ace in each location. Or a row of seven might be turned up to reveal 1-31-1945, the spectator's day of birth. More elegantly, four piles are cut off the deck, and the top four cards are turned up, appearing as a random failure, say, A, Q, 9, 7. With a wink, the performer turns the piles over to reveal their bottom cards: they are the A, Q, 9, and 7 of matching colors. These are easy-to-understand endings.

Consider Peirce's ending. There are two packets of cards, and the card at position j in one packet tells the position of the card of value j in the other. It is hard to parse but completely original. People have been doing such revelations for over 500 years, and we have never seen *anything* like this dyslexic effect. In Sections 5 and 6, we shine some mathematical and magical light in this corner. For the moment, let us review.

Peirce's trick involves an interminable amount of dealing for two fairly subtle results: never a red card shows, and the dyslexic revelation. His contributions need unpacking; we turn to this in Sections 3–8. To conclude this section, we give a much more hands-on description of Peirce's trick together with a standalone explanation of why it works. This is not needed to follow the rest of the chapter, and the reader may want to return to it when working through Peirce's trick with cards in hand.

2.2 Why the Trick Works

Peirce worked "mod 13," but all can be done with any prime number p of cards such that 2 is a primitive root of p (see Section 9 for more on primitive roots). Thus 5, 11, 13, 19, or 29 cards can be used, for example.

To explain the mathematical underpinnings of Peirce's card trick, we first consider in some detail what happens with cards ace through 10.

Analysis of a Two-Pile Shuffle. Let us start with ten cards, say the ace through 10 of spades. We will think of the ace as having the value 1. The cards are arranged face down in the order $1, 2, \ldots, 10$ from top to bottom (in hand, so to speak). We now deal the cards alternately into two face-up piles from left to right, with the left pile getting the first card (1) and the right pile getting the next card (2). After all ten cards have been dealt, the face-up piles look like:

9	10
7	8
5	6
3	4
1	2
Left	Right

We then take the left pile and place it on top of the right pile, and then *turn the combined pile over*, so that now the cards are all face down. The new pile now looks like:

Position	Value
1	2
2	4
3	6
4	8
5	10
6	1
7	3
8	5
9	7
10	9

Notice that the card in position i has the value $2i$ (mod 11) (How nice!). In general, if this operation is repeated k times, then the card in position i will have the value $2^k i$ (mod 11). In particular, if $k = 10$, then the cards will come back to their original order (since $2^{10} \equiv 1$ (mod 11)).

Now for Peirce's trick, we start with *two* sets of 11 cards, say, ace through jack of spades (the black cards), and ace through jack of hearts (the red cards). To begin with, we discard the jack of spades, since it will never be used (actually, it can be used at the very end). So at the beginning, we have the black cards face down in the order $1, 2, \ldots, 10$ from top to bottom (in hand, so to speak) and the red cards face down on the table in order $1, 2, \ldots, 10$, jack from top to bottom.

We now perform the shuffle described above. That is, we begin dealing the ten black cards face up into two alternating piles on the table, starting with the left pile first. However, when we reach card 10, instead of placing it on top of the right pile, we instead place it face up on the table starting a new black pile, and we take the top card of the red pile (which is an ace, or 1) and place it face up on top the right pile. So the piles now look like:

$$
\begin{array}{cc}
9 & 1 \\
7 & 8 \\
5 & 6 \\
3 & 4 \\
1 & 2
\end{array}
$$

As before, we now combine the left and right piles into one pile by placing the left pile on top of the right pile and turning combined pile upside down, so that the cards are all face down. Here is the situation after this first step:

Position	Value
1	2
2	4
3	6
4	8
5	1
6	1
7	3
8	5
9	7
10	9

The (slightly depleted) red pile now has the face-down cards in order $2, 3, \ldots, 10$ from top to bottom and the (new) black face-up pile has the single black card 10.

Now perform this step again on the combined pile. That is, we deal the cards face up into two piles alternatively, left to right with the left coming first, except when we come to the last card (which is a 9), we place it face up on the new black pile, and put the next card from the red pile (which is a 2) as the last card on the right pile. Now the left and right piles look like:

7	2
3	5
1	1
6	8
2	4
Left	*Right*

We then combine the two piles into one by placing the left pile on top of the right pile and turning the combined pile upside down, so that all the cards are face down. What we now have after this step is:

Position	*Value*
1	4
2	8
3	1
4	5
5	2
6	2
7	6
8	1
9	3
10	7

Let us now perform this "Peirce shuffle" altogether 10 times. We tabulate the combined face-down piles after each step below (the step is indicated on top in bold):

Position	**1**	**2**	**3**	**4**	**5**	**6**	**7**	**8**	**9**	**10**
1	2	4	8	5	1	2	3	4	5	6
2	4	8	5	1	2	3	4	5	6	7
3	6	1	2	4	8	5	1	2	3	4
4	8	5	1	2	3	4	5	6	7	8
5	1	2	3	4	5	6	7	8	9	10
6	1	2	4	8	5	1	2	3	4	5
7	3	6	1	2	4	8	5	1	2	3
8	5	1	2	3	4	5	6	7	8	9
9	7	3	6	1	2	4	8	5	1	2
10	9	7	3	6	1	2	4	8	5	1

In the meanwhile, the new black face-up pile has been building up. Turning it over so that all the cards are face down, it now looks like:

Position	Value
1	10
2	9
3	7
4	3
5	6
6	1
7	2
8	4
9	8
10	5

The first interesting claim is that the final red pile and the final black pile are **inverse permutations** of each other (check!). (See Section 4 for background on inverses.) Let us see why.

First consider the black pile. Its entries are formed by always taking the card in position 10 in the current list and placing it on the (new) black pile. But the card in the tenth position goes through the values $10, 2 \cdot 10, 2^2 \cdot 10, \ldots, 2^9 \cdot 10$ (mod 11) $= 10, 9, 7, 3, 6, 1, 2, 4, 8, 5$, as the piles are created. So this is the arrangement of the black pile. Each entry is twice the value (mod 11) of the value before it. So what is the rule that links the position i to the value? It is just this:

Rule 1: The card at position i in the black pile has the value 2^{4+i} (mod 11).
The 4 in the exponent was needed to adjust the cycle, so that the card in position 6 has value 1 $(= 2^{10}$ (mod 11)).

Now let's look at the red pile. The trick here is to represent the position as a power of 2 (mod 11). Thus, we could express the above table as:

Position	1	2	3	4	5	6	7	8	9	10
$1 \equiv 2^{10}$	2	4	8	5	1	2	3	4	5	6
$2 \equiv 2^1$	4	8	5	1	2	3	4	5	6	7
$3 \equiv 2^8$	6	1	2	4	8	5	1	2	3	4
$4 \equiv 2^2$	8	5	1	2	3	4	5	6	7	8
$5 \equiv 2^4$	1	2	3	4	5	6	7	8	9	10
$6 \equiv 2^9$	1	2	4	8	5	1	2	3	4	5
$7 \equiv 2^7$	3	6	1	2	4	8	5	1	2	3
$8 \equiv 2^3$	5	1	2	3	4	5	6	7	8	9
$9 \equiv 2^6$	7	3	6	1	2	4	8	5	1	2
$10 \equiv 2^5$	9	7	3	6	1	2	4	8	5	1

We saw that on the very first step, the card in position 10 (the black 10) was replaced by the first red card, 1. At step **2**, the card in position 10 (the black 9) was replaced by the next red card, 2. In general, once a card reaches position

10, it gets replaced at the next step by the next unused red card, which will be in position 5. This is why the successive cards in position 5 are 1, 2, 3, etc. Now consider the card in position 8. It started out with value $8 \equiv 2^3$ (mod 11). It would have hit the value 10 at step **2**, but 10 has already been replaced by 1. That is why the card in position 8 at step **2** is 1. The next card in position 8 in step **3** should have been $2 \cdot 10 \equiv 9$ (mod 11), but 9 has already been replaced by 2. That's why the card in position 8 at step **3** is 2. In general, once a 1 appears in a position in some step, then the subsequent cards in that position are 2, 3, ..., to the end. So the question is: What is the final value in position 2^j at the end? Well, the value in that position at step **k** would be 2^{j+k} (mod 11), if not for the red cards replacing the black cards. Thus, at step $4 - j$ (mod **10**), the value is 5. The next value at that position should have been 10, but 10 was replaced by 1. So how many more steps does it take us to get to the end from here? It takes just $10 - (4 - j) = 6 + j$ (mod 10) more steps to get to step **10**. Thus, we have the following rule.

Rule 2: The card at position 2^j (mod 10) in the red pile has the value $6 + j$ (mod 11).

We now need to show that these rules are reciprocal. To do this, we just bump up the index in rule 1 by 6 to get:

Rule 1′: The card at position $i + 6$ (mod 10) in the black pile has the value 2^i (mod 11).

This shows that the two permutations are inverses of each other.

The Peirce Pickup: Preliminary Analysis for the General Case. We begin with a pile of n face-down cards with the card in position i from the top having value i. We choose an arbitrary integer k *relatively prime* to n satisfying $1 \leq k < n$. We now deal the cards into k piles, starting from the left and turning the cards face up as we go. Define $r = n$ (mod k). Thus, $n = kt + r$ for some integer t. We now choose an arbitrary pile—say, pile s—and place it on top of pile $s + r$. Here the pile numbers start with 1 on the left and are always reduced modulo k. We then take this larger pile and place it on top of pile $s + 2r$. We have to be careful to count the positions of the missing piles when we are doing this pickup. We continue this process of picking up piles until all the piles have been combined into one pile of n face-up cards. We then turn the pile over, so that we now have a pile of n face-down cards. We then make a final cut so as to bring the card n to the bottom of the face-down stack. As usual, the question now is: What is the value of the card in position i from the top? The answer is this:

Rule 3′: The card at position i has the value ki (mod n).

As an example, take $n = 13$, $k = 5$, so that $r = 3$. After the piles are dealt, we have the situation:

$$11\ 12\ 13$$
$$6\ \ 7\ \ 8\ \ 9\ 10$$
$$1\ \ 2\ \ 3\ \ 4\ 5.$$

Let us choose $s = 2$ for the starting pile to pick up. It is placed on top of pile 5, and the combined pile is placed on pile $5 + 3 \equiv 3 \pmod{5}$, etc. In the combined pile, we then cut the deck, so that the 13 is on the bottom. The final pile, after being turned over to become face down, is:

Position	Value
1	5
2	10
3	2
4	7
5	12
6	4
7	9
8	1
9	6
10	11
11	3
12	8
13	13

As advertised, the card in position i has value $5i \pmod{13}$. Now, why does this work in general? Well, consider the ith pile before the piles have been collected. It has all the cards with value congruent to i modulo k arranged in increasing order from bottom to top. So the top card of this pile is either $kt + i$ or $kt + i - n$ depending on whether $i \le r$ or not. Hence, if we add k to the value of this top card, then in either case, we get the value $k(t + 1) + i - n = i - r$. But this is just the value of the bottom card of the $(i - r)$th pile, which is the pile we are putting on top of pile i. Thus, when going from the top card of pile i to the bottom card of pile $i - r$, we simply add $k \pmod n$ to the value of the top care of pile i. Since this holds for any pile-to-pile connection, then in the inverted combined list, the value of the card in position $s + 1$ is obtained from the value of the card at position s by adding $k \pmod n$. This implies that the value of the card at position i must have the form $ki + c$ for some c. Now, the final cut, which brings the card with value n to the bottom, is just a cyclic shift of the values. In particular, it forces the value of card n (on the bottom) to have the value n. Thus, this cyclic shift forces c to be zero! In other words, for the final stack, we have **Rule 3**, as stated above.

The same arguments show that the following more general rule holds for the Peirce Pickup.

Rule 3′: In the final stack of the Peirce Pickup using k piles with an arbitrary initial list of values, the value of the card in position i is the value of the card in position $ki \pmod n$ in the original list.

Finishing the Analysis. For the final part of the trick, we begin by placing the red card 11 (= jack of hearts) at the bottom of the red stack. We next choose (actually, the spectator chooses) an arbitrarily number k with $1 \leq k < 11$. We define $r = 11$ (mod k). Thus, for some t, $11 = kt + r$. We then perform the Peirce Pickup on this red stack, normalizing at the end, so that the card with value 11 is on the bottom. For example, suppose we choose $k = 4$. Thus, in this case, starting with pile 2, for example, we go from

Position	Value
1	6
2	7
3	4
4	8
5	10
6	5
7	3
8	9
9	2
10	1
11	11

by stacking to

$$2 \ 1 \ 11$$
$$10 \ 5 \ 3 \ 9$$
$$6 \ 7 \ 4 \ 8$$

by recombining, starting with pile 2, to

Position	Value
1	1
2	5
3	7
4	2
5	10
6	6
7	9
8	8
9	11
10	3
11	4

by inverting to

Position	Value
1	4
2	3
3	11
4	8
5	9
6	6
7	10
8	2
9	7
10	5
11	1

and by cutting the 11 to the bottom to

Position	Value
1	8
2	9
3	6
4	10
5	2
6	7
7	5
8	1
9	4
10	3
11	11

Now what is the rule for obtaining the value of the card at position i? For the starting red stack, the card at position 2^j had the value $6 + j$. By **Rule 3'**, we know that after performing the Peirce Pickup with $k = 4$, the card at position i now has the value originally held by the card at position $4i$ (mod 11). But by **Rule 2**, the card at position $4 \cdot 2^i = 2^{j+2}$ in the red stack has the value $6 + j + 2 = 8 + j$ (mod 11). This is just the value of the card in position j after completing the Peirce Pickup. Replacing the index $j + 2$ by j, this says that the card in position 2^j in the final red list has the value $6 + j$ (mod 11) (If we had chosen $k = 3$, for example, then we would have used $3 \cdot 2^j = 2^{j+8}$, since $2^8 \equiv 3$ (mod 11)).

Now for the final cut of the (long-ignored) black deck. Recall that it was (facedown):

Position	Value
1	10
2	9
3	7
4	3
5	6
6	1
7	2
8	4
9	8
10	5

and the card in position i has the value 2^{4+i}. Since the card in position 1 in the final red list has the value 8, we want to cut the black deck so that in it, the card in position 8 will have the value 1. This is necessary if the permutation are to be inverses. To do this, we simply cut the bottom two cards to the top, thus forming

Position	Value
1	8
2	5
3	10
4	9
5	7
6	3
7	6
8	1
9	2
10	4
(11)	(11)

If desired, the discarded black card 11 (the jack of spades) can be appended to this list with no harm. In this final black deck, the card in position i now has the value 2^{i+6}, since the indices were shifted by 2. Thus, making sure that some pair (here, 1 and 8) are "reciprocal" guarantees that all pairs are reciprocal. Of course, this is the inverse of the rule governing the arrangement in the final red list.

It should be clear how the arguments will go in the general case that the original number of cards n is prime (or a prime power), which has 2 as a primitive root. For this case, the rules for the red and black lists must be modified accordingly. In particular, they are as follows.

Rule 1 (general): The card in position i in the black pile has the value $2^{i+\frac{n-3}{2}}$ (mod n).

Rule 2 (general): The card in position 2^j in the red pile has the value $j + \frac{n+1}{2}$ (mod n).

3 Peirce's Cyclic Principle

The first part of Peirce's trick has cards repeatedly dealt into two piles, the last card switched for one of another color and the two piles combined so the new card goes into the center; the old card is set off to one side. Working with twelve cards, it turns out that on each repetition, the last card dealt is always one of the old cards, so all twelve cards are eventually replaced. As there are more and more new cards, it becomes more and more surprising that you don't hit one at the end of the deal.

This section abstracts Peirce's idea and shows how to make a standalone trick from it. Begin with the abstraction. You can replace "deal into two piles" with any repetitive procedure for any number of cards n as long as your procedure first repeats after n repetitions. Let us explain by example.

Example 1 (Down and under). *Take five cards, in order 1,2,3,4,5 from top down. Hold them face down as if dealing in a card game. Deal the top card onto the table, take off the next card, and put it under the current four in hand (put it under). Deal the next card face down onto the card, on the table, put the next under, then down, then under, then down, and put the last one down as well. The cards will be in order 2,4,5,3,1. Instead of switching the 2 for a different color, just turn it face up. You will find this recycles after exactly five repetitions. Further, the last card is always face down (before being turned over).*

Example 2 (Two-pile deal). *Consider next dealing 2n cards repeatedly into two piles. We will follow Peirce and start with the cards face down, turning them up as they are dealt, say, left/right, left/right, ... So the cards start out:*

$$\begin{array}{cc} \vdots & \vdots \\ 5 & 6 \\ 3 & 4 \\ 1 & 2 \end{array}$$

At the end, put the last card on the right pile, but leave it face down. Put the left pile on the right pile, and turn all the cards back over, so they are all face down (with one face up in the middle). This is one deal. How many deals does it take to recycle? The answer depends on n in an unknowable way. Indeed, Peirce's two-pile deal is what magicians call an "inverse perfect shuffle." Since the original bottom card winds up inside, it is an inverse in-shuffle. The number of repeats, call it r, to recycle satisfies $2^r \equiv 1$ (mod $2n + 1$). For example, when $2n = 12$ as

Figure 12.2. Some tarot cards.

for Peirce, we must find the smallest power of 2 so that $2^r \equiv 1$ *(mod 13). By brute force, the successive powers of 2 (mod 13) are 2,4,8,3,6,12,11,9,5,10,7,1. The first repeat comes after twelve doublings. We say "2 is a primitive root (mod 13)." It is natural to ask whether there are other values of n so that 2 is a primitive root (mod 2n+1.) It is unknown if this happens for infinitely many n. It does if the Generalized Riemann Hypothesis holds. The Riemann Hypothesis is perhaps the most famous problem in mathematics (with a $1,000,000 prize for its solution). Our simple card trick leads into deep waters.*

3.1 Some Mathematics

To learn more about perfect shuffles (and down/under shuffles), the reader can consult Diaconis and Graham [12, Chap. 7]. There are many other ways of mixing that repeat after exactly n. These are called "n-cycles," and it is not hard to see that there are $(n-1)!$ of them. However, if you pick a shuffle at random, it's not likely to be an n-cycle. Call the shuffle σ, and suppose it first repeats after $r(\sigma)$ iterations. In Erdös and Turán [24], it is shown that r is about $e^{(\log n)^2/2}$. More formally, for large n,

$$Pr\left\{\frac{\log r(\sigma) - \frac{(\log n)^2}{2}}{\sqrt{(\log n)^3/3}} \leq x\right\} \sim \frac{1}{\sqrt{2\pi}} \int_{-\infty}^{x} e^{-t^2/2}\, dt.$$

To say this in English: The log of the repetition number $r(\sigma)$ fluctuates about $(\log n)^2/2$ at scale $\sqrt{(\log n)^3/3}$ with fluctuations following a bell-shaped curve.

Permutations such as σ can easily have $r(\sigma)$ much larger than n. For example, when $n = 52$, the largest order is 180 180. If you look up permutations in an elementary group theory book, you will find that a permutation can be decomposed into cycles and that $r(\sigma)$ is the least common multiple of the different cycle lengths. This makes it easy to determine $r(\sigma)$. If $g(n)$ is the largest-possible order of a permutation of n ($g(n)$ is called Landau's function, with its own Wikipedia page), Landau showed that

$$\lim_{n\to\infty} \frac{\log g(n)}{\sqrt{n \log n}} = 1.$$

Thus, roughly, the largest order is $e^{\sqrt{n \log n}}$, which is much, much larger than $n = e^{\log n}$.

3.2 Some Magic

That's enough about the mathematics of Peirce's cyclic principle. How can a reasonable trick be formed from it? Here is a simple illustration, a variation of a trick called "turning the tarot" by Dave Arch [1].

Example 3 (A tarot trick).   *The performer shows seven tarot cards, drawn from the Major Arcana. (If you want to follow along, you can work with any seven face-down playing cards.) Explaining that some people believe that tarot cards can predict the future, the performer asks a spectator to select a lucky number between 1 and 7; suppose that 4 is named. The performer asks the spectator to mix the seven cards and then place cards singly from top to bottom of the packet, turning up the fourth. At this stage, a brief interpretation of the meaning of this card is given: A bit of homework is required to present this effect. The deal is repeated, the spectator putting cards singly from top to bottom and turning up the fourth. The performer continues the interpretation and adds, "If, as we go along, we have a repeat of an earlier card, the reading is off. If, by luck or fate, we get all the way through without a repeat, the reading has meaning." As continued, a fresh card is turned up each time. With a little practice, the successive interpretations can be linked together to cohere.*

We think it is clear that this simple cycle has been amplified into a solid piece of entertainment. The trick always works, because 7 is a prime, and any number between 1 and 7 has no common factors with 7 and so will first repeat after seven repetitions.

If you want to do the trick with n cards—for example, the full Major Arcana of the tarot deck has $n = 22$—the spectator must name a number having no common factor with n. The number of these has a name, $\phi(n)$, or *Euler's phi function*. For example, $\phi(22) = \phi(2) \times \phi(11) = 10$. The chance of success is $\phi(n)/(n-1)$. On average, for a typical n, this is about $6/\pi^2 \doteq 0.61$. In math, we are saying

$$\lim_{N \to \infty} \frac{1}{N} \sum_{n=1}^{N} \frac{\phi(n)}{n} = \frac{6}{\pi^2}.$$

To improve your chances, if a spectator names a number having a common factor with the deck size, you can say: "Deal your number of cards singly from top to bottom and turn up the *next* card." The chance that j or $j+1$ has no common factors with n is approximately

$$2 \cdot \frac{6}{\pi^2} - \prod_p \left(1 - \frac{2}{p^2}\right) \approx 0.87$$

(the product is over all primes p). We are just having fun mixing magic and mathematics. For practical performance, use a prime number for n.

Peirce amplified the cyclic principle, merging it with what we call his dyslexic principle. We explain and abstract this next.

4 Peirce's Dyslexic Principle

This section unpacks the strange ending of Peirce's card trick. Remember the setting: There are two rows of cards on the table. The value of the card in position j in the first row is the position of value j in the second row. Here is an example when $n = 5$:

First row   3   5   1   2   4
Second row   3   4   1   5   2.

To find the position of value 5 in the second row, look in the fifth position of the first row. The 4 there tells you there is a 5 in position 4 of the second row. This works for any value. It also works with the rows reversed: The 2 in position 5 of the second row tells us that the value of the card in position 2 of the first row is a 5. Switching rows switches value and position. What's going on?

While Peirce doesn't spell it out, his principle amounts to a classical statement: The two permutations involved are *inverses* of each other. To explain, think of a row of cards as a rule σ for assigning a value $\sigma(i)$ to position i. For the first row in the previous example, this is represented as

i	1	2	3	4	5
$\sigma(i)$	3	5	1	2	4

Here, 1 is assigned $\sigma(1) = 3$, and 5 is assigned $\sigma(5) = 4$. These assignments are called "permutations." Now, permutations can be multiplied. If σ and π are permutations, $\pi\sigma(i) = \pi(\sigma(i))$; first do σ and then do π. With σ as given previously, use π defined as

face-down i	1	2	3	4	5
$\pi(i)$	5	4	3	2	1

Then $\pi(\sigma(1)) = 3$, because σ takes 1 to 3 and π fixes 3:

i	1	2	3	4	5
$\pi\sigma$	3	1	5	4	2

Note that this is different from $\sigma\pi$. The *identity* permutation ι fixes everything:

i	1	2	3	4	5
$\iota(i)$	1	2	3	4	5

Two permutations π and σ are *inverses* if $\pi\sigma = \iota$. The reader can check that the two permutations used in our $n = 5$ example are inverses:

σ	3	5	1	2	4
π	3	4	1	5	2

We usually write σ^{-1} for the inverse of σ.

4.1 Our Finding

Peirce's dyslexic principle is equivalent to the two rows of cards being inverse to one another. Multiplication and inverse are the backbone of modern algebra and group theory. Peirce has let them in the back door as a card trick. We don't know if he intentionally did this (i.e., "Hmm, let's see, how can I make a card trick out of inverses?"). As far as we know, no one before or since has had such a crazy, original idea.

Peirce did much more: He found a completely nonobvious way of arriving at two inverse permutations. We hope that the reader finds the following as surprising as we did.

Consider Peirce's trick. He started with two sets of 12 cards in order A, 2, ..., Q (the red king is not really part of the picture). He repeatedly dealt the first set into two piles, discarding the last and switching it for the current top card of the second set. The discards from the first set were placed successively in a face-up pile on the table. After going through the two piles, he dealt them into two rows on the table. Amazingly,

the two rows are inverses.

Peirce's discovery suggests two questions, one mathematical, the other magical. How else can two inverse permutations be formed? How can this inverse relationship be exploited to make a good trick?

There is a charming, simple, surprising way to get inverses. Start with two rows of n cards, in order $1, 2, \ldots, n$, on the table. Suppose the top row is all red and the bottom row is all black. It doesn't matter whether all cards are face up or all are face down.

4.2 Inverses

Repeatedly take any red card in the top row, any black card in the bottom row, and switch their positions. Continue until all the red cards are in the bottom row. The two permutations are inverse:

$$
\begin{array}{ccccccccccc}
1\,2\,3\,4\,5 & & 1\,4\,3\,4\,5 & & 1\,4\,1\,4\,5 & & 2\,4\,1\,4\,5 & & 2\,4\,1\,5\,5 & & 2\,4\,1\,5\,3 \\
1\,2\,3\,4\,5 & \xrightarrow{} & 1\,2\,3\,2\,5 & \xrightarrow{} & 3\,2\,3\,2\,5 & \xrightarrow{} & 3\,1\,3\,2\,5 & \xrightarrow{} & 3\,1\,3\,2\,4 & \xrightarrow{} & 3\,1\,5\,2\,4
\end{array}
$$

You should check that the last two permutations are inverses. The spectators can be allowed to choose these random transpositions, or the performer can choose them, seemingly haphazardly, to ensure a given permutation in the first row. At least one of your authors still finds it viscerally surprising that any old way of switching the two rows will work.

Marty Isaacs [25] has suggested another way to get two inverse rows: Start with two rows of cards, the top row in order $1, 2, \ldots, n$, face up. The bottom row is in the same order, face down. Have the spectator switch two face-up cards, say, 2 and 7 in the face-up row. You now switch the cards at positions 2 and 7 in the bottom row. This may be continued for as many switches as you like. The two rows will always be inverse permutations. If both rows are face up, at any stage, values can be switched in either row with positions switched in the other. Finally, if our original method of producing inverses is used, as above, a few further such switches will keep the rows inverse.

Peirce's original procedure can be abstracted. Here is our main finding.

4.3 Our Rule

Consider *any* n-cycle σ (e.g., a repetitive procedure on n cards which first recycles after n repetitions). Take two packets of cards, the red packet and the black packet, originally in the same order: $1, 2, 3, \ldots, n$. Repeatedly perform σ to the red packet, but at a fixed step k, switch the current red card for the current top card of the black packet, setting the red cards face up in a pile at one side. After n steps, the two packets are in inverse orders.

We leave the proofs of both rules to our readers. The "cards-in-hand" method should prove completely convincing.

Turn next to the magical problem. How can inverses be exploited to make a good trick? Peirce just baldly revealed things by example. To see the problem, consider a possible slam-bang ending: With two piles of cards in inverse permutation order, deal one of the piles in a row face down on the table and the second pile in a row face up on the table. Suppose $n = 10$ and the two rows are as follows:

Face down	2	3	1	5	6	4	9	7	10	8
Face up	3	1	2	6	4	5	8	10	7	9

We want to put the face-up cards, starting with the leftmost 3, one at a time, onto their face-down mates in the top row. The two rows determine one another, so it should be easy (or at least possible). Okay, where should the 3 go? Well, the position of the $\boxed{3}$ in the top row is the value of the third card in the bottom row, a 2, so put the face-up 3 on the second card in the first row. Next to be placed is the $\boxed{1}$ the position of the 1 in the top row is the value of the first card in the bottom row. Oops, this has already been placed, but we remember it was a 3. So place the 1 in the third place of the top row. A similar problem occurs for placing the 2, the 4, and the 5. Your memory may be better than ours, but trying to flawlessly match the cards during a live performance seems foolish. Trouble mounts when the bottom permutation, call it σ, has a drop $\sigma(i) < i$, for then the information has been used up.

Okay, we can't use it for magic. What about math? Pick a permutation at random: How many drops does it have? This is a well-studied problem; indeed, one of us has written several papers on drops [3, 5–7], because they occur in the study of the mathematics of juggling. Permutations with a given number of drops are in bijection with permutations with a given number of descents. (A permutation σ has a descent at i if $\sigma(i+1) < \sigma(i)$, so 3 1 5 4 2 has three descents.) Euler studied descents, and your second author has written about them [8]. The bottom line is that a typical permutation has about $n/2 \pm \sqrt{n/12}$ drops, and the fluctuations follow a bell-shaped curve. This is too many drops for practical work.

Peirce himself offered an even more convoluted way to reveal the inverse relation. His manuscript contains "a second curiosity" (sections 643–645, or Peirce [36]), which is an amazing two-dimensional extension requiring the audience to understand a novel numbering system. It remains to be unpacked.

In what follows we offer our best effort to make a solid performable trick from Peirce's dyslexic principle.

Example 4 (Concentration, without memory). *Concentration—also known as pairs, match-up, pexeso, or shinkei suijaku—is a widely played family card game in which a deck of cards is laid out on the table. Two cards are turned over in each round, and the object is to turn up matching pairs (e.g., the two red aces or the two black jacks). If the pair turned up matches, you remove them and get a point. If not, they are turned face down again, and all players try to remember their positions for future rounds. The following magic-trick version makes for good entertainment.*

Effect. Show, say, the ace through king of hearts and the ace through king of spades, in order. Lay the hearts face down in a row on the table and the spades face up in a row below them (Figure 12.3).

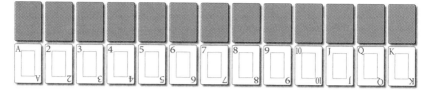

Figure 12.3. A layout of thirteen hearts and thirteen spades.

Figure 12.4. A layout of five hearts and five spades.

Turn your back on the proceedings, and instruct the spectator to take any face-down card in the top row and switch it with any face-up card in the bottom row. This is repeated until the bottom row is all face down and the top row is all face up. In practice, this should be done slowly, repeating the instructions, perhaps with commentary about trying to remember where the cards go. To finish, rapidly pick up the face-down cards, one by one, and place them on a face-up card. At the end, thirteen perfect matches are displayed—an ace on an ace, and so on.

Method. The trick (sort of) works itself. Of course, it will take a bit of practice to perform it in a rapid, efficient way. Here is how it goes: Suppose you are doing it with five cards, and when you turn around, you see the result in Figure 12.4.

As you go along, you will leave the face-up cards in place and put their matching face-up cards alongside. Where is the face-down 4? Look at position 4 in the top row; it's the ace. The face-down 4 is in position 1(!). Similarly, the face-down 5 is in position 3, the face-down 2 is in position 5, the face-down ace is in position 4, and the face-down 3 is in position 2. We find it surprising and pleasing when we work it. At the end, the cards appear as shown in Figure 12.5.

Figure 12.5. The layout after the switches.

The face-down cards half-cover their face-up mates; conclude by revealing the perfect matches.

How It Works. At the conclusion of the initial swapping phase—switching face-up and face-down pairs—the top and bottom rows make up an inverse pair of permutations. By Peirce's dyslexic principle, the position of the face-down value j is determined by the value of the face-up card at position j. That's it.

Performance Details. Any number n of cards can be used. Perhaps a complete suit is a good balance between difficulty and boredom. Two matching sets of alphabet cards could be used, or pictures of Disney characters, movie stars, abstract symbols, or ESP cards.

It is best to involve the audience's "concentration" along the way. After the initial switches, turn back to face the audience and ask, "Does anyone know where the ace is?" Or, pointing to a middle card, ask "Anyone remember what this card is?"

You don't have to match the cards left to right as described. You can have a spectator point to any face-up card and find its mate as you are questioning him.

We hope that Peirce would approve of our efforts to make a performable trick from his dyslexic principle. We give more developments in the following section.

5 More Dyslexia

A "one-row" version of Peirce's dyslexic principle can be developed by considering involutions. Work with an even number of cards, say, $2n$. Put them face up in order left to right, for example, with $2n = 10$:

$$1 \quad 2 \quad 3 \quad 4 \quad 5 \quad 6 \quad 7 \quad 8 \quad 9 \quad 10.$$

Now pick any pair of distinct cards. Turn them face down, and switch their positions. For example, if 3 and 7 are chosen, the row appears as

$$1 \quad 2 \quad \blacksquare \quad \blacksquare \quad 4 \quad 5 \quad 6 \quad \blacksquare \quad 8 \quad 9 \quad 10,$$

with the face-down cards being 7, 3 (left to right). Continue, each time picking a random pair of cards, turning them face-down and switching their positions. After n such moves, all the cards are face down and in a seemingly random arrangement. If you perform this as a trick, turn your back and have the spectator make the switches.

You may now turn around and have the cards "magically" sort themselves: Here is how it goes. As a presentation gambit, ask "Who knows where the 7 is?" or "Who can remember the position of any card now?" Ask the spectator to turn over the desired card. Say it is the 6 in position 4. Switch the 6 with the face-down card in position 6, and move that face-down card back to position 4, saying "We'll just put the 6 in its proper place." Keep going. Have the spectator choose another face-down card (other than the fourth), turn it up, and switch the two as before. After n such moves, half the cards are face up and in their correct positions. End the trick by turning over all the remaining face-down cards, revealing that the row is sorted correctly.

This is close enough to the surface that you may just see through it: Do you? We are reminded of the old line, "It's not much of a trick, but it makes you think—makes you think it's not much of a trick."

The trick doesn't have to be performed with cards. Indeed, what do "logarithms," "migrated," "spheroidal," and "multipronged" have in common? All are familiar words spelled with all letters distinct. Here is a little trick based on any one of these.

5.1 One-Handed Concentration

The performer removes eight letter tiles from a small bag and lays them out on the table. They spell

"I'll turn my back and ask you to make a mess by switching pairs: Pick up any two tiles, switch their positions, and turn them face down like this." Say *I* and *Y* are switched. The row becomes

The switched tiles are placed slightly below their original positions. The performer continues: "Jane, please choose two tiles, switch them and put them where they belong, just below. How about you, Patricia?" Keep going until all tiles are face down.

"Was anybody watching? Who remembers what wound up in the first position?" The performer points to the face-down tile in the original D position and has the spectators guess before turning it up. "Let's put it back where it started." Say the E tile is in the first position; switch it with the face-down tile at the original E position. At this stage, the row appears as

The performer moves to the second position, originally the Y: "Does anyone remember what's here?" Whatever it is, switch it with the face-down tile now at that letter's original position:

One more iteration—"Does anyone think they know any tile for sure?" Someone points out a face-down tile not in the first two positions; say, the C

shows up. Switch it with the face-down tile at the original C position. At this point, there is one final pair, which hasn't been touched. Without making a big deal of it, the performer casually switches these two face-down tiles: "That was hard work. I prefer magic! What was our original word?" The performer snaps his fingers and turns over the tiles in order to spell "D-Y-S-L-E-X-I-C" once more.

It isn't a great trick, but it has a certain charm. You can experiment with the length of the word used; it's more surprising to get more letters right but takes additional "clowning around time" to complete. Of course, you could do it with a row of cards, originally in order A 1 2 3 4 5 6 7 8 9 10, having four cards turned up in the original memory test and switching the last two. With cards, an opportunity to do a little manipulation presents itself. After four have been placed face up during the memory phase, gather the remaining six, managing to switch the two that need to be switched as you do so. Give the face-down packet of six a "false shuffle," and deal them back face down into the missing places. Finish as before. Does this help the mystery? Maybe; you have to decide. Instead of a false shuffle, you can mix the six by dealing them under and down and then replacing them in the correct order (a bit of study and practice being required to do this).

As ever, there are further possibilities: there *can* be repeated letters. Anagrams, such as "calipers/replicas" or "discounter/introduces/reductions," suggest a trick that starts out with one word and ends up with another. To go back to the start of this section, you could use "dyslexic."

As advertised, it is a one-row version of Peirce's dyslexic principle. Call the final permutation of $2n$ cards, when all are face down after first switching n pairs, π. So $\pi(1)$ is the card at position 1, and so on. The procedure forces $\pi\pi =$ identity: Applying π twice winds up leaving every card in its original place. We write $\pi = \pi^{-1}$ or have $\pi(i) = j$ if and only if $\pi(j) = i$. These are the *fixed-point free involutions*. We don't want to keep writing all that, so let's call them "big involutions" for this section.

How many big involutions are there? The answer is neat:

$$(2n-1)(2n-3)\cdots 5\cdot 3\cdot 1;$$

we write $(2n-1)!!$ for this "taking every other term" (the so-called *skip factorial*). When $2n = 4$, $(2n-1)!! = 3$. The three involutions are

$$2\,1\,4\,3 \qquad 3\,4\,1\,2 \qquad 4\,3\,2\,1.$$

When $2n = 6$, $(2n-1)!! = 5\cdot 3 = 15$. It is instructive to write these all out (at least once in one's life).

Whenever we meet a new object, we ask, "What does a typical one look like?" Ok, what does a typical big involution look like? This innocent question leads to complex destinations. To keep things civil, we will not explore this subject in detail but just get things started.

5.2 Descents

One obvious feature of a permutation is its up/down pattern. A permutation π has a descent at i if $\pi(i+1) < \pi(i)$. Thus, for $2n = 10$, the involution 2 1 9 6 10 4 8 7 3 5 has five descents, as indicated. Let $D(\sigma)$ be the number of descents. A big involution must have at least one descent (e.g., 4 5 6 1 2 3 has $D(\sigma) = 1$). As many as $2n - 1$ can occur; for example, 6 5 4 3 2 1 has $D(\sigma) = 5$. We can now ask: "Pick a big involution at random. How many descents does it typically have, and how are they distributed?" Does a typical σ have more or fewer descents on average than a typical permutation without any restrictions? A random permutation has $(2n - 1)/2$ descents on average. It is easy to see that for involutions, the average is larger,

$$\frac{2n-1}{2} + \frac{2n-1}{2n-2}$$

(so only about 1 larger). We won't dig in and do the work here, but it can be shown that a random big involution has about n descents with fluctuations of order $\sqrt{n}$ and, normalized by its mean and standard deviation, the fluctuations follow a bell-shaped curve. This is almost exactly the same as what happens for a random permutation. A careful proof and extension to the distribution of descents in other conjugacy classes is found in Kim and Lee [28].

The reader may discover that there is an extensive enumerative literature on descents in permutations, going back to Euler. Entry to this body of work is readily available [9, 41, 42, 44]. There is some parallel development for random big involutions in Guo and Zeng [37]. To open this door, think of a big involution as a "perfect matching" of $2n$ things. These can be diagrammed by drawing an arc between each switched pair. Thus, $\pi = 2\ 1\ 9\ 6\ 10\ 4\ 8\ 7\ 3\ 5$ appears as:

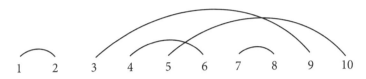

This representation suggests a host of questions, such as: How many arcs cross? The crossing number in the previous example is $c(\pi) = 2$. The answers are sometimes lovely. For example, the number of noncrossing matchings, $c(\pi) = 0$, is the Catalan number $\binom{2n}{n}/(n+1)$. A look at what is known and pointers to an extensive literature are available [4, 13, 14, 27]. More probabilistic results are in Kasraoui [26].

Perhaps the deepest result about random matchings is the Baik–Rains theorem [2], which determines the limiting distribution of the length of the longest increasing subsequence, $l(\pi)$. Returning to the previous sequence, $l(2\ 1\ 9\ 6\ 10\ 4\ 8\ 7\ 3\ 5) = 3$. Baik and Rains show that $l(\pi)$ is about $2\sqrt{2n}$, with fluctuations of size $(2n)^{1/6}$ having a remarkable Tracy–Widom distribution. See Stanley [42] for a detailed overview of these notable results.

Our bottom line for this section is that both the math and magic are interesting.

6 Peirce's Packet Pickup

Magic tricks devoted to dealing cards into piles and assembling the piles in various orders have been performed for at least 400 years. Charles Peirce introduced a truly new principle in this domain. He used his principle both as part of card tricks and as a way of teaching children basic arithmetic.

All of Peirce's dealing was done with the cards face down in the hand to start (say, in order $1, 2, 3, \ldots, n$ from the top down). The cards are dealt *face up* into k piles. The piles are picked up, in a careful order described in a moment, and then turned back face down. For example, ten cards dealt into five piles look like:

$$6\ 7\ 8\ 9\ 10$$
$$1\ 2\ 3\ 4\ 5.$$

Suppose the piles are picked up left to right, and all are turned back face down. The final order is 5, 10, 4, 9, 3, 8, 2, 7, 1, 6. If ten cards are dealt into two piles and assembled left to right, the order becomes 2, 4, 6, 8, 10, 1, 3, 5, 7, 9. To see how this can be used, consider twenty-five cards dealt into five piles and assembled in this way. If this action is repeated, you will find the cards back in their original order.

Here comes Peirce's contribution. In all of the previous examples, the total number of piles is a divisor of the total number of cards, so things worked out neatly. Peirce figured out a way to have things work out neatly with any number n of cards and any number k of piles, where k is relatively prime to n. As an illustration, suppose thirteen cards are dealt into five piles. They appear as:

$$11\ 12\ 13$$
$$6\ \ 7\ \ 8\ \ 9\ 10$$
$$1\ \ 2\ \ 3\ 4\ 5.$$

Peirce's rule says: Pick up the pile on which the last card is dealt. Suppose that this is j from the left. Drop it on the pile j to the right (going around the corner if needed); $j = 3$ in the arrangement above. Drop these two piles on the pile j to the right, and continue until there is only one big pile. It is important to count empty spaces as you go around. Thus, if the piles in the example are in places 1, 2, 3, 4, 5, you drop pile 3 on pile 1, both on pile 4, this packet on pile 2, and finally all on pile 5.

What's neat about it? It is that Peirce's pickup results in the card in position i is of value ki (mod n). In the illustration above, $n = 13$, $k = 5$, and the final arrangement is 5, 10, 2, 7, 12, 4, 9, 1, 6, 11, 3, 8, 13. That is, the top card (position 1) is 5 ($= 5 \times 1$). The next card is 10 ($= 5 \times 2$). The next is 2, because $= 5 \times 3 = 15 = 2$ (mod 13), and so on. There are all kinds of bells and whistles that can be added to this: starting the pickup with any pile, using the minimum of j and $k - j$, and others. Some of this may be found in Gardner [16, chap. 4] or Eisemann [20,21], but it is also fun to try your own variations. We have recorded many of ours in Section 7.

It is hard to know the exact history. Peirce said he developed the trick around 1860, and he certainly used his pickup in an early unpublished manuscript, "Familiar letters about the art of reasoning" (1890), reprinted in Peirce [19]. This was part of a many-sided effort that Peirce made to write a popular school arithmetic book. It was one of dozens of failed schemes to earn a living. He did receive some advances and produced a large number of partial chapters. Peirce used cyclic arithmetic as a way of teaching addition (mod n), and he introduced his dealing procedure as a way of teaching multiplication (mod n). In "Some amazing mazes," he analyzed repeated deals into different numbers of piles and gave a way of doing one final deal to return the deck to its original order. He also combined his deals with primitive roots to make a magic trick.

We give here a simpler magic trick, introduced in his "Familiar letters," which has not been presented before.

Example 5 (A simple Peirce trick). *For illustration, work (mod 13) with cards labeled A, 2, 3, ..., 10, J, Q, K. Deal into either five or eight piles, and perform a Peirce pickup. Then deal the cards face down around a circle:*

Ask a spectator, "What card would you like me to find?" Suppose she says, "The jack ($= 11$)." Starting from the king, count 11 clockwise to wind up at 3; turn it up. Then move counterclockwise three positions starting from the king, and show the

requested jack. In general, to find α, count forward α from the king, turn up β, then count β counterclockwise from the king, and turn up α. Briefly, if π(i) is the label at position i, with the king K = 0, then π(i) + π(−i) = 0 (mod 13). Hence, $5 + 8 = 3 + 10 = J + 2 = 6 + 7 = 1 + Q = 9 + 4 = 13$.

Peirce gives variations; ten cards dealt into three piles and properly picked up twice results in

This is reminiscent of Peirce's dyslexic principle, with a single pile of cards. We have given our best tricks with the principle in Sections 4 and 5. Peirce used card tricks many times in his efforts to liven up the teaching of arithmetic. It is surely worth looking further into his work.

In the next section, we present an original development of Peirce's pickup, showing how a quite nice magic trick can be based on his idea.

7 A Performable Peirce Trick

Peirce discovered a remarkable principle whereby after any number of shuffles, one further shuffle brings the deck back to its original order. His shuffles involve dealing a deck of cards into any number of piles and picking up the piles. He discovered that special pickup sequences permit analysis; this section presents an analysis of his shuffle. We begin with an example that makes a fine performable trick and then explain how the general principle works. As usual, the example is designed so that the reader can follow along with cards in hand. We urge you to go get a deck.

7.1 Together Again

The performer removes the thirteen spades from a regular deck of cards and arranges them in order from ace at the top through king at the bottom. "This is a story about the trials of couples in modern times." The performer removes the king and queen, turns them face up, and places them onto the rest of the face-down packet.

"The king and queen represent an ordinary couple, trying to get through life. Of course, life has its ups and downs. Would you cut the cards and complete the cut?" The packet of cards is spread face down. The king and queen are face up together. "Even though no one knew where they would wind up, at least they were together. Now here comes trouble. Would you, sir, give me any woman's first name?"

Whatever name is suggested, the cards are dealt face down into piles one at a time, from left to right as in a card game, into the number of piles equal to the number of letters in the name. For the example "Sue," three piles are dealt. Say the letters out loud as you deal, S-U-E, S-U-E, etc. As you deal, the face-up king and queen will be dealt into separate piles. "When Sue came into their lives, they separated."

Next, the piles must be picked up in a special order. The general rule will be given later. If you are following along—and you should be!—the piles are called 1, 2, 3 from left to right, and pile 3 should be placed on pile 1, with pile 2 on top of these.

Now point to a woman and ask her to provide a man's name: say it's "Barry." Deal the cards into five piles, one for each letter of the name. Pick up by placing pile 3 on pile 1, pile 5 on these, 2 on these, and 4 on top of all. These deals can be continued. You can ask how many children a couple has and deal that number of piles, or simply ask for a small number. To get to the finish, we will stop here.

To finish, a deal must be made which brings the king and queen together. The number of piles will depend on the configuration but is easy to find. Spread the cards, and count how many are between the king and queen. In our example, it will be five or six, depending on the initial cut. Add 1 to this number and deal that many piles. If you like, you can name an appropriate word out loud (e.g., "divorce") for a deal into seven piles. In this case, your final deal will be into six or seven piles. If six, pick up pile 1, place it on 2, the whole on 3, then on piles 4, 5, and 6 in turn. If you deal seven piles, pick up pile 7, place it on 6, the whole on 5, then 4, then 3, then 2, and 1.

In either case, your final pickup and deal will have brought the king/queen pair together. Spread the cards, and cut the face-up pair to the top. The final patter goes as follows: "These trials and tribulations have brought the couple back together. What about the rest of their world? How did they fare through all this chaos?" Here, the rest of the packet is spread face up and shown to be in perfect order, from ace through jack.

7.2 Some Analysis

To explain how the trick works (and how it can be adapted to other deck sizes), an excursion into cyclic arithmetic is necessary. We begin by explaining a general rule for picking up piles that makes further analysis simple. This is our version of Peirce's rule (he dealt cards face up). Our development benefits from contributions of Bob Page, a California card man.

Peirce's pickup. Let n cards be dealt into p piles. The deal is left to right, as in a card game. Hence, if thirteen cards numbered from 0 to 12 are dealt into five piles, they would wind up as

$$
\begin{array}{l}
10\ 11\ 12 \\
5\ \ 6\ \ 7\ 8\ 9 \\
0\ \ 1\ \ 2\ 3\ 4.
\end{array}
$$

Suppose that the number of cards is not evenly divisible by the number of piles. Then there will be some piles that are "short." Call the number of short piles s. In the example here, $s = 2$. The piles are gathered by choosing a base pile, consecutively picking up piles at distance s, moving clockwise around a circle, and placing on the base pile.

Suppose thirteen cards are dealt into five piles. If the five piles are numbered 1 through 5 from left to right and the leftmost pile is chosen as base, the pickup proceeds by placing pile 3 on pile 1 (remember there are $s = 2$ short piles), then pile 5 on these, pile 2 next, and finally pile 4 on top. The deck will be in final arrangement:

$$
\begin{array}{ccccccccccccc}
8 & 3 & 11 & 6 & 1 & 9 & 4 & 12 & 7 & 2 & 10 & 5 & 0 \\
\text{top} & & & & & & & & & & & & \text{bottom}
\end{array}
$$

Starting with a different base pile merely results in a cyclic shift. Note that the arrangement has consecutive cards differing by 5 (mod 13). The next theorem shows how this works in general.

Theorem 1. *If n cards are dealt into p piles, with n and p having no common factors, and the piles are picked up by putting every sth pile on the first pile, where s is the remainder when n is divided by p, then the card originally at position j winds up at position*

$$
\frac{-j}{p} - 1 \quad (\mathrm{mod}\ n).
$$

To explain the notation, consider what $-1/p$ (mod n) means. This is a number such that

$$
p \cdot \left(\frac{-1}{p} \right) \equiv -1.
$$

It is a basic fact of elementary number theory that any p larger than 0 and less than n has such an inverse. For example, if n is 13 and p is 5, $-1/p = 5$, because $5 \times 5 = 25 = 26 - 1 \equiv -1$ (mod 13). The inverses of all possible values of p

TABLE 12.1.
Values of $-1/p$ (mod 13) when $n = 13$

p	1	2	3	4	5	6	7	8	9	10	11	12
$-1/p$	12	6	4	3	5	2	11	8	10	9	7	1

when $n = 13$ are given in Table 12.1. There is no simple formula for $-1/p$. It can be found efficiently by using the Euclidean algorithm, but for small n a bit of trial and error usually suffices.

Going back to the example: Thirteen cards numbered from 0 to 12 are dealt into five piles and assembled via the Peirce pickup into one pile with card 0 at the bottom. The cards are in order

$$8\ 3\ 11\ 6\ 1\ 9\ 4\ 12\ 7\ 2\ 10\ 5\ 0.$$

The theorem states that card j goes to $5j - 1$ (mod 13), because $-1/5 = 5$. Thus, the card in position 0 winds up at $5 \times 0 - 1 = -1 = 12$ (mod 13). The card at position 1 winds up in position $4 = 5 \times 1 - 1$ (remember, the top card is in position 0). The card in position 8 winds up in $5 \times 8 - 1 = 39 = 0$ (mod 13).

The theorem has corollaries that yield card tricks. The first explains the trick, showing how many piles the final deal must be to bring back the original order.

Corollary 1.   *If n cards are dealt into p_1, then $p_2, \ldots$, and then p_k piles, then the final position of a card at position j is*

$$-1 + \frac{1}{p_k} - \frac{1}{p_{k-1}p_k} + \ldots + \frac{(-1)^k}{p_2 \cdots p_{k-1}p_k} + \frac{(-1)^k j}{p_1 p_2 \cdots p_{k-1}p_k} \quad (\text{mod } n).$$

If one final deal into

$$\frac{(-1)^{k+1}}{p_1 p_2 \cdots p_k} \quad (\text{mod } n)$$

piles is made, then the deck will return to its original order.

Example 6.   *In the trick used to illustrate this section, deals into three and then five piles were made. Now $3 \times 5 = 15 = 2$ (mod 13), and $-1/2 = 6$ from 12.1.*

It is perfectly possible to do the calculation required to bring the trick to a successful conclusion in your head. This allows the trick to be worked with all cards face down. The face-up cards allow the determination without calculation.

Some further ideas are contained in the practical suggestions at the end of this section.

The next corollary abstracts Peirce's original discovery. It is followed by two tricks using the structure captured in the corollary.

Corollary 2. *Suppose that* $n = a \cdot b$. *If* n *and* p *have no common factors and* p *piles are dealt and picked up according to Peirce's pickup, then cards at position* $j, j + a, j + 2a, \ldots$, *are permuted among each other, for each* $j = 0, 1, 2, \ldots, b - 1$.

To explain, consider a twelve-card deck made up of four clubs, four hearts, and four spades. Suppose the original order is

If the cards are dealt into five (or seven) piles and picked up properly, every fourth card will be of the same suit. Twelve can also be represented as 2×6, and this order would be preserved as well. Thus, if the cards are arranged in suit order as above, or alternate as picture-spot-picture-spot, etc., the order is still preserved.

One way to capitalize on Peirce's discovery is to use numbers like 25 or 49, squares of primes, which can be dealt into any number of piles other than the primes themselves and still preserve the order. In this form, the principle was discovered by the English card man, Roy Walton. His trick was published in the 1967 volume of the privately published magic journal, *Pallbearer's Review*. The Walton trick, along with our own variation, is a fine example of how the ideas of this section can be made into magic.

7.3 Draw Poker Plus

In this trick a deck of twenty-five cards is repeatedly mixed by dealing into piles of varying sizes. At the finish, five hands of poker are dealt, and it is discovered that every hand is a pat hand, (e.g., a straight, flush, full house, four-of-a-kind, or straight flush).

The working is more or less automatic. Remove any five pat poker hands, and arrange them five apart: A twenty-five-card deck is stacked 1, 2, 3, 4, 5, 1, 2, 3, 4, 5, ..., where 1 represents any card from the first hand, etc. The deck can be freely cut. Walton allowed the pack to be dealt into a spectator's choice of two, three, four, or six piles. The piles are assembled by picking up the one at the leftmost position and placing it on top of the next pile to the left, placing these two packets on the next pile to the left, and so on, until one final packet has been formed. Walton's discovery is that every fifth card will still be from the same poker hand. The pack can be cut and dealt repeatedly; whenever you wish, deal five hands. Each will contain a pat poker hand.

To connect with Corollary 2, observe that when twenty-five cards are dealt into two, three, four, or six piles, the first pile has an extra card, and the remaining piles are short. The Peirce pickup agrees with Walton's pickup, so Corollary 2 is in force. Of course, we now know how to preserve things if seven, eight, nine, … piles are used.

Over the years, we have developed a different use of Walton's idea. This is a nice example where less is more. Let's all do the trick that follows next.

7.4 Draw Poker Minus

In effect, a packet of cards is dealt into piles to mix them. Finally, they are dealt into five hands. The performer explains that of course, one of the hands must be best, but which one? He lets the spectator freely choose any of the hands. The others are shown, and while some fairly good poker hands turn up, inevitably the hand chosen beats them all.

This trick begins as in draw poker plus, explained previously. Five pat poker hands are removed and stacked every fifth. Have the spectator cut. Spread the cards face up, casually asking that a subliminal impression be formed. Ask for a small number of piles: If two, three, four, or six, deal and pick up, asking the spectator to observe carefully to ensure proper cardtable procedure is used. Have the cards cut. Ask for another number of piles, deal, and pick up. Say that you hope that the spectator is subconsciously following along. Continue until the number of piles requested is five, and deal into those five piles.

Point out that one of the hands wins (most probably). Ask the spectator to place her hand on one. This is a good time to build up the tension. Ask if she feels sure about her choice or wants to change her mind. Finally, the unchosen four hands are formed into a pile with patter like: "Do you want to trade for this hand? No? How about this one?" The twenty remaining cards are in a pile on the table. Pick them up by removing the top card and using it as a scoop to lift the others. This shifts one card from the top to the bottom. This should be done surreptitiously, with a distracting remark such as, "Have I come near the cards you chose, or interfered with your selection in any way?"

Now fan off the top five cards of the twenty-card packet. This will no longer be a pat hand, though it may be a pretty good hand. Continue to fan off packets of five cards to find the best hand among the four in the pile. Remind the spectator that the cards were cut freely and mixed at her discretion. Ask that the selected hand be turned up one card at a time, and make the most of it.

Some Practical Details. The hands chosen may be flushes, straights, or full houses. Avoid four-of-a-kinds, since there is a small chance (1 in 5) that transferring the card from top to bottom won't break the hand.

Instead of openly transferring the top card, those who are adept can use slight-of-hand. For example, if the first card dealt off the twenty-card packet is dealt from the bottom, all will still go well.

There are many variants possible. Craig Snyder has suggested using twenty-five tarot cards. These can be repeatedly cut and dealt, which can be made to look like part of a fortune-telling ritual. A final deal into five piles is made, and a hand selected. This will be one of the five pat hands originally set. With a bit of thought, it is easy to decide which hand—if all else fails, you can try marking the cards—and you can have five prepared "readings" (preferably inscribed on parchment) put into five different pockets. Remove the appropriate reading, and go to town drawing out the drama, turn up the five chosen cards, and reveal their interpretation.

Another potential use involves an ESP deck. This consists of twenty-five cards depicting symbols, such as $\bigcirc$, $+$, $\int\int\int$, $\square$, $*$, each repeated five times. These can be arranged so that every fifth card is the same design. Then, when five hands are dealt and the spectator selects one, it consists of just one symbol throughout. Again, with a bit of preparation, you can determine which symbol the hand contains. Turn your back, and have one card freely selected from the chosen hand and concentrated upon. Reveal it in a dramatic manner.

Alternatively, have several spectators select a hand and then a card within that hand, and read their minds all at once. It is also possible to arrange the cards initially so that every fifth card is different. Then the final chosen hand of five cards has no repeated values; this seems less easy to turn into a miracle.

Our next section returns to Peirce's last novel magical principle.

8 Peirce's Primitive Principle

Suppose that the number of cards p is a prime, like 11 or 13. A basic fact of number theory is that there are always some numbers a such that $a, a^2, \ldots, a^{p-1}$ are all distinct (mod p). Such an a is called a *primitive root* of p. For example, when $p = 11$, $a = 2$ is a primitive root:

j	0	1	2	3	4	5	6	7	8	9
a^j	1	2	4	8	5	10	9	7	3	6

A classical theorem of Fermat (Fermat's little theorem) states that for any nonzero a, $a^{p-1} \equiv 1$ (mod p), so this is as far as it goes. When $p = 11$, $a = 3$ is *not* a primitive root: The successive powers of 3 mod 11 are 1, 3, 9, 5, 4. There are $\phi(p-1)$ distinct primitive roots mod p, where ϕ is Euler's ϕ function, the number of a relatively prime to $p-1$ (so $\phi(10) = 4$), but there is no "formula" for finding one. Indeed, it is unknown whether 2 is a primitive root for infinitely many primes; see the Wikipedia entry for the Artin conjecture. Peirce gave his own proof of the existence of primitive roots as part of his "Some amazing mazes" card routine. He also gave a card-dealing proof of Fermat's little theorem.

If a is a primitive root mod p, then p cards dealt into a piles, and picked up à la Peirce, will recycle after exactly $p-1$ repetitions. This occurs at the start of Peirce's trick with $p = 13$, $a = 2$. However, Peirce used primitive roots in a much more subtle way. Suppose the $p-1$ values $1, 2 \ldots, p-1$ are arranged with a^j in position j ($j = 0$ being the first position) through $p-2$. When $p = 13$ and $a = 2$, the values are in order

$$A\ 2\ 4\ 8\ 3\ 6\ Q\ J\ 9\ 5\ 10\ 7.$$

Now, deal into b piles and pick up à la Peirce. This takes the card originally at position j to position bj. Suppose that $b = a^k$ and the cards are in order $a^0, a^1, a^2, \ldots$. After the deal, they are in order $a^k, a^{k+1}, a^{k+2}, \ldots, a^{k+p-1}$. The *exponents* behave in a simple way, all shifting by k. Dealing again, say into c piles with $c = a^l$, results in an additional shift by l. If after a series of deals, the total shift is by m, one additional deal into d piles, with $d = a^{-m}$, will bring the cards back to order.

Peirce used this in an ingenious way. He worked with two piles, the hearts in order $1, a, a^2, \ldots$, and the spades in *inverse* order. If after any number of deals of the hearts, the final arrangement consists of a shift by m (in the exponent), merely cutting m cards from the bottom of the spades pile to the top results in the two orderings being inverses. We call this use of primitive roots, in connection with inverses, *Peirce's primitive principle*.

The procedure of going between $b = a^j$ and j back to b is a number-theoretic version of logarithms. Turning multiplication into addition has well-known benefits. In recent years, it forms the basis of a raft of cryptographic schemes [40]. It is easy, given a and p, to find a^j (mod p). It seems difficult to go backward: Given a^j (mod p), find j. This is the problem of finding logarithms in finite fields, and extensive, very sophisticated theoretical work has been developed to replace simple trial and error. The best algorithms still require $2^{p^{1/3}}$ steps, ensuring the security of the associated crypto schemes.

The use of primitive roots to perform computations (and prove theorems) in number theory is classical, going back to Euler. Lists of primitive roots for small primes were compiled and tables for going from a^j to j have been published. We think that Peirce wanted to make a card trick that would illustrate these ideas, going well beyond the mechanism of almost all other magic tricks. The primitive element theorem underlies our tricks with de Bruijn sequences, and we recommend Diaconis and Graham [12, chaps. 2–4] for further developments.

9 Summing Up

Peirce's magical contribution has been broken into pieces:

- the cyclic principle,
- the dealing principle,
- the dyslexic principle, and
- the primitive principle.

Each of these has depth and absolute originality. We have tried to show how these can be broken off, modified, extended, and adapted to make solid, entertaining magic tricks. Peirce *combined* them. This section shows that the whole is more than (and less than) the sum of its parts.

Let's first dispose of "less than." The bottom line is that Peirce's "Some amazing mazes" is a poor, essentially unperformable trick. Martin Gardner called it "surely the most complicated and fantastic card trick ever invented."

Now for "more than." Peirce managed to *combine* his principles into one seamless whole. He begins with two packets in standard order. After repeated dealing—his cyclic principle, which has some mild entertainment built in—the two packets are in inverse order, demonstrating his dyslexic principle. But they are not in just any order: Peirce has arranged that they are in "primitive root order." This is because 2 is a primitive root mod 13, and the last card dealt has double the value of its position; this is his primitive principle. Next, Peirce mixes the cards by repeatedly dealing into spectator-chosen numbers of piles. He picks up carefully—his dealing principle. Finally, by making a single simple cut in the other packet—his primitive principle fully applied—he again has the two packets in inverse order. The finale is a display of the dyslexic principle. It is astounding to layer these concepts in such a fluid fashion.

Let us go back to a much larger picture: The impossible task of showing how his magic illuminates his huge body of work. Of course, this can be dangerous business. Consider evaluating Linus Pauling, who won two Nobel Prizes, on the basis of his strange, late-in-life fixation with vitamin C. We knew Pauling then, and many of his peers regarded him as a crackpot.

We feel Peirce's magic work is enlightening. At a distance, and even after a first, second, and third look, his card trick is an unperformable nightmare: tedious dealing that goes on and on coupled with a weak, ungraspable effect that is sure to leave the audience confused. But Gardner [16] also expressed its underlying value: "I cannot recommend it for entertaining friends unless they have a passion for number theory, but for a teacher who wants to 'motivate' student interest in congruence arithmetic, it is superb."

Writing now in 2017, we have been trying to unpack Peirce over a 60-year period, having first come across his work in 1957. Long years of study have revealed fascinating new ideas there, completely original and with every possibility of generating marvelous new magic tricks. It is at least plausible that Peirce's work in *any* of the myriad subjects on which he wrote has similar complexity and uniqueness, and innumerable scholars have taken up the challenge of this research. In our survey of the Peirce literature, we find it has developed its own language and themes. We suggest a return to the original Peirce corpus: take an essay, and try to follow along and make your own sense of it.

10 A Peirce Magic Letter

We don't know much about Peirce's interactions with magic. Introducing "Some amazing mazes" Pearce [36] wrote:

> About 1860 I cooked up a melange of effects of most of the elementary principles of cyclic arithmetic; and ever since at the end of some evening's card play, I have occasionally exhibited it in the form of a "trick" (though there is really no trick about the phenomenon) with the uniform result of interesting and surprising all the company, albeit their mathematical powers have ranged from a bare sufficiency for an altruistic tolerance of cards up to those of some of the mightiest mathematicians of the age, who assuredly with a little reflection could have unraveled the marvel.

Peirce (1839–1914) was 21 in 1860. At about this time, there were regular magic performances in Boston theaters, attracting performers, such as John Henry Anderson, Signor Blitz, and Andrew MacAllister. (Much more detail can be found in Moulton [29].) One could find magic books in libraries and printshops. These were often compendia of sports and games with some magic thrown in. Titles such as *The Boy's Own Book of Sports and Games* (1859), *The Modern Cabinet of Arts* (1846, 1856) by Thornton, *Wyman's Book of Magic* (1851), and *The Whole Art of Legerdemain* (1830, 1852) were common (see Figure 12.6).

One of the better books, *The Magician's Own Book* (1857), incorporated recent American tricks. There are some self-working tricks of the type Peirce favored. These include "the pairs repaired" (*Mutus nomen dedit cocis,* or the Latin card trick), "the 21 card trick," "the clock trick," and tricks with a prearranged pack. All of these are pretty tame compared to what Peirce was cooking up. Many mathematical card tricks go back hundreds of years—see Heeffer [15] for a wonderful history—but Peirce's inventions are much, much deeper.

By remarkable good luck, a long letter from Peirce to his friend Chauncey Wright in September 1865, in which he described card tricks, has recently turned up [31]. Wright was one of Peirce's gurus, a central figure in the birth of Cantabrigian-American philosophy, alongside Peirce, William James, and Oliver Wendell Holmes. Fascinating details are in Menand [38], including their shared interest in card tricks.

Figure 12.6. Pages from *The Magician's Own Book* (1857).

After apologizing for not returning a borrowed book, Peirce commences:

I have invented a little trick at cards. Take a pack containing a multiple of four cards, an equal number of each suit. Arrange them in regular order in their suits. Milk the pack three times. That is, take alternate cards from the face and back of the pack and put them in a pile on the table, back up, until the cards are exhausted. Do this three times. Then, holding the cards back up, count out four into the other hand so as to reverse their order. Then count four *underneath* these so as *not* to reverse their order. Then four *above*, so as to reverse their order again and so on till the pack is exhausted. Then turn the pack over and deal out in four hands.

This combination of milk and over/under shuffles is original. The careful study by Monge [45] concerning over/under shuffles is here being extended to encompass dealing off packets and reversing some of the cards along the way. Peirce begins some mathematical analysis for milk shuffles: Where does card j end up, and how many repeats to recycle?

In a postscript, Peirce describes a few extensions and variations. Once of them is a little "story trick." In brief, remove the ace of hearts (the priest) and two couples, the queen and jack of hearts and the queen and jack of spades. Arrange them at the top of a face-up pack as A, Q, J, Q, J; the ace is five cards down. "The couples went to see a priest to arrange a joint wedding, but then the complexities of life intervened." Place different cards, here denoted X, between the couples as

A, X, Q, X, J, X, X,
Q, X, X, X, J,

and then cut six cards from the top to the bottom of the pack. "The storms of life continued." Turn the deck face down, and proceed as follows: Deal the top card on the table, the next under the pack, the next on the table (onto the first card dealt there), then under, then down, and so forth. At the finish—"when life settled down the hymeneal altar was found intact"—spread the deck face up to show the couples and priest together.

Peirce called this story "a three-volume novel" and went on to an even more elaborate variation. Unknowingly, we applied some of Peirce's principles in a similar story; see "Together again" in Section 7. We may hope for further adventures of Peirce in our shared magicians' land and perhaps some supporting material from Chauncey Wright.

Acknowledgments

This material was given at the Conference on Logic and Literary Form (UC Berkeley, April 14–15, 2017) and at the MOVES Conference on Recreational Mathematics (CUNY, August 6–8, 2017). We thank David Eisenbud and Cindy Lawrence. Steve Freeman, Susan Holmes, Marty Isaacs, Richard Stanley, and Steve Stigler have contributed in essential ways. The authors gratefully acknowledge the careful reading of an early draft done by Steve Butler as well as his help in producing our figures.

Supported in part by National Science Foundation award DMS 1608182.

References

[1] Dave Arch. Turning the tarot. *Syzygy* **3** (1996) 245.

[2] J. Baik and E. Rains. The asymptotics of monotone subsequences of involutions. *Duke Math. J.* **109** (2001) 205–281.

[3] J. Buhler, D. Eisenbud, R. Graham, and C. Wright. Juggling drops and descents. *Amer. Math. Monthly* **101** (1994) 507–519.

[4] W. Chen, E. Deng, R. Du, R. Stanley, and C. Yan. Crossings and nestings of matchings and partitions. *Trans. Amer. Math. Soc.* **359** (2007) 1555–1575.

[5] F. Chung, A. Claesson, M. Dukes, and R. L. Graham. Descent polynomials for permutations with bounded drop size. *European J. Combinatorics* **31** (2010) 1853–1867.

[6] F. Chung and R. L. Graham. The drop polynomial of a weighted digraph. *J. Comb. Th. Ser. A* **126** (2017) 62–82.

[7] F. Chung and R. L. Graham. The digraph drop polynomial. In *Connections in Discrete Mathematics*, S. Butler, J. Cooper, and G. Hurlbert, eds., Cambridge University Press, Cambridge, 2018, 86––103.

[8] F. Chung and R. L. Graham, Inversion-descent polynomials for restricted permutations. *J. Comb. Th. Ser. A* **120** (2013) 366–378.

[9] A. Borodin, P. Diaconis, and J. Fulman. On adding a list of numbers (and other one-dependent determinantal processes). *Bull. Amer. Math. Soc* (N.S.) **47** (2010) 639–670.

[10] P. Diaconis and L. Saloff-Coste. Convolution powers of complex functions on $\mathbb{Z}$. *Math. Nachr.* **287** (2014) 1106–1130.

[11] P. Diaconis and B. Skyrms. *Ten Great Ideas about Chance*. Princeton University Press, Princeton, NJ, 2017.

[12] P. Diaconis and R. L. Graham. *Magical Mathematics: The Mathematical Ideas That Animate Great Magic Tricks,* Foreword by Martin Gardner. Princeton University Press, Princeton, NJ, 2012.

[13] B. Chern, P. Diaconis, D. Kane, and R. Rhoades. Closed expressions for averages of set partition statistics. *Res. Math. Sci.* **1** (2014) SpringerOpen Journal (online).

[14] B. Chern, P. Diaconis, D. Kane, and R. Rhoades. Central limit theorems for some set partition statistics. *Adv. Appl. Math.* **70** (2015) 92–105.

[15] Albrecht Heeffer. Récréations mathématiques: A study of its authorship, sources and influence. *Gibecière* **1** (2006) Conjuring Arts Research Center.

[16] Martin Gardner. *Fractal Music, Hypercards and More … Mathematical Recreations from SCIENTIFIC AMERICAN Magazine.* W. H. Freeman and Co., 1992.

[17] Martin Gardner. *The Whys of a Philosophical Scrivener.* St. Martin's Press, 1999.

[18] Martin Gardner. *Logic Machines and Diagrams.* McGraw-Hill, New York, 1958.

[19] Charles S. Peirce. *The New Elements of Mathematics, Vol. IV,* Carolyn Eisele, ed. Mouton Publishers, The Hague and Paris, and Humanities Press, Atlantic Highlands, NJ, 1976.

[20] K. Eisemann. Number-theoretic analysis and extensions of "the most complicated and fantastic card trick ever invented." *Amer. Math. Monthly* **91** (1984) 284–289.

[21] K. Eisemann. A self-dual card trick based on congruences and k-shuffles. *Amer. Math. Monthly* **93** (1986) 201–205.

[22] S. Minch. *The Collected Works of Alex Elmsley, Vol. I.* L & L Publishing, Tahoma, CA, 1991.

[23] S. Minch. *The Collected Works of Alex Elmsley, Vol. II.* L & L Publishing, Tahoma, CA, 1994.

[24] P. Erdős and P. Turán. On some problems of a statistical group-theory. (III). *Acta Math. Acad. Sci. Hungar.* **18** (1967) 309–320.

[25] Marty Isaacs (personal communication, June 2017).

[26] A. Kasraoui. On the limiting distribution of some numbers of crossings in set partitions. ArXiv 1301.6540 (2013).

[27] N. Khare, R. Lorentz, and C. Yan. Moments of matching statistics. *J. Comb.* **8** (2017) 1–27.

[28] G. Kim and S. Lee. Central limit theorem for descents in conjugacy classes of S_n. ArXiv 1803.10457v1 (2018).

[29] H. J. Moulton. *Houdini's History of Magic in Boston, 1792–1915.* Meyerbooks, Glenwood, IL, 1983.

[30] P. Meier and S. Zabell. Benjamin Peirce and the Howland will. *J. Amer. Statist Assoc.* **75** (1980) 497–506.

[31] Charles S. Peirce to Chauncy Wright, 2 September 1865, Chauncey Wright Papers, Box 1, American Philosophical Society, Philadelphia.

[32] Charles S. Peirce. *The Essential Peirce, Selected Philosophical Writings, Volume 1 (1867–1893)*, N. Houser and Ch. Kloesel, eds. Indiana University Press, Bloomington and Indianapolis, IN, 1992.

[33] Charles S. Peirce. *The Essential Peirce, Selected Philosophical Writings, Volume 2 (1893–1913)*. Peirce Edition Project. Indiana University Press, Bloomington and Indianapolis, IN, 1998.

[34] Charles S. Peirce. *Collected Papers of Charles Sanders Peirce*, Vols. 1–6, 1931–1935, Charles Hartshorne and Paul Weiss, eds.; Vols. 7–8, 1958, Arthur W. Burks, ed. Harvard University Press, Cambridge, MA, 1958.

[35] Charles S. Peirce. Some amazing mazes. *Monist* **18** (1908) 227–241.

[36] Charles S. Peirce. Some amazing mazes: A second curiosity. *Monist* **19** (1909) 36–45.

[37] V.J.W. Guo and J. Zeng. The Eulerian distribution on involutions is indeed unimodal, *J. Comb. Th. Ser. A* **113** (2006) 1061–1071.

[38] L. Menand. *The Metaphysical Club: A Story of Ideas in America.* Farrar, Straus and Giroux, New York, 2001.

[39] J. Kaag. *American Philosophy: A Love Story.* Farrar, Straus and Giroux, New York, 2016.

[40] A. Odlyzko. Discrete logarithms in finite fields and their cryptographic significance. *Lecture Notes Comput. Sci.* **209** (1985) 224–314.

[41] T. K. Petersen. *Eulerian Numbers.* Birkhäuser/Springer, New York, 2015.

[42] R. P. Stanley. *Enumerative Combinatorics, Volume 1.* Cambridge University Press, Cambridge, 2012.

[43] S. M. Stigler. Mathematical statistics in the early states. *Ann. Statist.* **6** (1978) 239–265.

[44] Y. Zhuang. Eulerian polynomials and descent statistics. *Adv. Appl. Math.* **90** (2017) 86–144.

[45] G. Monge. Réflexions sur un tour de cartes. Mem. Math. Phys. Acad. de Sciences Paris (1773) 390–412.

13

CAN YOU WIN KHALOU IN FIVE MOVES?

Brian Hopkins

Khalou is a puzzle app, written in 2013 by Aurélien Wenger, that features a 4×4 array of stones, each with a black side and a white side. By selecting certain patterns of three or four stones, those stones flip to display the other color. In particular, the allowed three-stone patterns are elbow-shaped (any 2×2 block omitting one position). The four-stone patterns include complete rows and complete columns. The goal is to make all the stones show white.

As discussed in Section 2, this game has similarities with both Lights Out and a switching game created by Elwyn Berlekamp. Points are awarded in Khalou depending on both the elapsed time and the number of steps—after five steps, the message "Each position has to be solved in a maximum of 5 moves" appears, and no points are possible (i.e., you can keep playing, but even if you achieve the all-white goal, you will receive no points). See Figure 13.1 for an example of an initial state.

I spent enough time playing the game that I could solve most initial states in at most five steps, but not always. This made me wonder whether the statement about the maximum number of steps was correct. In the language used with Rubik's Cube [7], is God's number for Khalou really five? We will settle that question in Section 3. Staying with the Rubik's Cube analogy, the value of God's number changes, depending on what is allowed as a move [6]. In Section 4, we explore how the maximum number of steps required in Khalou changes as the number and shape of moves change. For example, thinking as a game designer, what not-too-large collection of moves could guarantee a solution in at most four steps?

1 Comparisons and an Upper Bound

There are two similar recreations that have been considered in the literature.

First, in the late 1960s, Elwyn Berlekamp built a switching game at Bell Labs, consisting of a 10×10 array of lights. On the back, 100 switches allowed any configuration of lights to be turned on. On the front (available to the

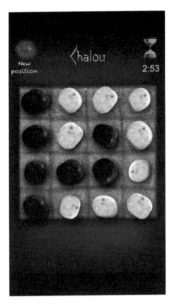

Figure 13.1. An initial state of Khalou (which happens to have a two-step solution).

player) were twenty switches, one for each row and column—flipping the switch toggled the entire row or column between on and off. It is not always possible to turn off all the lights (there are 2^{100} possible initial states and only 2^{20} possible states for the twenty switches), so the task is to leave as few lights on as possible. The challenging question is this: Among all these minimal positions with the least possible number of lights on, what is the greatest possible number of lights on? Fishburn and Sloane [3] used binary linear codes and solved "by hand" the analogous questions for square arrays up to 9×9; they used a computer to claim that the solution to Berlekamp's question is 34. However, Carlson and Starlaski [2], students in the Freshman Summer Institute 2002 at the California Institute of Technology, proved (using a computer) that the correct answer is 35.

Second, in 1995, Tiger Electronics released the game Lights Out with a 5×5 array of buttons. Pushing a button makes it and its immediate row and column neighbors toggle between on and off. Depending on a button's position, a move could change three, four, or five buttons. Anderson and Feil [1] used linear algebra to show that there are 2^{23} "winnable" configurations (i.e., for 1/4 of the possible 2^{25} configurations, there is a sequence of buttons to push resulting in an all-off array).

Many researchers have written about these games. Lights Out, especially, has been generalized in many ways, both commercially and mathematically, such as by using cellular automata [8]. See the survey [4] and recent article [5] for additional results and citations.

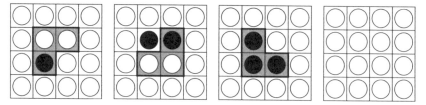

Figure 13.2. Three elbow moves from a single black stone to the all-white state.

Khalou may be considered a mix of these games, having both the universal row and column moves and the local elbow moves (although in just a 4 × 4 array, the distinction is primarily philosophical). However, unlike Berlekamp's switching game and Lights Out, every initial state of Khalou can be solved: To show this, it is enough to demonstrate a sequence of moves that flip any given stone with all other stones remaining the same.

As shown in Figure 13.2, three elbow moves change a single black stone to the desired all-white array. Notice that the twelve edge stones were not affected by these three steps. Also, the other three interior stones could have been any combination of white and black and would have been returned to their original colors. Thus, Figure 13.2 shows not just the solution sequence for a single black stone, but also a sequence that flips that one stone from black to white with no net change to the other stones. This gives a very crude bound for the number of moves necessary for any initial state: A state with k black stones can be solved in at most $3k$ steps by using this three-step procedure for each black stone.

2 Methodology

States of white/black and off/on can be modeled by modulo 2 arithmetic. All situations described here share the fact that doing an operation twice (flipping a switch, pushing a button, selecing a pattern) is equivalent to doing nothing. Similarly, the order of operations will make no difference.

For Khalou, let us represent the moves as 4 × 4 binary matrices. Here is the collection of moves M_1 with several matrices listed explicitly:

$$M_1 = \left\{ \begin{pmatrix} 1&1&0&0 \\ 1&0&0&0 \\ 0&0&0&0 \\ 0&0&0&0 \end{pmatrix}, \begin{pmatrix} 1&0&0&0 \\ 1&1&0&0 \\ 0&0&0&0 \\ 0&0&0&0 \end{pmatrix}, \begin{pmatrix} 0&1&0&0 \\ 1&1&0&0 \\ 0&0&0&0 \\ 0&0&0&0 \end{pmatrix}, \begin{pmatrix} 1&1&0&0 \\ 0&1&0&0 \\ 0&0&0&0 \\ 0&0&0&0 \end{pmatrix}, \begin{pmatrix} 0&1&1&0 \\ 0&1&0&0 \\ 0&0&0&0 \\ 0&0&0&0 \end{pmatrix}, \ldots , \right.$$

$$\left. \begin{pmatrix} 1&1&1&1 \\ 0&0&0&0 \\ 0&0&0&0 \\ 0&0&0&0 \end{pmatrix}, \begin{pmatrix} 0&0&0&0 \\ 1&1&1&1 \\ 0&0&0&0 \\ 0&0&0&0 \end{pmatrix}, \ldots , \begin{pmatrix} 1&0&0&0 \\ 1&0&0&0 \\ 1&0&0&0 \\ 1&0&0&0 \end{pmatrix}, \begin{pmatrix} 0&1&0&0 \\ 0&1&0&0 \\ 0&1&0&0 \\ 0&1&0&0 \end{pmatrix}, \ldots \right\}.$$

TABLE 13.1.
Number of states i steps from winning with 44 moves consisting of elbows, rows, and columns

i	0	1	2	3	4	5	6
$|M_i|$	1	44	898	9240	35116	20096	141

Our approach may seem backward compared to the game descriptions. We start with the winning all-white state M_0, the matrix of all zeros. The matrices of M_1 represent states that are one move away from winning. Define

$$M_2 = \{a + b \mid a, b \in M_1\} \setminus (M_0 \cup M_1).$$

These matrices represent states that are exactly two steps away from winning; the set-minus operation in the above equation removes any configurations that are closer to the winning state. (It actually suffices to remove just the M_0 matrix, since nothing in M_1 has the form $a + b$ for $a, b \in M_1$.) In general, for $n \geq 2$, let

$$M_n = \{a + b \mid a \in M_{n-1}, b \in M_1\} \setminus \bigcup_{i=0}^{n-1} M_i,$$

the set of matrices representing the states that require exactly n steps to reach M_0, the winning state. (Again, removing $\bigcup_{i=0}^{n-2} M_i$ would suffice.) These sets can be computed easily in *Mathematica*, for instance.

For example, the sequence shown in Figure 13.2 for solving a state with one black stone corresponds to

$$\begin{pmatrix} 0\ 0\ 0\ 0 \\ 0\ 1\ 1\ 0 \\ 0\ 1\ 0\ 0 \\ 0\ 0\ 0\ 0 \end{pmatrix} + \begin{pmatrix} 0\ 0\ 0\ 0 \\ 0\ 0\ 1\ 0 \\ 0\ 1\ 1\ 0 \\ 0\ 0\ 0\ 0 \end{pmatrix} + \begin{pmatrix} 0\ 0\ 0\ 0 \\ 0\ 1\ 0\ 0 \\ 0\ 1\ 1\ 0 \\ 0\ 0\ 0\ 0 \end{pmatrix} = \begin{pmatrix} 0\ 0\ 0\ 0 \\ 0\ 0\ 0\ 0 \\ 0\ 1\ 0\ 0 \\ 0\ 0\ 0\ 0 \end{pmatrix}.$$

With some argument that no two matrices in M_1 could have this sum (e.g., show that any matrix in M_2 has at least two 1s), this computation shows that a matrix with a single 1 is in M_3.

The Khalou claim is true if $\sum_{i=0}^{5} |M_i| = 2^{16}$.

3 Solving Khalou

Suppose M_1 consists of the 36 elbow moves, four row moves, and four column moves. Table 13.1 gives the sizes of the resulting M_i. So in fact, about 0.2% of initial states require six steps to win! The reader can confirm that the state with

TABLE 13.2.
Number of states i steps from winning with the 46 moves consisting of elbows, rows, columns, and two diagonals

i	0	1	2	3	4	5
$|M_i|$	1	46	987	10680	39067	14755

TABLE 13.3.
Number of states with k black stones that are i steps from winning with the 46 Khalou moves

i\k	0	1	2	3	4	5	6	7	8	9	10	11	12	13	14	15	16
0	1	0	0	0	0	0	0	0	0	0	0	0	0	0	0	0	0
1	0	0	0	36	10	0	0	0	0	0	0	0	0	0	0	0	0
2	0	0	82	84	176	108	356	168	13	0	0	0	0	0	0	0	0
3	0	16	38	320	744	2024	2002	2128	1572	1224	532	72	8	0	0	0	0
4	0	0	0	120	833	2008	4820	7344	8611	7092	4498	2608	834	212	86	0	1
5	0	0	0	0	57	228	830	1800	2674	3124	2978	1688	978	348	34	16	0

Note: Row sums match the entries of Table 13.2.

a black stone in each of the four corners and all other stones white requires six steps.

However, I realized that the Khalou claim is correct with the actual moves allowed in the game. Lost in translation somewhere, I had missed two additional moves: flipping stones along the two diagonals, corresponding to the matrices

$$\begin{pmatrix} 1 & 0 & 0 & 0 \\ 0 & 1 & 0 & 0 \\ 0 & 0 & 1 & 0 \\ 0 & 0 & 0 & 1 \end{pmatrix}, \begin{pmatrix} 0 & 0 & 0 & 1 \\ 0 & 0 & 1 & 0 \\ 0 & 1 & 0 & 0 \\ 1 & 0 & 0 & 0 \end{pmatrix}.$$

With these moves included in M_1, the claim in true; see Table 13.2.

The reader can confirm that the four corner states can now be solved in five steps, including a diagonal move. The values in Table 13.2 also show that Khalou is well designed in the sense that about 22.5% of initial states require the maximum five steps, with another 59.6% requiring four steps.

Another attractive aspect of the game are the exceptions to the general trend that the minimum number of steps required to win increases with the number of initial black stones. Table 13.3 breaks down the states counted in Table 13.2 by number k of black stones.

4 Changing the Moves

Notice that including the two diagonal flips made a set of 46 moves that allowed every state to be solved in at most five steps; without them, the remaining 44 moves required six steps to solve some initial states. What more can we say about the relation between the number of allowed moves and the minimum number of steps required to solve any initial state?

There are two extreme cases. If we allow flipping any individual stone, then those 16 moves allow any state to be solved in at most 16 steps—just select each black stone to flip it to white. If we allow each of the 2^{16} states to be a single move (never mind how such would be entered), then those 2^{16} moves allow any state to be solved in exactly one step.

As a first exploration, starting from the 44 elbows, rows, and columns, is it necessary to include the two diagonals to make a collection of moves with a maximum of five steps to solve any initial state, or could one additional move suffice? Testing shows that neither diagonal alone suffices. Among seven highly symmetric possibilities, the X move with matrix

$$\begin{pmatrix} 1 & 0 & 0 & 1 \\ 0 & 1 & 1 & 0 \\ 0 & 1 & 1 & 0 \\ 1 & 0 & 0 & 1 \end{pmatrix}$$

combining the two diagonal moves is the only 45th move that allows for every possible solution in five steps. Table 13.4 suggests this could be considered a slight improvement over the 46 Khalou moves: In addition to being one move simpler, it is more challenging on average, having a few more initial states requiring four and five steps.

In the spirit of having fewer possible moves (although more than flipping the 16 individual stones), let us consider subsets of the moves discussed so far. Allowing just row and column moves is the 4×4 version of Berlekamp's game. Again, the 2^8 move combinations cannot cover the 2^{16} possible states. In fact, the eight moves determine a 7-dimensional subspace of states: Flipping the first row, for instance, is the same as flipping the three other rows and all four columns. The maximum number of black stones in a minimum state is four [3].

By the discussion near Figure 13.2 for flipping a single stone, we know that the 36 elbow moves do suffice to solve all possible initial states. The minimal number of steps is, of course, far less than the crude $3k$ estimate for a state with k black stones. The details are given in Table 13.5.

In the same way that the set of rows and columns are dependent, the set of 36 elbow moves are also redundant. We could use three families of elbows, leaving out the ones that do not use the lower-right corner of a 2×2 square, say, corresponding to the first and fifth matrices in the M_1 list given above. The cost

TABLE 13.4.
Number of states that are i steps from winning with the 45 moves consisting of elbows, rows, columns, and the X

i	0	1	2	3	4	5
$|M_i|$	1	45	942	10118	39201	15229

TABLE 13.5.
Number of states that are i steps from winning with the 36 elbow moves

i	0	1	2	3	4	5	6	7	8
$|M_i|$	1	36	582	4716	17106	25740	15047	2276	32

TABLE 13.6.
Number of states that are i steps from winning with 27 elbow moves, excluding one family

i	0	1	2	3	4	5	6	7	8	9
$|M_i|$	1	27	351	2599	10502	21485	20910	8649	1004	8

TABLE 13.7.
Number of states that are i steps from winning with the 60 moves consisting of elbows and dominos

i	0	1	2	3	4	5	6
$|M_i|$	1	60	1289	10666	30411	21182	1927

for this reduction from 36 to 27 moves is one potential additional step for just eight initial states, as detailed in Table 13.6. It is clear that no one family of nine elbow moves could solve all states, as they could not affect, say, a black stone in the lower-right corner of the array. The reader is invited to verify that no two families of elbow moves can solve all states, either.

There is an art to selecting moves that reduce the maximum number of steps required to solve any initial state. Consider the set of 24 domino moves. By themselves, they cannot reach every state (convince yourself that, starting from an all-white board, they will always leave an even number of black stones). Adding them to the 36 elbows gives 60 total moves that can be used to solve every state, but as shown in Table 13.7, they can require six steps, no better than the 44 moves initially considered here. It is not at all hard to increase the number of moves in ways that do not reduce the maximum number of steps required to solve any initial state.

TABLE 13.8.
Number of states that are i steps from winning with the 127 moves from Khalou and three additional Tetris pieces

i	0	1	2	3	4
$\|M_i\|$	1	127	4981	38882	21545

TABLE 13.9.
Various collections of moves, their number, and the associated maximum number of steps needed

Collection of moves	Number	Maximum
Each single stone	16	16
Three families of elbows	27	9
All elbows	36	8
Rows, columns, elbows	44	6
Rows, columns, elbows, X	45	5
Khalou moves	46	5
Elbows and dominos	60	6
Khalou plus some Tetris	127	4
All states	65536	1

What moves can be added to the Khalou moves so that four moves suffice? After much experimentation, I found a collection inspired by an older game, Tetris. Add all positions, rotations, and reflections of the moves called in the literature L (48 of these), Q (nine), and T (24), corresponding to the matrices

$$\begin{pmatrix} 1\,0\,0\,0 \\ 1\,0\,0\,0 \\ 1\,1\,0\,0 \\ 0\,0\,0\,0 \end{pmatrix}, \begin{pmatrix} 1\,1\,0\,0 \\ 1\,1\,0\,0 \\ 0\,0\,0\,0 \\ 0\,0\,0\,0 \end{pmatrix}, \begin{pmatrix} 1\,1\,1\,0 \\ 0\,1\,0\,0 \\ 0\,0\,0\,0 \\ 0\,0\,0\,0 \end{pmatrix},$$

respectively (including Z does not help). The resulting set of 127 moves is much larger than the Khalou collection, but hopefully familiar to players. As shown in Table 13.8, this set does reach every state in at most four steps, with 32.9% requiring that maximum number of steps.

Table 13.9 summarizes our examples exploring the relation between number of moves and maximum number of steps required to solve any initial state.

References

[1] M. Anderson and T. Feil. Turning lights out with linear algebra. *Math. Mag.* **71** (1998) 300–303.

[2] J. Carlson and D. Starlaski. The correct solution to Berlekamp's switching game. *Discrete Math.* **287** (2004) 145–150.

[3] P. Fishburn and N. Sloane. The solution to Berlekamp's switching game. *Discrete Math.* **74** (1989) 263–290.

[4] R. Fleischer and J. Yu. A survey of the game "Lights Out!." In *Space-Efficient Data Structures, Streams, and Algorithms*, A. Brodnik, A. López-Ortiz, V. Raman, A. Viola, eds. Springer, Berlin, 2013, 176–198.

[5] M. Kreh. "Lights Out" and variants. *Amer. Math. Monthly* **124** (2017) 937–950.

[6] T. Rokicki. Towards God's number for Rubik's Cube in the quarter-turn metric. *College Math. J.* **45** (2014) 242–253.

[7] T. Rokicki, H. Kociemba, M. Davidson, and J. Dethridge. The diameter of the Rubik's Cube group is twenty. *SIAM J. Discrete Math.* **27** (2013) 1082–1105.

[8] K. Sutner. Linear cellular automata and the Garden-of-Eden. *Math. Intelligencer* **11**, no. 2 (1989) 49–53.

14

ON PARTITIONS INTO SQUARES OF DISTINCT INTEGERS WHOSE RECIPROCALS SUM TO 1

Max A. Alekseyev

Integers 11 and 24 share an interesting property: each can be partitioned into distinct positive integers whose reciprocals sum to 1. Indeed, $11 = 2 + 3 + 6$ and $1/2 + 1/3 + 1/6 = 1$, and $24 = 2 + 4 + 6 + 12$, where again $1/2 + 1/4 + 1/6 + 1/12 = 1$. Sums of reciprocals of distinct positive integers are often referred to as *Egyptian fractions* [3], and from known Egyptian fractions of 1, we can easily construct other numbers with the same property. The smallest such numbers are 1, 11, 24, and 30 (sequence A052428 in the OEIS [4]), and they tend to appear rather sparsely among small integers. So it may come as a surprise that *any* number greater than 77 has this property. This was proved in 1963 by Graham [1], who further conjectured that any sufficiently large integer can be partitioned into *squares* of distinct positive integers whose reciprocals sum to 1 [2, section D11]. Examples of such partitions can again be obtained from known examples of Egyptian fractions of 1:

$$1 = \tfrac{1}{1} \qquad\qquad 1^2 = 1,$$

$$1 = \tfrac{1}{2} + \tfrac{1}{3} + \tfrac{1}{6} \qquad\qquad 2^2 + 3^2 + 6^2 = 49,$$

$$1 = \tfrac{1}{2} + \tfrac{1}{4} + \tfrac{1}{6} + \tfrac{1}{12} \qquad\qquad 2^2 + 4^2 + 6^2 + 12^2 = 200,$$

$$1 = \tfrac{1}{2} + \tfrac{1}{3} + \tfrac{1}{10} + \tfrac{1}{15} \qquad\qquad 2^2 + 3^2 + 10^2 + 15^2 = 338.$$

In fact, 1, 49, 200, and 338 are the smallest such numbers (sequence A297895 in the OEIS [4]), and they seemingly appear even more sparsely. Nevertheless, in this chapter, we prove Graham's conjecture and establish the exact bound for the existence of such partitions.

Call a positive integer m *representable* if there exists a set of positive integers $X = \{x_1, x_2, \ldots, x_k\}$ such that

$$1 = \frac{1}{x_1} + \cdots + \frac{1}{x_n} \quad \text{and} \quad m = x_1^2 + \cdots + x_n^2.$$

Let us further say that X is a *representation* of m. For example, 200 is representable, since it has representation $\{2, 4, 6, 12\}$.

Our main result is the following theorem.

Theorem 1. The largest integer that is not representable is 8542.

We provide a proof of Theorem 1, generalizing the original approach of Graham [1] based on constructing representations of larger numbers from those of smaller ones. More generally, let us refer to such a construction as a *translation* of representations and introduce a class of translations that acts on restricted representations. This provides us with yet another proof of Theorem 1.

Since this approach requires computation of representations of certain small numbers, let us start with a discussion of an algorithm that generates representations of a given number. The same algorithm is also used to prove that 8542 is not representable.

1 Computing Representations

Our goal is to design an efficient exhaustive-search algorithm for a representation of a given integer m. We start with proving bounds for the search, using the power mean inequality.

Recall that the qth power mean of positive numbers $x_1, \ldots, x_k$ is defined as

$$A_q(x_1, \ldots, x_k) = \left(\frac{x_1^q + \cdots + x_k^q}{k} \right)^{\frac{1}{q}}.$$

In particular, $A_q(x_1, \ldots, x_k)$ represents the harmonic, geometric, or arithmetic mean when $q = -1, 0, 1$, respectively.[1] The power mean inequality, generalizing the arithmetic mean–geometric mean (AM-GM) inequality, states that $A_q(x_1, \ldots, x_k) \leq A_{q'}(x_1, \ldots, x_k)$ whenever $q \leq q'$.

The following lemma will be crucial for our algorithm.

Lemma 1. For a positive integer d and a finite set of positive integers X, let

$$s = \sum_{x \in X} \frac{1}{x} \quad \text{and} \quad n = \sum_{x \in X} x^d. \tag{1}$$

---

[1] Formally speaking, the geometric mean equals the limit of $A_q(x_1, \ldots, x_k)$ as $q \to 0$.

Then

$$|X| \le s \sqrt[d+1]{\frac{n}{s}}, \tag{2}$$

and

$$\left\lceil \frac{1}{s} \right\rceil \le \min X \le \left\lfloor \min\left\{ \sqrt[d+1]{\frac{n}{s}}, \sqrt[d]{n} \right\} \right\rfloor. \tag{3}$$

Proof. Suppose that $X = \{x_1, \ldots, x_k\}$, where $x_1 < x_2 < \cdots < x_k$, and so $|X| = k$ and $\min X = x_1$.

From (1) and the power mean inequality, it follows that

$$\frac{k}{s} = A_{-1}(x_1, \ldots, x_k) \le A_d(x_1, \ldots, x_k) = \sqrt[d]{\frac{n}{k}},$$

which further implies (2).

Since $x_1 = \min X$, we have

$$\frac{1}{x_1} \le s = \frac{1}{x_1} + \cdots + \frac{1}{x_k} \le \frac{k}{x_1},$$

implying that

$$\frac{1}{s} \le x_1 \le \frac{k}{s}.$$

Similarly, from $x_1^d \le x_1^d + \cdots + x_k^d = n$, we obtain $x_1 \le \sqrt[d]{n}$. Finally, using (2), we get

$$x_1 \le \frac{k}{s} \le \sqrt[d+1]{\frac{n}{s}},$$

which completes the proof of (3). □

Lemma 1 for $d = 2$ enables us to search for a representation $X = \{x_1 < x_2 < \cdots < x_k\}$ of a given integer m using backtracking as follows. Clearly, X should satisfy the equalities (1) for $s = 1$ and $n = m$, and so we let $s_1 = 1$ and $n_1 = m$. Then the value of $x_1 = \min X$ lies in the range given by (3) for $s = s_1$ and $n = n_1$. For each candidate value of x_1 in this range, we compute $s_2 = s_1 - \frac{1}{x_1}$ and $n_2 = n_1 - x_1^2$, representing the sum of reciprocals and squares, respectively, of the elements of $X \setminus \{x_1\}$. Then (3) for $s = s_2$ and $n = n_2$ defines a range for $x_2 = \min(X \setminus \{x_1\})$ (additionally we require $x_2 \ge x_1 + 1$), and so on. The procedure

Algorithm 1. For a given rational number s and integers t, n, function CONSTRUCTX(t, s, n) constructs a set X of positive integers such that $\min X \geq t$, $\sum_{x \in X} x^{-1} = s$, and $\sum_{x \in X} x^2 = n$; or returns the empty set $\emptyset$ if no such X exists.

```
function ConstructX(t, s, n)
    if s ≤ 0 or n ≤ 0 then
        return ∅
    end if
    L := ⌈max { 1/s , t }⌉
    U := ⌊min { ∛(n/s) , √n }⌋
    for x := L, L + 1, ..., U do
        s_new := s − 1/x
        n_new := n − x²
        if s_new = 0 and n_new = 0 then
            return {x}
        end if
        X := ConstructX(x + 1, s_new, n_new)
        if X ≠ ∅ then
            return X ∪ {x}
        end if
    end for
    return ∅
end function
```

stops when $s_{k+1} = 0$ and $n_{k+1} = 0$ for some k, implying that $1 = s_1 = \sum_{i=1}^{k} \frac{1}{x_i}$ and $m = n_1 = \sum_{i=1}^{k} x_i^2$, that is, $\{x_1, \ldots, x_k\}$ is a representation of m. However, if all candidate values have been explored without finding a representation, then no such representation exists (i.e., m is not representable).

Note that the bounds in (3) do not depend on $|X|$, and thus we do not need to know the size of X in advance. Furthermore, the inequality (2) guarantees that the algorithm always terminates and either produces a representation of m or establishes that none exist.

Algorithm 1 presents a pseudocode of the described algorithm as the recursive function CONSTRUCTX(t, s, n). To construct a representation of m, one needs to call CONSTRUCTX$(1, 1, m)$. For $m = 8542$, this function returns the empty set and thus implies the following statement.

Lemma 2. The number 8542 is not representable.

It is easy to modify Algorithm 1 to search for representations with additional restrictions on the elements (e.g., with certain numbers forbidden). We will see a need for such representations below.

2 Proof of Theorem 1

For a set S, we say that a representation X is S-*avoiding* if $X \cap S = \emptyset$, that is, X contains no elements from S.

Graham [1] introduced two functions:

$$f_0(X) = \{2\} \cup 2X \qquad \text{and} \qquad f_3(X) = \{3, 7, 78, 91\} \cup 2X, \qquad (4)$$

defined on $\{39\}$-avoiding representations, where the set $2X$ is obtained from X by multiplying each element by 2 (i.e., $2X = \{2x \,:\, x \in X\}$). Indeed, since a representation of $m > 1$ cannot contain 1, the sets $\{2\}$ and $2X$ are disjoint, implying that

$$\sum_{y \in f_0(X)} \frac{1}{y} = \frac{1}{2} + \sum_{x \in X} \frac{1}{2x} = \frac{1}{2} + \frac{1}{2}\sum_{x \in X} \frac{1}{x} = 1.$$

Similarly, the sets $\{3, 7, 78, 91\}$ and $2X$ are disjoint, since $2X$ consists of even numbers and $2 \cdot 39 = 78 \notin 2X$ (as X is $\{39\}$-avoiding), implying that

$$\sum_{y \in f_3(X)} \frac{1}{y} = \frac{1}{3} + \frac{1}{7} + \frac{1}{78} + \frac{1}{91} + \sum_{x \in X} \frac{1}{2x} = \frac{1}{2} + \frac{1}{2}\sum_{x \in X} \frac{1}{x} = 1.$$

Furthermore, if X is a representation of an integer m, then $\sum_{y \in 2X} y^2 = \sum_{x \in X}(2x)^2 = 4m$, implying that $f_0(X)$ and $f_3(X)$ are representations of integers $g_0(m) = 2^2 + 4m = 4m + 4$ and $g_3(m) = 3^2 + 7^2 + 78^2 + 91^2 + 4m = 4m + 14423$, respectively. Trivially, for any integer m, we have $g_0(m) \equiv 0 \pmod 4$ and $g_3(m) \equiv 3 \pmod 4$, which explain the choice of indices in the function names. Finally, one can easily check that neither $f_0(X)$ nor $f_3(X)$ contains 39, and thus they map $\{39\}$-avoiding representations to $\{39\}$-avoiding representations.

While the functions $f_0(X)$ and $f_3(X)$ were sufficient for the problem addressed by Graham, we will need two more functions, $f_1(X)$ and $f_2(X)$, such that the corresponding functions $g_i(m)$ $(i = 0, 1, 2, 3)$ form a complete residue system modulo 4. It turns out that such functions cannot be defined on $\{39\}$-avoiding representations, which leads us to a further restriction of the domain. Specifically, it is convenient to deal with the $\{21, 39\}$-avoiding representations, on which we define

$$f_1(X) = \{5, 7, 9, 45, 78, 91\} \cup 2X \qquad \text{and} \qquad f_2(X) = \{3, 7, 42\} \cup 2X. \qquad (5)$$

One can easily see that the functions f_1 and f_2 map a $\{21, 39\}$-avoiding representation X of an integer m to a $\{21, 39\}$-representation of integers $g_1(m) = 4m + 16545$ and $g_2(m) = 4m + 1822$, respectively. The functions f_0 and f_3 can also be viewed as mappings on $\{21, 39\}$-avoiding representations. As planned, we have $g_i(m) \equiv i \pmod 4$ for each $i = 0, 1, 2, 3$, which will play a key role in the proof of Theorem 1 below.

Lemma 3. Let m be an integer such that $8543 \le m \le 54533$. Then

- m is representable; and
- m has a $\{21, 39\}$-avoiding representation unless $m \in E$, where

$$E = \{ 8552, 8697, 8774, 8823, 8897, 8942, 9258, 9381, 9439, 9497 \}.$$

Proof. The proof is established computationally. We provide representations of all representable integers $m \le 54533$ in a supplementary file (see Section 4), where the listed representation of each $m \ge 8543$, $m \notin E$ is $\{21, 39\}$-avoiding. □

We are now ready to prove Theorem 1.

Proof of Theorem 1. Thanks to Lemma 2, it remains to prove that every number greater than 8542 is representable. First, Lemma 3 implies that every number m in the range $8543 \le m \le 9497$ is representable. For larger m, we will use induction on m to prove that every integer $m \ge 9498$ has a $\{21, 39\}$-avoiding representation.

Again by Lemma 3, we have that every number m in the range $9498 \le m \le 54533$ has a $\{21, 39\}$-avoiding representation.

Consider $m > 54533$, and assume that all integers in the interval $[9498, m-1]$ have $\{21, 39\}$-avoiding representations. Let $i = m \bmod 4$, and so $i \in \{0, 1, 2, 3\}$. Then there exists an integer m' such that $m = g_i(m')$ and thus $m' \ge \frac{m-16545}{4} > \frac{54533-16545}{4} = 9497$. By the induction assumption, m' has a $\{21, 39\}$-avoiding representation X. Then $f_i(X)$ is a $\{21, 39\}$-avoiding representation of m, which concludes the proof. □

3 t-Translations

The functions defined in equations (4)–(5) inspire us to consider a broader class of functions that map representations of small integers to those of larger ones. For an integer t, let us call an integer m *t-representable* if it has a representation X with $\min X \ge t$, called a *t-representation* of m. Clearly, an integer $m > 1$ is representable if and only if it is 2-representable. A t-representation (if it exists) of a given integer m can be constructed with Algorithm 1 by calling CONSTRUCTX$(t, 1, m)$.

Let t be a positive integer. A tuple of positive integers $r = (k; y_1, \ldots, y_l)$ is called a *t-translation* if

$$1 - \frac{1}{k} = \frac{1}{y_1} + \cdots + \frac{1}{y_l},$$

$t \le y_1 < y_2 < \cdots < y_l$, and for every $i \in \{1, 2, \ldots, l\}$, we have either $y_i < tk$ or $k \nmid y_i$.

With every t-translation $r = (k; y_1, \ldots, y_l)$, we associate two parameters: *scale* $\mathrm{sc}(r) = k^2$ and *shift* $\mathrm{sh}(r) = y_1^2 + \cdots + y_l^2$.

Lemma 4. Let m be a t-representable integer, and r be a t-translation. Then the number $\mathrm{sc}(r) \cdot m + \mathrm{sh}(r)$ is t-representable.

Proof. Suppose that $\{x_1, \ldots, x_n\}$ is a t-representation of m. If $r = (k; y_1, \ldots, y_l)$ is a t-translation, then

$$\mathrm{sc}(r) \cdot m + \mathrm{sh}(r) = y_1^2 + \cdots + y_l^2 + (kx_1)^2 + \cdots + (kx_n)^2,$$

and

$$\frac{1}{y_1} + \cdots + \frac{1}{y_l} + \frac{1}{kx_1} + \cdots + \frac{1}{kx_n} = \left(1 - \frac{1}{k}\right) + \frac{1}{k} = 1.$$

Notice that $\{y_1, \ldots, y_l\} \cap \{kx_1, \ldots, kx_n\} = \emptyset$, since for any $i \in \{1, 2, \ldots, l\}$ and $j \in \{1, 2, \ldots, n\}$, we have either $y_i < tk \le kx_j$ or $k \nmid y_i$ (i.e., $y_i \ne kx_j$). Hence, the set $Y = \{y_1, \ldots, y_l, kx_1, \ldots, kx_n\}$ forms a representation of $\mathrm{sc}(r) \cdot m + \mathrm{sh}(r)$. Furthermore, it is easy to see that $\min Y \ge t$, i.e., Y is a t-representation. □

The function $f_0(X)$ defined in (4) corresponds to a 2-translation $(2; 2)$; however, $(2; 3, 7, 78, 91)$ corresponding to the function $f_3(X)$ is not a 2-translation because of the presence of $78 = 2 \cdot 39$. As we will see below, it is preferable to have the scale small, ideally equal $2^2 = 4$, which is possible for the only 2-translation $(2; 2)$.

At the same time, for $t = 6$, we can construct a set of 6-translations, such as

$$\{(2; 9, 10, 11, 15, 21, 33, 45, 55, 77), \quad (2; 6, 7, 9, 21, 45, 105), \tag{6}$$
$$(2; 7, 9, 10, 15, 21, 45, 105); \quad (2; 6, 9, 11, 21, 33, 45, 55, 77)\},$$

where the translations have scale $2^2 = 4$ and shifts $\{13036, 13657, 13946, 12747\}$.

A set of translations S is called *complete* if the set

$$\{ \mathrm{sh}(r) \pmod{\mathrm{sc}(r)} : r \in S \}$$

forms a complete residue system; that is, for any integer m there exists $r \in S$ such that $m \equiv \mathrm{sh}(r) \pmod{\mathrm{sc}(r)}$. It can be easily verified that the 6-translations in (6) form a complete set of translations.

The following theorem generalizes Theorem 1.

Theorem 2. Let S be a complete set of t-translations with maximum scale q and maximum shift s. If numbers $n+1, n+2, \ldots, qn+s$ are t-representable, then so is any number greater than n.

Proof. Suppose that all numbers $n+1, n+2, \ldots, qn+s$ are t-representable. Let us will prove by induction that so is any number $m > qn+s$.

Assume that all numbers from $n+1$ to $m-1$ are t-representable. Since S is complete, there exists a t-translation $r \in S$ such that $\text{sh}(r) \equiv m \pmod{\text{sc}(r)}$. Note that $\text{sh}(r) \leq s$ and $\text{sc}(r) \leq q$.

For a number $m' = \frac{m - \text{sh}(r)}{\text{sc}(r)}$, we have $m' < m$, and $m' > \frac{(qn+s)-s}{q} = n$, implying by induction that m' is t-representable. Then by Lemma 4, the number $m = \text{sc}(r) \cdot m' + \text{sh}(r)$ is t-representable. $\qquad\square$

We can observe that small 6-representable numbers tend to appear very sparsely, with the smallest ones being 2579, 3633, 3735, 3868 (sequence A303400 in the OEIS [4]). Nevertheless, using the complete set of 6-translations in (6), we can prove the following theorem.

Theorem 3. The largest integer that is not 6-representable is 15707.

Proof. First, we computationally establish that 15707 is not 6-representable. Thanks to the complete set of 6-translations given in (6), by Theorem 2, it remains to show that all integers m in the interval $15708 \leq m \leq 4 \cdot 15707 + 13946 = 76774$ are 6-representable, which we again establish computationally (see Section 4). $\qquad\square$

We have also computed similar bounds for other $t \leq 8$ (sequence A297896 in the OEIS [4]), although we do not know whether such bounds exist for all t.

Theorem 3 together with Lemma 3 provides yet another proof of Theorem 1.

4 Supplementary Files

The following supplementary files support this study:

- https://oeis.org/A297895/a297895.txt contains representations of all representable integers $m \leq 54533$, which are $\{21, 39\}$-avoiding if $m \geq 8543$ and $m \notin E$ (see Lemma 3);
- https://oeis.org/A303400/a303400.txt contains 6-representations of all 6-representable integers up to 76774, which include all integers in the interval $[15708, 76774]$.

References

[1] R. L. Graham. A theorem on partitions. *J. Australian Mathematical Soc.* **3** no. 4 (1963) 435–441.

[2] R. Guy. *Unsolved Problems in Number Theory, Volume 1.* Springer Science and Business Media, New York, 2013.

[3] R. Knott. Egyptian fractions. Online at http://www.maths.surrey.ac.uk/hosted-sites/R.Knott/Fractions/egyptian.html, 2017. Last accessed June 2018.

[4] The OEIS Foundation. *The Online Encyclopedia of Integer Sequences.* Online at http://oeis.org, 2018. Last accessed June 2018.

15

PUZZLES, PARITY MAPS, AND PLENTY OF SOLUTIONS

David Nacin

KenKen puzzles were invented in 2004 by Japanese math teacher Tetsuya Miyamoto, with the intention of improving the arithmetic skills of his students. He originally named them Kashikoku-Naru puzzles, which translates to "a puzzle that makes you smarter." The current name of KenKen translates loosely to "cleverness squared." These puzzles are also referred to as Mathdoku or Calcudoku, often by those who wish to publish them without having the rights to the better-known but copyrighted name. KenKen puzzles now appear regularly in the *New York Times* and in many other publications worldwide.

The objective of KenKen is to fill in the cells of an n-by-n grid with symbols from a set of size n in an attempt to form a *Latin square*. This means that each symbol must appear exactly once in each row and column, a constraint that also occurs in Sudoku and many other puzzle variants. In KenKen, there are additional constraints in the form of heavily outlined regions of cells that are called *cages*, at least some of which contain clues. These clues consist of both an operation and a target. When the operation is applied to the entries in each of the cells in some order, the output must equal this target. Most commonly, the set $\{1, 2, \cdots, n\}$ is used, together with the operations of addition, subtraction, multiplication, and division (see Figure 15.1). For $n \geq 2$, this set is not closed under any of these operations, so the targets often fall outside the original set.

One way to make our targets fall within our set is to construct puzzles using a finite group. Here, we will study a collection of these puzzles over one such group, see when their solution is unique, and find how many solutions they can have when it is not.

1 A Collection of Puzzles over the Group $\mathbb{Z}_4$

Our story begins with a side note in a 2010 blog entry by Brian Hayes [1]. There, he included a sample puzzle over the set of complex numbers $\{1, i, -1, -i\}$, shown here in Figure 15.2. He stated that he did not have a deductive method for solving such puzzles and that he was not sure that the solution was unique,

1	2 ˣ³⁶	3	4 ⁺¹¹
2	3	4	1 ⁺⁶
3	4 ˣ¹⁶	1 /²	2
4	1	2	3

Figure 15.1. A sample solved puzzle.

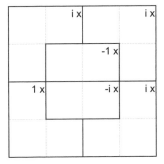

Figure 15.2. The original bit-player puzzle.

saying that it was at best a "jokenken." The solution is indeed unique, and the reader may wish to take a moment to attempt to solve this original puzzle before reading further.

In this case, our set happens to be closed under the single operation given in the clues of this puzzle. This is a closed finite subset of the group of nonzero complex numbers under multiplication, and hence is a subgroup. As i (or similarly, $-i$) generates the group, it must be isomorphic to any cyclic group of order four. We can therefore map this puzzle over to the group of integers modulo four under addition, which we will refer to here as $\mathbb{Z}_4$. Any pairing of a generator from each defines an isomorphism, so we have two possible choices. Picking the map sending i to 1 allows us to convert to the puzzle shown in Figure 15.3. We can then drop the operation from our clues, understanding it to be the group product, as we can with KenKen over any finite group.

The cage pattern for this puzzle may at first seem somewhat arbitrary, but aside from the vertical and horizontal symmetry, it has another important feature. It contains some cages with an odd number of cells. If all the cages contained an even number of cells, then we could add two to each entry of a

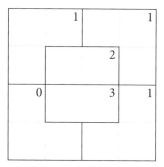

Figure 15.3. A $\mathbb{Z}_4$ version of the bit-player puzzle.

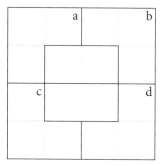

Figure 15.4. The general $\mathbb{Z}_4$ puzzle, $P\left(\left[\begin{smallmatrix} a & b \\ c & d \end{smallmatrix}\right]\right)$.

solution. This operation would leave the sum in each cage unchanged modulo four, creating a different solution to the same puzzle. Thus having some odd-sized cages is necessary for unique solutions to be possible.

If we take the collection of all possible cage patterns with the same symmetry conditions, some odd sized cages, and no size one cages, then there are only a handful of choices. It has been shown in Nacin [4] that each of these are equivalent as puzzles to the one shown in Figure 15.4. Let us refer to this puzzle by $P = P\left(\left[\begin{smallmatrix} a & b \\ c & d \end{smallmatrix}\right]\right)$ for $a, b, c, d \in \mathbb{Z}_4$, and only consider puzzles of this type throughout the rest of this chapter.

Notice that the four-cell center cage has not been given a clue here. The sum of any row must be $0 + 1 + 2 + 3$, since each row contains all four entries. The sum of the whole board must then equal $4 \times (0 + 1 + 2 + 3)$, which is equivalent to zero modulo four. For the whole board to sum to zero, the sum of the four center entries must equal the negative of the sum of the four given clues. Therefore, for a solvable puzzle, we can always find the unique possible value for a clue in this center cage.

By leaving out this center clue, we only remove the possibility of purposely using it to create puzzles with no solutions, something that we assume is not our intent. Similarly, we would gain no information by splitting the center four entries into two cages with two cells each. The sum of any two adjacent clues in the center cage can be determined directly from the two clues on the same half of the board, together with the fact that the sum of the entries on any half of the board is $2 \times (0 + 1 + 2 + 3)$, or zero modulo four. An analysis of the puzzle $P\left(\left[\begin{smallmatrix} a & b \\ c & d \end{smallmatrix}\right]\right)$ thus contains the full theory for a variety of cage patterns.

A clue-based criterion has been determined for when puzzles of this type have a unique solution [4], but here, let us aim to answer a more involved question. Based on the clues alone, how can we tell exactly how many solutions a given puzzle has, and in each case, is there a way to clearly see why? We seek a total classification, together with an explanation, for the number of solutions to any given puzzle.

2 Group Actions on the Set of All Puzzles

How do we begin classifying puzzles by their number of solutions? One strategy is to consider all possible clue combinations, solve and count the solutions for each puzzle, and then compile all our information accordingly into one large chart. To avoid spending time on puzzles with no solutions, another strategy is to take all Latin squares over $\{0, 1, 2, 3\}$, use each one to generate a puzzle by setting the clues to fit, and then continue in the same fashion. These strategies both leave much to be desired. They involve considering $4^4 = 256$ or 576 cases, respectively [2]. The outcome would hardly be in a digestible form, and even if we recognize some defining quality for each group of puzzles having the same number of solutions, the method gives no insight into what is causing any puzzle to have that number of solutions. Let us therefore take a different approach here.

Rotating any puzzle by $90°$ changes where the four clues appear, but leaves the cage pattern unchanged. The puzzle $P\left(\left[\begin{smallmatrix} 0 & 0 \\ 1 & 2 \end{smallmatrix}\right]\right)$ has the same number of solutions as $P\left(\left[\begin{smallmatrix} 1 & 0 \\ 2 & 0 \end{smallmatrix}\right]\right)$, whatever that number is, because turning a puzzle on its side before solving it does not affect how many solutions we get. We can actually apply any number of $90°$ rotations and vertical, horizontal, or diagonal reflections to a puzzle without changing the number of solutions. Instead of studying all possible puzzles, we can examine one puzzle from each equivalence class up to these operations. To better apply this strategy, we introduce the notation of group actions.

A group action of the group G on the set X is a map

$$G \times X \to X$$

satisfying

$$a(bx) = (ab)x$$

and

$$ex = x$$

for any a and b in G and any x in X. Here e is used to represent the identity in G. Define the orbit of x in X to be the set

$$orb(x) = \{gx : g \in G\},$$

and the stabilizer to be the set

$$stab(x) = \{g \in G : gx = x\}.$$

For any g in G, define the fixed points of g to be the set

$$X^g = \{x \in X : gx = x\}.$$

Allow the dihedral group

$$D_4 \cong \langle r, s \mid r^4, s^2, sr = r^3 s \rangle$$

to act on the set of all puzzles by having r rotate the clues of a puzzle clockwise $90°$ and s flip the clues of a puzzle about a vertical line of symmetry. To keep things in line with our group action notation, apply these elements in D_4 from right to left. Thus rs first flips a puzzle and then rotates it. Some examples are:

$$rP\left(\left[\begin{smallmatrix} 0 & 1 \\ 3 & 2 \end{smallmatrix}\right]\right) = P\left(\left[\begin{smallmatrix} 3 & 0 \\ 2 & 1 \end{smallmatrix}\right]\right), \qquad sP\left(\left[\begin{smallmatrix} 0 & 1 \\ 3 & 2 \end{smallmatrix}\right]\right) = P\left(\left[\begin{smallmatrix} 1 & 0 \\ 2 & 3 \end{smallmatrix}\right]\right),$$

$$rsP\left(\left[\begin{smallmatrix} 0 & 1 \\ 3 & 2 \end{smallmatrix}\right]\right) = P\left(\left[\begin{smallmatrix} 2 & 1 \\ 3 & 0 \end{smallmatrix}\right]\right), \qquad srP\left(\left[\begin{smallmatrix} 0 & 1 \\ 3 & 2 \end{smallmatrix}\right]\right) = P\left(\left[\begin{smallmatrix} 0 & 3 \\ 1 & 2 \end{smallmatrix}\right]\right),$$

$$stab(P\left(\left[\begin{smallmatrix} 1 & 2 \\ 2 & 3 \end{smallmatrix}\right]\right)) = \{e, r^3 s\}, \qquad stab(P\left(\left[\begin{smallmatrix} 1 & 2 \\ 1 & 2 \end{smallmatrix}\right]\right)) = \{e, r^2 s\},$$

$$X^s = \{P\left(\left[\begin{smallmatrix} a & a \\ b & b \end{smallmatrix}\right]\right) : a, b \in \mathbb{Z}_4\}, \qquad X^{r^2} = \{P\left(\left[\begin{smallmatrix} a & b \\ b & a \end{smallmatrix}\right]\right) : a, b \in \mathbb{Z}_4\}.$$

We also have

$$orb\left(P\left(\left[\begin{smallmatrix} 1 & 2 \\ 2 & 3 \end{smallmatrix}\right]\right)\right) = \{P\left(\left[\begin{smallmatrix} 1 & 2 \\ 2 & 3 \end{smallmatrix}\right]\right), P\left(\left[\begin{smallmatrix} 2 & 1 \\ 3 & 2 \end{smallmatrix}\right]\right), P\left(\left[\begin{smallmatrix} 3 & 2 \\ 2 & 1 \end{smallmatrix}\right]\right), P\left(\left[\begin{smallmatrix} 2 & 3 \\ 1 & 2 \end{smallmatrix}\right]\right)\},$$

and

$$orb\left(P\left(\left[\begin{smallmatrix} 1 & 1 \\ 2 & 3 \end{smallmatrix}\right]\right)\right) = \left\{ \begin{matrix} P\left(\left[\begin{smallmatrix} 1 & 1 \\ 2 & 3 \end{smallmatrix}\right]\right), P\left(\left[\begin{smallmatrix} 2 & 1 \\ 3 & 1 \end{smallmatrix}\right]\right), P\left(\left[\begin{smallmatrix} 3 & 2 \\ 1 & 1 \end{smallmatrix}\right]\right), P\left(\left[\begin{smallmatrix} 1 & 3 \\ 1 & 2 \end{smallmatrix}\right]\right), \\ P\left(\left[\begin{smallmatrix} 1 & 1 \\ 3 & 2 \end{smallmatrix}\right]\right), P\left(\left[\begin{smallmatrix} 3 & 1 \\ 2 & 1 \end{smallmatrix}\right]\right), P\left(\left[\begin{smallmatrix} 2 & 3 \\ 1 & 1 \end{smallmatrix}\right]\right), P\left(\left[\begin{smallmatrix} 1 & 2 \\ 1 & 3 \end{smallmatrix}\right]\right) \end{matrix} \right\}.$$

Setting $x \sim y$ if $x \in orb(y)$ produces an equivalence relation and thus partitions the set of all puzzles. As elements in each orbit share the same number of solutions, we only need to study one representative from each orbit to complete our classification. To continue, we introduce two general theorems about group actions: the Orbit-Stabilizer theorem, which will tell us the size of any given orbit based on the stabilizer, and the Cauchy-Frobenius lemma, which tells us the number of distinct orbits. The latter is often inappropriately attributed to William Burnside, who did write about it but made no claims of having first discovered the result himself [5]. The first tells us that for any $x \in X$,

$$|orb(x)| \cdot |stab(x)| = |G|,$$

and the second tells us that the number $|X \backslash G|$ of distinct orbits is given by

$$|X \backslash G| = \frac{1}{|G|} \sum_{g \in G} |X^g|.$$

To count the number of orbits under the D_4 action, we first must compute $|X^g|$ for the various $g \in D_4$. For r and r^3, we get $|X^g| = 4^1$, as all clues must be equal for the puzzle to remain the same after a 90° rotation in either direction. Puzzles in X^g for g equal to r^2, s, or r^2s must have two clues, each equal to two others, giving us $|X^g| = 4^2$. For the two diagonal reflections rs and r^3s, two clues lie on the line of reflection, and we get $|X^g| = 4^3$, as the other two clues must be equal. Finally, every puzzle is fixed under e, so $|X^e| = 4^4$. Our theorem now tells us that we have

$$\frac{1}{|D_4|}(2 \times 4^1 + 3 \times 4^2 + 2 \times 4^3 + 4^4) = \frac{440}{8} = 55$$

distinct orbits. Studying one puzzle from each of these fifty-five orbits would give us all the information we need for a classification, though this still leaves a large number of cases to sort through. Instead, let us seek to find a different group action that still preserves the number of solutions, in hopes of finding a coarser partition.

If ϕ is any automorphism of $\mathbb{Z}_4$, we can also apply ϕ to the set of all puzzles by letting

$$\phi\left(P\left(\left[\begin{smallmatrix} a & b \\ c & d \end{smallmatrix}\right]\right)\right) = P\left(\left[\begin{smallmatrix} \phi(a) & \phi(b) \\ \phi(c) & \phi(d) \end{smallmatrix}\right]\right).$$

We can similarly apply ϕ pointwise to any Latin square. Because ϕ preserves the product, when L is a solution to P, $\phi(L)$ will be a solution to $\phi(P)$. Because ϕ is injective on the set of Latin squares, this tells us that $\phi(P)$ has at least as many

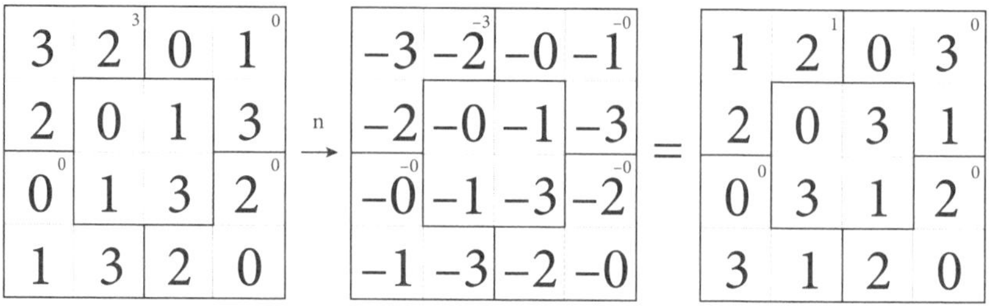

$$
\begin{array}{|cc|cc|}
\hline
3 & 2 & 0 & 1 \\
2 & 0 & 1 & 3 \\
0 & 1 & 3 & 2 \\
1 & 3 & 2 & 0 \\
\hline
\end{array}
\quad \xrightarrow{n}\quad
\begin{array}{|cc|cc|}
\hline
-3 & -2 & -0 & -1 \\
-2 & -0 & -1 & -3 \\
-0 & -1 & -3 & -2 \\
-1 & -3 & -2 & -0 \\
\hline
\end{array}
\quad = \quad
\begin{array}{|cc|cc|}
\hline
1 & 2 & 0 & 3 \\
2 & 0 & 3 & 1 \\
0 & 3 & 1 & 2 \\
3 & 1 & 2 & 0 \\
\hline
\end{array}
$$

Figure 15.5. Negating the entries of a solution to $P\left(\left[\begin{smallmatrix} 3 & 0 \\ 0 & 0 \end{smallmatrix}\right]\right)$ to construct a solution to $nP\left(\left[\begin{smallmatrix} 3 & 0 \\ 0 & 0 \end{smallmatrix}\right]\right)$.

solutions as does P. In fact, it has the same number of solutions, because if $\phi(P)$ had more solutions for some ϕ and P, we could apply $\phi^{-1} \circ \phi$ to P and create a puzzle that has strictly more solutions than itself.

We can therefore take the group of all automorphisms of $\mathbb{Z}_4$ and form a new group action that also preserves the number of solutions on the set of all puzzles. Unfortunately, this group is not very large. Any automorphism of a cyclic group must map generators to generators, and thus the group $\mathbb{Z}_4$ only has one nontrivial automorphism. This is the negation map $n(x) = -x$, also expressible as the permutation $(1, 3)$ of the set $\{0, 1, 2, 3\}$, since this automorphism switches odd elements and fixes even ones. The automorphism group is equal to $\langle\, n \mid n^2 \,\rangle \cong \mathbb{Z}_2$. See Figure 15.5 for an example of negating a solution of a puzzle P to construct a solution to nP. We can allow this group to act pointwise on the clues of our puzzles, but using such a small group to construct another action will not help reduce our overall problem by very much. Thankfully, there are other ways in which useful pointwise-defined maps can arise.

Consider the map $a \colon \mathbb{Z}_4 \to \mathbb{Z}_4$, which adds one modulo four to each input, also expressible as the permutation $(0, 1, 2, 3)$ of $\{0, 1, 2, 3\}$. Though this certainly is not an automorphism, we can still extend this to a map on the set of all puzzles. Applying it pointwise to the four clues, define

$$
aP\left(\left[\begin{smallmatrix} b & c \\ d & e \end{smallmatrix}\right]\right) = P\left(\left[\begin{smallmatrix} b+1 & c+1 \\ d+1 & e+1 \end{smallmatrix}\right]\right).
$$

If we add 3 to each of the entries of a solution to P, the sums in our outer cages each increase by $3 + 3 + 3 \equiv 1$ modulo four, giving us a solution to aP. One example of this is given in Figure 15.6. Thus the puzzle aP has at least as many solutions as P. The map a cannot increase this number either. Otherwise, a^4 would have more solutions than P, although the two are actually the same puzzle. Like negation, this map also preserves the number of solutions.

Though we could construct group actions using $\langle n \mid n^2 \rangle \cong \mathbb{Z}_2$ or $\langle a \mid a^4 \rangle \cong \mathbb{Z}_4$ we choose instead to define an action of a larger group containing both.

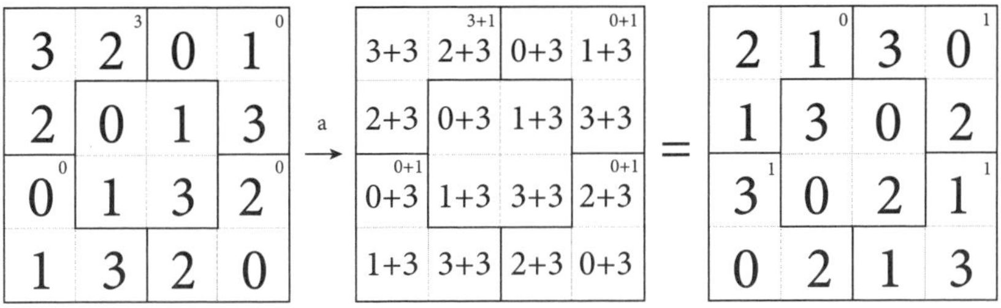

$$
\begin{array}{|cccc|}
\hline
3 & 2 & 0 & 1 \\
2 & 0 & 1 & 3 \\
0 & 1 & 3 & 2 \\
1 & 3 & 2 & 0 \\
\hline
\end{array}
\quad \xrightarrow{a} \quad
\begin{array}{|cccc|}
\hline
3+3 & 2+3 & 0+3 & 1+3 \\
2+3 & 0+3 & 1+3 & 3+3 \\
0+3 & 1+3 & 3+3 & 2+3 \\
1+3 & 3+3 & 2+3 & 0+3 \\
\hline
\end{array}
\quad = \quad
\begin{array}{|cccc|}
\hline
2 & 1 & 3 & 0 \\
1 & 3 & 0 & 2 \\
3 & 0 & 2 & 1 \\
0 & 2 & 1 & 3 \\
\hline
\end{array}
$$

Figure 15.6. Adding 3 to the entries of a solution to $P\left(\left[\begin{smallmatrix} 3 & 0 \\ 0 & 0 \end{smallmatrix}\right]\right)$ to construct a solution to $aP\left(\left[\begin{smallmatrix} 3 & 0 \\ 0 & 0 \end{smallmatrix}\right]\right)$.

Note that as permutations of $\{0, 1, 2, 3\}$, both a^4 and n^2 equal the identity, and $na = a^3 n$. These are precisely the relations we get from D_4 with a in place of rotation and n in place of reflection. We can now construct a different D_4 group action on the set of all puzzles using the group

$$D_4 \cong \langle a, n \mid a^4, n^2, na = a^3 n \rangle.$$

As both a and n preserve the number of solutions, and each element here is a composition of these, this D_4 action preserves the number of solutions as well.

For some examples with our new group action, we have:

$$aP\left(\left[\begin{smallmatrix} 0 & 1 \\ 2 & 2 \end{smallmatrix}\right]\right) = P\left(\left[\begin{smallmatrix} 1 & 2 \\ 3 & 3 \end{smallmatrix}\right]\right) \qquad nP\left(\left[\begin{smallmatrix} 0 & 1 \\ 2 & 2 \end{smallmatrix}\right]\right) = P\left(\left[\begin{smallmatrix} 0 & 3 \\ 2 & 2 \end{smallmatrix}\right]\right),$$

$$anP\left(\left[\begin{smallmatrix} 0 & 1 \\ 3 & 2 \end{smallmatrix}\right]\right) = P\left(\left[\begin{smallmatrix} 1 & 0 \\ 2 & 3 \end{smallmatrix}\right]\right), \qquad naP\left(\left[\begin{smallmatrix} 0 & 1 \\ 3 & 2 \end{smallmatrix}\right]\right) = P\left(\left[\begin{smallmatrix} 3 & 2 \\ 0 & 1 \end{smallmatrix}\right]\right),$$

$$stab\left(P\left(\left[\begin{smallmatrix} 0 & 1 \\ 2 & 2 \end{smallmatrix}\right]\right)\right) = \{e\} \qquad stab\left(P\left(\left[\begin{smallmatrix} 1 & 1 \\ 1 & 3 \end{smallmatrix}\right]\right)\right) = \{e, a^2 n\},$$

$$X^{an} = \emptyset, \qquad X^{a^2 n} = \{P\left(\left[\begin{smallmatrix} b & c \\ d & e \end{smallmatrix}\right]\right) : b, c, d, e \in \{1, 3\}\},$$

We also have that

$$orb\left(P\left(\left[\begin{smallmatrix} 1 & 3 \\ 3 & 3 \end{smallmatrix}\right]\right)\right) = \left\{P\left(\left[\begin{smallmatrix} 1 & 3 \\ 3 & 3 \end{smallmatrix}\right]\right), P\left(\left[\begin{smallmatrix} 2 & 0 \\ 0 & 0 \end{smallmatrix}\right]\right), P\left(\left[\begin{smallmatrix} 3 & 1 \\ 1 & 1 \end{smallmatrix}\right]\right), P\left(\left[\begin{smallmatrix} 0 & 2 \\ 2 & 2 \end{smallmatrix}\right]\right)\right\},$$

and

$$orb\left(P\left(\left[\begin{smallmatrix} 1 & 2 \\ 2 & 3 \end{smallmatrix}\right]\right)\right) = \left\{ \begin{array}{l} P\left(\left[\begin{smallmatrix} 1 & 2 \\ 2 & 3 \end{smallmatrix}\right]\right), P\left(\left[\begin{smallmatrix} 2 & 3 \\ 3 & 0 \end{smallmatrix}\right]\right), P\left(\left[\begin{smallmatrix} 3 & 0 \\ 0 & 1 \end{smallmatrix}\right]\right), P\left(\left[\begin{smallmatrix} 0 & 1 \\ 1 & 2 \end{smallmatrix}\right]\right), \\ P\left(\left[\begin{smallmatrix} 3 & 2 \\ 2 & 1 \end{smallmatrix}\right]\right), P\left(\left[\begin{smallmatrix} 0 & 3 \\ 3 & 2 \end{smallmatrix}\right]\right), P\left(\left[\begin{smallmatrix} 1 & 0 \\ 0 & 3 \end{smallmatrix}\right]\right), P\left(\left[\begin{smallmatrix} 2 & 1 \\ 1 & 0 \end{smallmatrix}\right]\right) \end{array} \right\}.$$

In this group, the permutations a, a^2, a^3, an, and $a^3 n$ leave no element of $\mathbb{Z}_4$ fixed and therefore can fix no puzzles. To remain fixed under n, each clue in

TABLE 15.1.
$|X^g|$ for the sixty-four $g \in D_4 \times D_4$

	e'	a	a^2	a^3	n	an	a^{2n}	a^3n
e	4^4	0	0	0	2^4	0	2^4	0
r	4^1	4^1	4^1	4^1	4^1	4^1	4^1	4^1
r^2	4^2	0	4^2	0	4^2	4^2	4^2	4^2
r^3	4^1	4^1	4^1	4^1	4^1	4^1	4^1	4^1
s	4^2	0	4^2	0	4^2	4^2	4^2	4^2
rs	4^3	0	0	0	$2^2 \cdot 4^1$	0	$2^2 \cdot 4^1$	0
r^2s	4^2	0	4^2	0	4^2	4^2	4^2	4^2
r^3s	4^3	0	0	0	$2^2 \cdot 4^1$	0	$2^2 \cdot 4^1$	0

a puzzle must be even, giving us $|X^n| = 2^4$. Since a^{2n} switches evens and fixes odds, we have $|X^{a^2n}| = 2^4$ as well. Finally as the identity fixes all 4^4 puzzles, we have

$$\frac{1}{|D_4|}(2 \times 2^4 + 1 \times 4^4) = \frac{288}{8} = 36.$$

This is an improvement, but again there are more orbits than desirable. We now combine all we have done by constructing a larger group containing all the transformations we have considered.

Consider how our two separate D_4 actions relate to each other. The elements in each individual D_4 do not generally commute, but two elements from different D_4 actions will. For example, if we add and then rotate or reflect, we get the same puzzle as if we performed the rotations and reflections to our puzzle before performing the addition. The same goes for negation. Thus, we can construct a $D_4 \times D_4$ group action of the order sixty-four group

$$G = \langle r, s \mid r^4, n^2, sr = r^3s \rangle \times \langle a, n \mid a^4, n^2, na = a^3n \rangle$$

on the set of all puzzles.

The fixed-point calculations here are a bit more involved, so let us organize each of the sixty-four $|X^g|$ into Table 15.1. The exponents have been left in intentionally, as this may give some insight into each of the individual calculations. For example, for the fixed points of r^3sn, we want to find the puzzles that are fixed under negation before or after diagonal reflection. The clues on the diagonal are fixed by the reflection, so to stay the same, they must be fixed by negation. That means they must both be even, which can happen in 2^2 different ways. The clues off the diagonal line of reflection must equal the negation of each other. Thus, one is determined by the other, and we can pick these in 4^1 ways. This allows us to arrive at $2^2 \cdot 4^1$ for the number of fixed points of r^3sn.

Plenty of Solutions • 231

TABLE 15.2.
Orbit and stabilizer information from the minimal puzzle in each orbit

x	$stab(x)$	$|stab(x)|$	$|orb(x)|$
$P\left(\begin{bmatrix}0&0\\0&0\end{bmatrix}\right)$	$\langle n,r,s\rangle$	16	4
$P\left(\begin{bmatrix}0&0\\1&0\end{bmatrix}\right)$	$\langle rs\rangle$	2	32
$P\left(\begin{bmatrix}0&0\\2&0\end{bmatrix}\right)$	$\langle n,rs\rangle$	4	16
$P\left(\begin{bmatrix}0&0\\1&1\end{bmatrix}\right)$	$\langle s,anrr\rangle$	4	16
$P\left(\begin{bmatrix}0&0\\2&2\end{bmatrix}\right)$	$\langle s,n,aarr\rangle$	8	8
$P\left(\begin{bmatrix}0&1\\1&0\end{bmatrix}\right)$	$\langle rs,sr,ans\rangle$	8	8
$P\left(\begin{bmatrix}0&2\\2&0\end{bmatrix}\right)$	$\langle rs,sr,aas,n\rangle$	16	4
$P\left(\begin{bmatrix}0&0\\2&1\end{bmatrix}\right)$	$\langle e\rangle$	1	64
$P\left(\begin{bmatrix}0&0\\3&1\end{bmatrix}\right)$	$\langle ns\rangle$	2	32
$P\left(\begin{bmatrix}0&1\\2&0\end{bmatrix}\right)$	$\langle rs\rangle$	2	32
$P\left(\begin{bmatrix}0&1\\3&0\end{bmatrix}\right)$	$\langle nrr,rs\rangle$	4	16
$P\left(\begin{bmatrix}0&1\\3&2\end{bmatrix}\right)$	$\langle nsr,ar\rangle$	8	8
$P\left(\begin{bmatrix}0&1\\2&3\end{bmatrix}\right)$	$\langle aas,anrr\rangle$	4	16

The Cauchy-Frobenius lemma allows us to add each of the terms in this table and divide by $|G| = 64$ to find the number of distinct orbits, which gives $\frac{832}{64} = 13$. Puzzles in each orbit all have the same number of solutions, and we have finally arrived at a reasonable number of orbits to consider.

We only need to study one puzzle from each orbit to count solutions. We choose a representative of each orbit form the stabilizer, starting with the top-left entry and circling clockwise, and then apply the Orbit-Stabilizer theorem to find the size of each orbit from the stabilizer. The results are compiled in Table 15.2. Horizontal lines separate the puzzles by the number of distinct clues, a number that happens to be invariant under all sixty-four transformations.

3 Parity Maps and Graphs

Let us turn our attention back to the original puzzle from bit-player, which happens to fall into the $P\left(\begin{bmatrix}0&0\\1&0\end{bmatrix}\right)$ orbit. Though we can count solutions to puzzles in this orbit using any representative, here let us choose to stick with the original puzzle. We will need a method of referring to specific entries in any given puzzle P. Refer to the entry in row i and column j as $P_{i,j}$, and refer to the four entries $P_{2,2}, P_{2,3}, P_{3,2}$ and $P_{3,3}$ as our *inner entries* (Figure 15.7).

Since our top clues in $P\left(\begin{bmatrix}1&1\\0&1\end{bmatrix}\right)$ are both equal to one, and the top eight cells must sum to zero, we know that $P_{2,2} + P_{2,3} \equiv 2$. As $P_{2,2}$ and $P_{2,3}$ must be distinct,

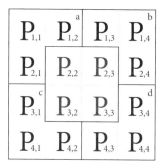

Figure 15.7. The entries of $P\left(\left[\begin{smallmatrix} a & b \\ c & d \end{smallmatrix}\right]\right)$, with the four inner entries highlighted in green.

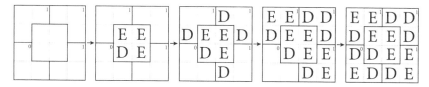

Figure 15.8. Making a parity map for $P\left(\left[\begin{smallmatrix} 1 & 1 \\ 0 & 1 \end{smallmatrix}\right]\right)$.

the only possible conclusion is that they must be zero and two in some order. Summing the rightmost eight cells similarly reveals that $P_{2,3}$ and $P_{3,3}$ must also be zero and two in some order. As the bottom half must sum to zero, we know that

$$P_{3,2} + P_{3,3} + 1 + 0 \equiv 0,$$

so $P_{3,2}$ must be odd. We can now begin to fill in what we know about each cell so far, forming a map of the parity of each entry. We use "E" for even cells and "D" to represent odd cells, in an attempt to avoid any confusion between the letter "O" and number zero. The parity of $P_{1,3}, P_{4,3}, P_{2,1}$, and $P_{2,4}$ is determined by the fact that our solution must be a Latin square. We can now use that information, together with the parities of the cage clues, to determine the parity of all entries. The result is shown in Figure 15.8.

The value of any entry determines the value of the entries with equal parity in the same row or same column. We can track this dependency by turning the map into a parity graph using our sixteen cells as vertices. We connect two distinct cells with an edge if they have the same parity and are in the same row or column. With this definition, the value of any cell immediately determines the value of all cells in the same cycle. For $P\left(\left[\begin{smallmatrix} 1 & 1 \\ 0 & 1 \end{smallmatrix}\right]\right)$, we get the graph shown in Figure 15.9. Use blue edges to connect even cells and red for odd, only so the cell parity matches the parity of the number of letters in the given color.

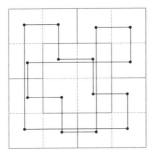

Figure 15.9. The parity graph for $P\left(\left[\begin{smallmatrix}1&1\\0&1\end{smallmatrix}\right]\right)$.

There are two possibilities for placing odds into the cells of the red cycle, but at most only one can fit. This is because the upper-right cage contains only entries from this cycle, our two options differ by two in each cell, and $2+2+2$ is not congruent to zero in $\mathbb{Z}_4$. Once this cycle is determined, we can use it together with the bottom-left-cage clue to find the value of $P_{4,1}$. That entry then determines the entries in the blue cycle completely. We have shown that this puzzle, and hence each puzzle in the same orbit, has at most one solution.

Our methods have shown that at most one Latin square fits the clues of our puzzle, but they do not guarantee that one exists that fits them all at the same time. We could produce a solution, but instead, let us try to generalize the same techniques used for $P\left(\left[\begin{smallmatrix}1&1\\0&1\end{smallmatrix}\right]\right)$ to other puzzles, getting the best possible upper bound for the number of solutions in each case.

Taking a step back, consider a KenKen puzzle over an arbitrary finite group, H. Suppose that the product of two adjacent cells equals the element c in H. The entries for those cells must satisfy the equation $ab=c$. For each a in H, $a^{-1}c$ is the only possibility for b, but to form a Latin square, we need $b \neq a$. Thus, the total number of possibilities equals

$$|H| - |\{a \in H : a^2 = c\}|.$$

For our puzzles over $\mathbb{Z}_4$, the squares are just the doubles, equal to the collection $\{0, 2\}$ of even numbers. Thus there are four possibilities when c is odd, but only two possibilities for a and b when c is even. If c is zero, these must be one and three, and if c is two, they must be zero and two.

This is particularly useful for us, when considering any two adjacent inner entries. Using the labeling shown in Figure 15.7, we know that

$$P_{2,2} + P_{2,3} + a + b \equiv$$
$$P_{3,2} + P_{3,3} + c + d \equiv$$
$$P_{2,2} + P_{3,2} + a + c \equiv$$
$$P_{2,3} + P_{3,3} + b + d \equiv 0,$$

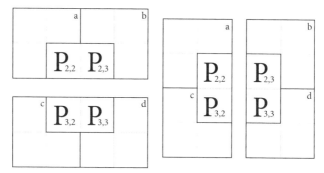

Figure 15.10. The four halves of $P\left(\left[\begin{smallmatrix} a & b \\ c & d \end{smallmatrix}\right]\right)$. In any solution, the entries of each must sum to zero.

because each of the four halves shown in Figure 15.10 must sum to zero. Given the value or parity of any one of these four inner entries, we have enough equations to determine the value or parity of the rest.

For example, with $P\left(\left[\begin{smallmatrix} 1 & 1 \\ 2 & 3 \end{smallmatrix}\right]\right)$, we get that $P_{2,2} + P_{2,3} + 2$, $P_{3,2} + P_{3,3} + 1$, $P_{2,2} + P_{3,2} + 3$, and $P_{2,3} + P_{3,3}$ are all zero. Notice that on their own, the equations $P_{3,2} + P_{3,3} + 1 \equiv 0$ and $P_{2,2} + P_{2,3} + 3 \equiv 0$ each allow for four solutions. However, $P_{2,2} + P_{2,3} + 2$ can only be zero if $P_{2,2}$ and $P_{2,3}$ are distinct evens, and $P_{3,2} + P_{3,3}$ can only be zero if they are distinct odds. This reveals the parity of all four inner entries, though the parity of any one inner entry together with our four equations would reveal the same. Why did two of the equations give us more information? They came from the sides of the board with clues of the same parity.

Suppose our puzzle has adjacent clues of the same parity. Then the equation we get from summing the entries on that side of the board is of the form $x + y + e \equiv 0$, where x and y are inner entries, and e is even. If e is two, then x and y must be even, and if e is zero, they must be odd. We have now shown all parts of the following lemma.

Lemma 1 (Inner Entry Lemma). *Let L be any solution to the puzzle $P = P\left(\left[\begin{smallmatrix} a & b \\ c & d \end{smallmatrix}\right]\right)$.*

1. *The parity of any one inner entry of L determines the parity of all four inner entries.*
2. *The value of any one inner entry of L determines the value of all four inner entries.*
3. *If P has adjacent clues that sum to zero, then the inner entries on that side of the board in L are distinct odds.*
4. *If P has adjacent clues that sum to two, then the inner entries on that side of the board in L are distinct evens.*

5. *If P has any two adjacent clues of equal parity, then the parity of all inner entries of L is determined.*

The final part of this lemma is applicable to the majority of our puzzles. Turning to the orbit representatives in Table 15.2, we see that only three out of our thirteen orbits lack a pair of adjacent clues of the same parity. Picking any side (the right, for example), we can go through the representatives in the table and use this lemma to see that $P_{2,3}$ and $P_{3,3}$ must be odd in any solution to $P\left(\left[\begin{smallmatrix} 0 & 0 \\ 0 & 0 \end{smallmatrix}\right]\right)$, $P\left(\left[\begin{smallmatrix} 0 & 0 \\ 1 & 0 \end{smallmatrix}\right]\right)$, $P\left(\left[\begin{smallmatrix} 0 & 0 \\ 2 & 0 \end{smallmatrix}\right]\right)$, and $P\left(\left[\begin{smallmatrix} 0 & 1 \\ 2 & 3 \end{smallmatrix}\right]\right)$, and even in any solution to $P\left(\left[\begin{smallmatrix} 0 & 0 \\ 2 & 2 \end{smallmatrix}\right]\right)$ and $P\left(\left[\begin{smallmatrix} 0 & 2 \\ 2 & 0 \end{smallmatrix}\right]\right)$. We will revisit each of these orbits in turn after developing one more result.

Seeking to replicate what we did for $P\left(\left[\begin{smallmatrix} 1 & 1 \\ 0 & 1 \end{smallmatrix}\right]\right)$, define a *parity map* for any puzzle P to be an assignment of one parity to each cell in a board, so that each parity appears twice in every row and column. A *parity graph* G is a graph constructed from a parity map, where two cells are connected by an edge if they are in the same row or column and have the same parity. Edges are colored red or blue if their cells are odd or even, respectively. A solution *satisfies a parity graph or map* if the parity of all entries matches that of the graph or map.

A puzzle with multiple solutions may have solutions satisfying different parity graphs, but no individual solution can satisfy more than one such graph. Therefore, by summing the number of possible solutions fitting each possible parity graph, we can get an upper bound for the number of solutions to any puzzle. We will need to establish some results for parity graphs.

Lemma 2 (Parity Graph Lemma). *Let G be a parity graph for a puzzle $P = P\left(\left[\begin{smallmatrix} a & b \\ c & d \end{smallmatrix}\right]\right)$.*

1. *Any parity graph is either composed of a single odd and single even cycle, or two odd and two even cycles.*
2. *If two cells are in the same cycle, their entries in any solution satisfying that graph must be equal if the distance between them is even. If the distance is odd, then they differ by 2 modulo four.*
3. *The value of any cell in a solution satisfying a parity graph determines the value of all cells in the same cycle.*
4. *If a cycle in a parity graph contains two cells from a three-cell cage, then the value of the third cell in that cage is determined in any solution satisfying that graph.*
5. *At most one solution can satisfy a parity graph composed of a single even and a single odd cycle.*

Since every row and column contains exactly two entries of each parity, every cell will be connected to exactly two other cells in the parity graph. Therefore, our graph will be a union of cycles of even size, and that size must be greater

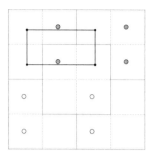

Figure 15.11. A cycle of size four implies that the four opposite parity entries in the same two rows, and the four opposite parity entries in the same two columns, must each form separate cycles of size four.

than two. As there are eight entries of each parity, the only possibilities are two cycles of size four, or one of size eight. In the case of two cycles of size four, each cycle is contained in two rows and columns, forcing a single cycle of opposite parity to be in the same two rows, and another in the same two columns. Thus if one parity has cycles of size four, then the other does as well. An example of this is shown in Figure 15.11.

The next two statements in Lemma 2 follow from the definition and the fact that any solution must form a Latin square. As elements are connected to same parity elements of different value, we can simply trace our way around the cycle to find the value of any cell.

For the fourth statement, consider the case where a cage contains two adjacent entries from the same cycle. Distinct even entries will sum to two, distinct odd entries will sum to zero, and either way we can subtract this sum from the clue to find the third entry in the cage. A cage may contain nonadjacent entries from the same cycle as well. In this situation, the same argument can be applied if the distance between them along the cycle is odd. If this distance is even, then those entries have the same value. As $0 + 0 \equiv 2 + 2$ and $1 + 1 \equiv 3 + 3$ modulo four, for either parity, their sum is determined, and we can subtract that sum from the cage clue to find the value of the third entry. An example is given in Figure 15.12, where the upper-right cage contains elements in the same cycle of even distance, and the bottom-left cage contains elements in the same cycle of odd distance. In both cages, we can determine the remaining entry exactly.

Parity graphs are not always composed of a single odd cycle and single even cycle, but when they are, all entries are determined. Each of the four outer cages must contain at least two elements from the same cycle. Due to the shape of these cages, the same cycle cannot occupy two cells in all four of the cages. There must be one cage with two open vertices and one filled, and another with two filled and one open. We can then use previous parts of Lemma 2 to determine

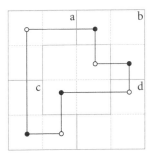

Figure 15.12. In this example, regardless of whether the open or filled vertices represent the values zero or two, we must still have $P_{1,4} = b$ and $P_{3,1} = c - 2$.

all entries in both cycles. This completes the proof of Lemma 2, and we can now proceed to see what these results say about puzzles in different orbits.

4 The Thirteen Orbits

Now that we have generalized the tools originally developed for the orbit of $P\left(\left[\begin{smallmatrix} 0 & 0 \\ 1 & 0 \end{smallmatrix}\right]\right)$, we can apply these techniques to the other twelve orbits. The strategy is as follows. First, let us take one puzzle for each orbit and consider the set of all possible parity graphs. Then use each parity graph to find an upper bound for the total number of solutions to all puzzles in that orbit. Finally, use these bounds, together with the size of each orbit and the total number of Latin squares, to attempt to reveal the exact number of solutions in every case.

Let us begin with the five orbits where it is impossible to even produce a parity graph, and then move to cases where there is more freedom, ending with the three orbits where even the parity of the inner entries cannot be determined.

4.1 Impossible Parity Graphs

The puzzles $P\left(\left[\begin{smallmatrix} 0 & 0 \\ 2 & 0 \end{smallmatrix}\right]\right)$, $P\left(\left[\begin{smallmatrix} 0 & 0 \\ 2 & 2 \end{smallmatrix}\right]\right)$, and $P\left(\left[\begin{smallmatrix} 0 & 0 \\ 2 & 1 \end{smallmatrix}\right]\right)$ each have adjacent clues of zero on the top half of the board. Because those clues sum to zero, Lemma 1 implies that in any solution, $P_{2,2}$ and $P_{2,3}$ must be odd. They also have adjacent clues summing to two on the left half of the board, and thus the same lemma implies that $P_{2,2}$ and $P_{3,2}$ must be even. As $P_{2,2}$ cannot be both odd and even, there can be no solutions for puzzles of these three types.

Both the top and bottom sets of clues sum to zero in $P\left(\left[\begin{smallmatrix} 0 & 0 \\ 3 & 1 \end{smallmatrix}\right]\right)$ and $P\left(\left[\begin{smallmatrix} 0 & 0 \\ 0 & 0 \end{smallmatrix}\right]\right)$, and thus Lemma 1 tells us that all four inner entries must be odd. For our solution to form a Latin square, this means that the corner entries of $P_{1,1}, P_{1,4}, P_{4,1}$, and $P_{4,4}$ must also be odd. Thus, each of our cages would need to contain two

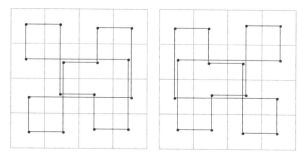

Figure 15.13. The two parity graphs for $P\left(\left[\begin{smallmatrix} 0 & 0 \\ 1 & 1 \end{smallmatrix}\right]\right)$.

evens and one odd, implying that all of our clues must be odd numbers, another contradiction.

Together, these two arguments show that five of the orbits listed in Table 15.2 contain only puzzles that cannot be solved. As we investigate further, we will see that these are the only cases with no solutions.

4.2 The Orbit of $P\left(\left[\begin{smallmatrix} 0 & 0 \\ 1 & 1 \end{smallmatrix}\right]\right)$

Applying Lemma 1 here, we find that in any solution, $P_{2,2}$ and $P_{2,3}$ must both be odd, and $P_{3,2}$ and $P_{3,3}$ must both be even. Together with our Latin square condition, this determines the parity of all entries in the two middle rows. This allows us to complete a parity map in two possible ways, leading to the two parity graphs shown in Figure 15.13. These each have a single even and odd cycle, so by Lemma 2, each can be completed in at most one way. Hence $P\left(\left[\begin{smallmatrix} 0 & 0 \\ 1 & 1 \end{smallmatrix}\right]\right)$, and any puzzle in the same orbit, has a maximum of two solutions.

4.3 The Orbit of $P\left(\left[\begin{smallmatrix} 0 & 1 \\ 2 & 0 \end{smallmatrix}\right]\right)$

Using Lemma 1, we see that solutions to $P\left(\left[\begin{smallmatrix} 0 & 1 \\ 2 & 0 \end{smallmatrix}\right]\right)$ must have $P_{2,2}, P_{3,2}$, and $P_{3,3}$ all even, and $P_{2,3}$ odd. The Latin square condition then forces $P_{3,1}, P_{3,4}, P_{1,2}$, and $P_{4,2}$ to all be odd. At this point, the parities of $P_{1,3}$ and $P_{2,4}$, which cannot both be odd, determine the parity of the other entries. Thus we get the three parity graphs shown in Figure 15.14. Each graph has a single even and odd cycle, so by Lemma 2, these graphs are each attached to at most one solution. Thus $P\left(\left[\begin{smallmatrix} 0 & 1 \\ 2 & 0 \end{smallmatrix}\right]\right)$ and other puzzles in its orbit have at most three solutions.

4.4 The Orbit of $P\left(\left[\begin{smallmatrix} 0 & 2 \\ 2 & 0 \end{smallmatrix}\right]\right)$

For this case, Lemma 1 shows that the inner entries are all even in any solution to $P\left(\left[\begin{smallmatrix} 0 & 2 \\ 2 & 0 \end{smallmatrix}\right]\right)$. This allows us to complete the parity graph in only the way shown in Figure 15.15.

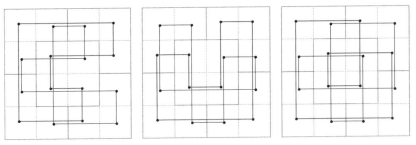

Figure 15.14. The parity graphs for $P\left(\left[\begin{smallmatrix}0 & 1\\ 2 & 0\end{smallmatrix}\right]\right)$.

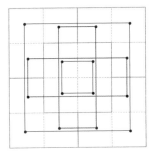

Figure 15.15. The parity graph for $P\left(\left[\begin{smallmatrix}0 & 2\\ 2 & 0\end{smallmatrix}\right]\right)$.

We have two cycles of each parity here, though the value of any inner entry, together with any two non-inner entries from different cycles, is enough to determine all other entries. There are two possibilities each for those values, so we arrive at an upper bound of eight solutions for puzzles in this orbit.

4.5 The Orbit of $P\left(\left[\begin{smallmatrix}0 & 1\\ 2 & 3\end{smallmatrix}\right]\right)$

As the clues on the left half and right half of $P\left(\left[\begin{smallmatrix}0 & 1\\ 2 & 3\end{smallmatrix}\right]\right)$ sum to two and zero, respectively, Lemma 1 tells us that in any solution, $P_{2,2}$ and $P_{3,2}$ must be even, and $P_{2,3}$ and $P_{3,3}$ must be odd. This determines the parity of all entries in the middle two columns. At this point, the parities in the right column are determined by those in the left. The parity of $P_{1,2}$ and $P_{4,2}$, together with the parity of the cage clues, imply that the sums $P_{1,1} + P_{2,1}$ and $P_{3,1} + P_{4,1}$ must both be odd. This leads to the four possibilities shown in Figure 15.16.

The first two graphs shown in the figure have only a single cycle of each parity and thus by Lemma 2, at most one solution each. The next two each have at most four solutions, as in both cases, the values of the entries in the cycles of either parity are determined by the values of the entries in the cycles of the

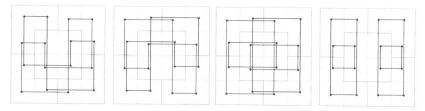

Figure 15.16. The parity graphs for $P\left(\left[\begin{smallmatrix} 0 & 1 \\ 2 & 3 \end{smallmatrix}\right]\right)$.

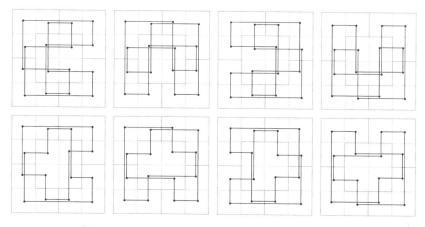

Figure 15.17. Opposite corner cells have different parity.

other parity. Adding all these possibilities together shows that puzzles in this orbit have at most ten solutions.

4.6 The Orbits of $P\left(\left[\begin{smallmatrix} 0 & 1 \\ 1 & 0 \end{smallmatrix}\right]\right)$, $P\left(\left[\begin{smallmatrix} 0 & 1 \\ 3 & 0 \end{smallmatrix}\right]\right)$, and $P\left(\left[\begin{smallmatrix} 0 & 1 \\ 3 & 2 \end{smallmatrix}\right]\right)$

In these three final orbits, the parity of the inner entries of solutions is not determined, because no two adjacent clues have the same parity. At best we can use Lemma 2 to conclude that the inner entries are either of the form $\left[\begin{smallmatrix} E & D \\ D & E \end{smallmatrix}\right]$ or $\left[\begin{smallmatrix} D & E \\ E & D \end{smallmatrix}\right]$. This implies that $P_{2,1}$ and $P_{2,4}$ must have opposite parity. The same is true for $P_{3,1}$ and $P_{3,4}$, $P_{1,2}$ and $P_{4,2}$, and $P_{1,3}$ and $P_{4,3}$.

This gives us thirty-two possible cases, though some have no valid configurations for the corners, whereas others have multiple possibilities. A short counting argument gives a total of sixteen possible parity maps. Let us form the corresponding parity graphs and then divide them into the three groups shown in Figures 15.17, 15.18, and 15.19 based on the relationship between the cycles and any two opposite corner cages.

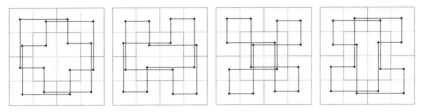

Figure 15.18. Opposite corners are equal.

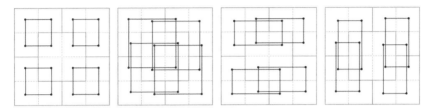

Figure 15.19. Two cycles for each parity.

Which of these parity graphs occur as solutions to $P\left(\left[\begin{smallmatrix} 0 & 1 \\ 1 & 0 \end{smallmatrix}\right]\right)$, $P\left(\left[\begin{smallmatrix} 0 & 1 \\ 3 & 0 \end{smallmatrix}\right]\right)$, and $P\left(\left[\begin{smallmatrix} 0 & 1 \\ 3 & 2 \end{smallmatrix}\right]\right)$ individually? Counting the distance between cells in the cycles of the parity graphs shown in Figure 15.17, we can see that the cage clues of opposite corners must be distinct, though of the same parity. This means that these eight graphs can only arise from solutions to $P\left(\left[\begin{smallmatrix} 0 & 1 \\ 3 & 2 \end{smallmatrix}\right]\right)$. The four parity graphs in Figure 15.18 require the cage clues in opposite corners to be equal. Thus, these can only arise from solutions to $P\left(\left[\begin{smallmatrix} 0 & 1 \\ 1 & 0 \end{smallmatrix}\right]\right)$. Finally, we have the four parity graphs shown in Figure 15.19. Though the parity of the cage clues is still determined here, our cycles allow for each cage clue to be either entry of that parity, independent of the other three cage clues. This adds four to the total number of solutions for each of $P\left(\left[\begin{smallmatrix} 0 & 1 \\ 3 & 2 \end{smallmatrix}\right]\right)$, $P\left(\left[\begin{smallmatrix} 0 & 1 \\ 1 & 0 \end{smallmatrix}\right]\right)$, and $P\left(\left[\begin{smallmatrix} 0 & 1 \\ 3 & 0 \end{smallmatrix}\right]\right)$.

Adding up the count for all three cases, we have at most twelve possible solutions to $P\left(\left[\begin{smallmatrix} 0 & 1 \\ 3 & 2 \end{smallmatrix}\right]\right)$, eight solutions to $P\left(\left[\begin{smallmatrix} 0 & 1 \\ 1 & 0 \end{smallmatrix}\right]\right)$, and four solutions to $P\left(\left[\begin{smallmatrix} 0 & 1 \\ 3 & 0 \end{smallmatrix}\right]\right)$.

4.7 Putting It All Together

Let $N\left(\left[\begin{smallmatrix} a & b \\ c & d \end{smallmatrix}\right]\right)$ be the exact number of solutions to $P\left(\left[\begin{smallmatrix} a & b \\ c & d \end{smallmatrix}\right]\right)$. We have determined through parity-graph arguments in the preceding subsections that the following inequalities all hold:

$$N\left(\left[\begin{smallmatrix} 0 & 0 \\ 0 & 1 \end{smallmatrix}\right]\right) \leq 1, \; N\left(\left[\begin{smallmatrix} 0 & 0 \\ 1 & 1 \end{smallmatrix}\right]\right) \leq 2, \; N\left(\left[\begin{smallmatrix} 0 & 1 \\ 1 & 0 \end{smallmatrix}\right]\right) \leq 8, \quad N\left(\left[\begin{smallmatrix} 0 & 2 \\ 2 & 0 \end{smallmatrix}\right]\right) \leq 8,$$

$$N\left(\left[\begin{smallmatrix} 0 & 1 \\ 2 & 0 \end{smallmatrix}\right]\right) \leq 3, \; N\left(\left[\begin{smallmatrix} 0 & 1 \\ 3 & 0 \end{smallmatrix}\right]\right) \leq 4, \; N\left(\left[\begin{smallmatrix} 0 & 2 \\ 3 & 1 \end{smallmatrix}\right]\right) \leq 10, \; N\left(\left[\begin{smallmatrix} 0 & 1 \\ 3 & 2 \end{smallmatrix}\right]\right) \leq 12.$$

(1)

TABLE 15.3.

The number of solutions for one puzzle of each orbit with solutions

x	$|orb(x)|$	Solutions	x	$|orb(x)|$	Solutions
$P\left(\left[\begin{smallmatrix} 0 & 0 \\ 1 & 0 \end{smallmatrix}\right]\right)$	32	1	$P\left(\left[\begin{smallmatrix} 0 & 1 \\ 2 & 0 \end{smallmatrix}\right]\right)$	32	3
$P\left(\left[\begin{smallmatrix} 0 & 0 \\ 1 & 1 \end{smallmatrix}\right]\right)$	16	2	$P\left(\left[\begin{smallmatrix} 0 & 1 \\ 3 & 0 \end{smallmatrix}\right]\right)$	16	4
$P\left(\left[\begin{smallmatrix} 0 & 1 \\ 1 & 0 \end{smallmatrix}\right]\right)$	8	8	$P\left(\left[\begin{smallmatrix} 0 & 1 \\ 3 & 2 \end{smallmatrix}\right]\right)$	8	12
$P\left(\left[\begin{smallmatrix} 0 & 2 \\ 2 & 0 \end{smallmatrix}\right]\right)$	4	8	$P\left(\left[\begin{smallmatrix} 0 & 1 \\ 2 & 3 \end{smallmatrix}\right]\right)$	16	10

Every Latin square of order four is the solution to at most one puzzle, and all puzzles in the same orbit have the same number of solutions. If we take the sum of the size of each orbit times the number of solutions to puzzles in that orbit, we must arrive at the total number of Latin squares:

$$32N\left(\left[\begin{smallmatrix} 0 & 0 \\ 0 & 1 \end{smallmatrix}\right]\right) + 16N\left(\left[\begin{smallmatrix} 0 & 0 \\ 1 & 1 \end{smallmatrix}\right]\right) + 8N\left(\left[\begin{smallmatrix} 0 & 1 \\ 1 & 0 \end{smallmatrix}\right]\right) + 4N\left(\left[\begin{smallmatrix} 0 & 2 \\ 2 & 0 \end{smallmatrix}\right]\right) +$$

$$32N\left(\left[\begin{smallmatrix} 0 & 1 \\ 2 & 0 \end{smallmatrix}\right]\right) + 16N\left(\left[\begin{smallmatrix} 0 & 1 \\ 3 & 0 \end{smallmatrix}\right]\right) + 8N\left(\left[\begin{smallmatrix} 0 & 1 \\ 3 & 2 \end{smallmatrix}\right]\right) + 16N\left(\left[\begin{smallmatrix} 0 & 1 \\ 2 & 3 \end{smallmatrix}\right]\right) = 576. \tag{2}$$

What is the largest that the left-hand side of equation (2) could be? Suppose that each of the inequalities shown in equation (1) is as large as possible. Then by adding everything together, we get

$$32(1) + 16(2) + 8(8) + 4(8) + 32(3) + 16(4) + 8(12) + 16(10) = 576.$$

We must achieve a total of 576, and the only way that we can get to 576 is if each of the upper bounds is tight. Therefore, each $N\left(\left[\begin{smallmatrix} a & b \\ c & d \end{smallmatrix}\right]\right)$ must actually achieve the upper bound attained from our parity graph calculations. We have found the exact value of each of our $N\left(\left[\begin{smallmatrix} a & b \\ c & d \end{smallmatrix}\right]\right)$.

All solvable puzzles fall into the orbit of one of these eight cases. Our results are compiled in Table 15.3.

Instead of continuing to use a fixed representative to describe each orbit, we can use the number of distinct clues together with the relative parity and position of those clues. These are all properties that happen to be invariant under our group action. This allows us to gather all of our results into a single classification theorem without making any references to orbit representatives or the machinery used to find these results.

Theorem 1 (A classification for $\mathbb{Z}_4$ puzzles). *Let P be a puzzle with n distinct clues.*

- *For $n = 1$, there are no solutions.*

- *Suppose that n = 2 and one clue appears three times. The puzzle has a unique solution if the repeated and unique clues are of different parity. Otherwise, there is no solution.*
- *Suppose that n = 2 and both clues appear twice. If any two adjacent pairs of clues are distinct, then the puzzle has eight solutions. Otherwise, there are no solutions if the clues have the same parity, and two solutions if they have different parity.*
- *If n = 3 there are no solutions if the repeated clues are adjacent. Otherwise, there are four solutions if the parity of the repeated clue differs from the other two, and three solutions if the repeated clue shares the same parity as one of the others.*
- *For n = 4 there are ten solutions if there exist adjacent clues of the same parity, and twelve solutions otherwise.*

5 Conclusions, Exercises, and Ideas for Further Study

Without solving a single puzzle, we used graphs to provide upper bounds for each orbit, and used the total number of Latin squares to show that these bounds are tight. This is certainly not the only way to arrive at our classification. By finding the best-possible lower bound for each orbit, either by brute force or other means, we could have used the same total to show those bounds are tight as well.

One way to get lower bounds for each orbit is by allowing the group to act not just on the set of all possible puzzles, but also on the set of all Latin squares over $\{0, 1, 2, 3\}$. The stabilizer of each puzzle will then map solutions of that puzzle to solutions for the same puzzle. The partitions that arise from considering the orbits of solutions under this action can be used to produce lower bounds without having to write out each individual solution.

We did not need to look outside the techniques of this chapter to find the number of Latin squares of order four, as parity graphs provide an excellent way of computing this and also reveal an additional fact. Exactly half of all squares decompose into a single cycle of each parity, and half have two cycles for each parity. Finding the total number of each type of graph, and noting that each cycle can be assigned values in two ways, provides a simple method for arriving at the number 576.

It is possible to extend some of our techniques to larger groups. In the five-by-five case, even with strict constraints on cage size and the same symmetry conditions we have here, we get a large collection of possible cage patterns. There are multiple patterns allowing for unique solutions over $\mathbb{Z}_5$, and we get very different classification results in these different cases.

In the six-by-six case, the number of possible cage patterns is huge unless we choose the absolute tightest of constraints. Though the results may be considered somewhat less interesting when each classification is only specific to one of a large list of equally valid cage patterns, working over $\mathbb{Z}_6$ has some benefits. Unlike with $\mathbb{Z}_5$, here we again have nontrivial subgroups for which we can ask which coset individual entries must fall into. Considering the possibilities for each cell modulo three gives us a structure similar to the parity maps for $\mathbb{Z}_4$, and we can form three-color coset graphs in the same fashion. We can also consider the values of entries modulo two to form a parity map, and considering individual rows and columns leads to a two-color parity hypergraph. Together, these give us many options for examining the solutions to any single puzzle.

There are still many puzzle variants left to examine in the four-by-four case. One can loosen or drop the symmetry conditions we insisted on, which greatly opens up the possibilities for cage patterns. Without these restrictions, the number of choices becomes large, though we still need some cages of odd size if we want puzzles with unique solutions. We can require rotational symmetry of 90° or 180° for the cage patterns; insist on reflective symmetry along a single vertical, horizontal, or diagonal axis; or require no such symmetry at all.

We can define a new type of puzzle, where the goal is to determine only the value of each entry modulo two, while still keeping the clues modulo four. Here we get multiple orbits with unique solutions, as $P\left(\left[\begin{smallmatrix} 0 & 2 \\ 2 & 0 \end{smallmatrix}\right]\right)$ can only be completed in one way. We can reduce things further by only providing the parity of each clue. This is equivalent to constructing a $\mathbb{Z}_2$ puzzle, where the goal is to fill in a square with 0s and 1s so that each number appears exactly twice in each row and column.

It is possible to drop the idea of groups entirely and consider puzzles over the ring $\mathbb{Z}_4$ under multiplication. Are unique solutions possible over the cage pattern we have here? What happens over other cage patterns? At this point, we are dealing with $\mathbb{Z}_4$ as a ring, so there is no reason not to mix additive and multiplicative clues. Alternately, we can choose to use neither. We can allow our operation to be any semigroup of order four, providing a large number of choices for us to work with [3]. This brings up one more question: How much structure do we need from an operation before unique solutions become impossible? If our binary operation is constant, all clues from cages with two or more entries become that constant. How many products need to differ from that constant in a semigroup before we get unique solutions over some cage pattern with cage sizes greater than one?

References

[1] B. Hayes. KenKen friendly numbers. Online at http://bit-player.org/2010/kenken -friendly-numbers. Last accessed June 2018.

[2] OEIS Foundation. Number of Latin squares of order n; or labeled quasigroups. Online at http://oeis.org/A002860. Last accessed June 2018.

[3] OEIS Foundation. Number of semigroups of order n, considered to be equivalent when they are isomorphic or anti-isomorphic (by reversal of the operator). Online at http://oeis.org/A001423. Last accessed June 2018.

[4] D. Nacin. On a complex KenKen problem. *College Math. J.* **48** no. 4 (2017) 274–282.

[5] P. M. Neumann. A lemma that is not Burnside's. *Mathematical Scientist* **4** no. 2 (1979) 133–141.

PART IV

◇◇◇◇◇◇◇◇◇◇◇◇◇◇◇◇◇◇◇◇◇◇◇◇◇◇◇◇◇◇◇◇◇◇◇◇◇

Geometry and Topology

16

SHOULD WE CALL THEM FLEXA-BANDS?

Yossi Elran and Ann Schwartz

Flexagons—flexible polygons—are folded paper toys first discovered in 1939. One of their most fascinating characteristics is their surprising depth. Although folded into a flat polygon, a flexagon can be manipulated to reveal new surfaces. A hexagon that is red on one side and blue on the other can become one that is blue on one side and yellow on the other.

A half-twisted band is produced by half twisting a strip of paper or similar material n times and joining together its ends. The most famous such band is the Möbius strip. Half-twisted bands have intriguing properties. In particular, when cut lengthwise down the middle, many surprising new topologies are formed; for example, they may have one side or two, and they may even be knotted. These objects are sometimes referred to as paradromic bands, Möbius strips, Möbius bands, or Möbius-like strips or bands. There is some ambiguity in the use of these words, so we will refer to them as *half-twisted bands*.

Our aim in this chapter is to acquaint the reader with some basic half-twisted bands and flexagons, restricting our discussion to those made out of straight strips of paper. We show that flexagons and half-twisted bands share some interesting properties and suggest a notation that captures these similarities. The chapter is organized as follows. In the first section, we discuss half-twisted and knotted bands, and present our suggested notation. In the second section, we discuss flexagons in general, and in the third section, we show that flexagons and half-twisted bands are one and the same, and present three examples of straight-strip flexagons and their half-twisted band twins.

1 Half-Twisted Bands

1.1 Odds and Evens

A strip of paper has two sides, front and back. Connecting its ends produces a band. The band also has two sides, an inside and an outside, and no creases or twists. Cutting lengthwise along the strip's center line creates two new bands,

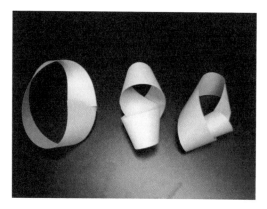

Figure 16.1. The 1-, 2-, and 3-bands.

each half the width of the original. Making a half-twist before joining the ends of a strip generates the famous *Möbius band*, also called the *Möbius strip*, which is one-sided and cannot be oriented. Other one-sided bands can be obtained by making a greater *odd* number of half-twists. Two-sided bands, like the trivial untwisted one, can be obtained by making an *even* number of half-twists in the strip before joining the ends. We will call the number of half-twists the *order* of the band. For example, a 3-half-twisted band has three half-twists. From here on, we omit the words "half-twists" when referring to half-twisted bands, unless necessary for clarification purposes. Therefore, the 3-band means a band with three half-twists in it. Figure 16.1 shows the 1-, 2-, and 3-bands.

1.2 Knotted Bands

If we want to make three or more half-twists in a strip of paper, there is another way to do this—by tying a knot or knots! For example, the 3-band can be made either in the normal way (by bringing the two ends of a strip together as if making an untwisted band and then making three half-twists at one of the ends) or by tying an overhand knot in the strip. The two resulting 3-bands are not the same. It is important to note that we count half-twists on the strip after it has been closed as a band and smoothed out, otherwise ambiguity would arise, for example when one half-twist is twisted clockwise and another anticlockwise. The number of distinguishable *n*-bands of the same order grows as *n* increases. When *n* is greater than 3, we can make *n*-bands that are both half-twisted and knotted. That is, the *n*-bands contain additional half-twists that are not part of the knots. Strictly speaking, twisted, knotted bands are neither Möbius bands nor knotted surfaces. For our purposes, we suggest refining the band classification to include knotted bands by adding "knot numbers," a classification used in knot theory. It is important to understand, though, that we will be applying the notion of a knot from knots in strings (which is what mathematicians typically mean by "knot") to knots in surfaces. The distinction is important, because strings have neither sides nor edges, whereas paper strips ("surfaces") have both. Still, once we know the terminology for string knots, we can use the same notation for our paper-strip knots. For readers interested in other mathematical notations in this context, used, for example, to discuss twisting in DNA strands, we refer them to Banchoff and White [2].

1.3 Knot Numbers: A Very Brief Primer

A mathematical knot is an abstraction of the knots we create using string [1]. Mathematical knots are always drawn as closed loops. That is, the open ends of the knotted string are joined together. The simplest possible knot is the unknot, a closed loop of string without a knot in it. A simple, closed loop of string can be tangled and twisted so that it looks quite different, but as long as the string is not cut open, knotted and joined again, it is still just an unknot. The simple and the tangled unknot are equivalent.

Figure 16.2 shows two illustrations of the next simplest knot, the trefoil knot. These illustrations are called *knot diagrams*. In the diagrams, we see some junctions where the string crosses over itself. These junctions are called *crossings*. Notice that the number of crossings is different in the two diagrams. The image on the right in Figure 16.2 has three crossings, while the one on the left has six. Why is this? Recall that any knot in real life can be manipulated by twisting or tangling, or even just by placing part of the string above or below another. Although these manipulations do not change the true essence of the knot, since the tangled version can be untangled to its simplest state, it does create a potentially infinite number of knot diagrams. However, there is always a simplest representation, with a minimal number of crossings (called the knot's *crossing number*). The unknot has a crossing number of 0. The trefoil knot's crossing number is 3, meaning that however much you try, you will not be able to untangle the knot so that it has less than three crossings. Hence, the image on the right in Figure 16.2 shows the simplest representation of the trefoil knot, while that on the left depicts a tangled version.

Note that two different knot diagrams can also be drawn if we project a trefoil knot from different perspectives. The fact that the same knot has different diagrams is the cause of much anguish among mathematicians, who want a reliable classification scheme for knots. To make things worse, two (or more) knots can be joined together by cutting both knots and joining the pairs of ends, resulting in what is called a *composite knot*. This resembles the multiplication

Figure 16.2. The trefoil knot from two different perspectives.

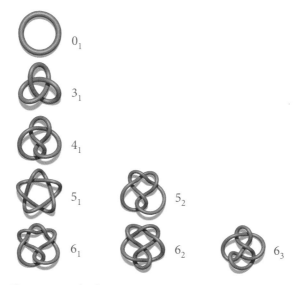

0_1

3_1

4_1

5_1 5_2

6_1 6_2 6_3

Figure 16.3. The first eight entries in the Rolfsen knot table.

of prime numbers to get composite numbers. There are an infinite number of knots, each of which can have an infinite number of knot diagrams; hence mathematicians proposed some classification schemes. The most common notation is from a table known as the Rolfsen or Alexander-Briggs-Rolfsen knot table (Figure 16.3) [11]. The Rolfsen table lists only *prime knots*, which are those that cannot be decomposed into simpler knots. Each prime knot is listed in Rolfsen's

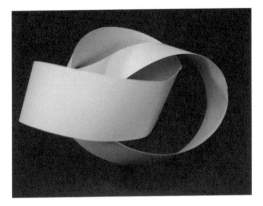

Figure 16.4. The 3_1-3-trefoil band.

table using only its (minimal) crossing number knot diagram. It is assigned a *knot number*, which is a label composed of a number with a subscript. The number is the knot's crossing number (each row in the table represents a different crossing number). The subscript, which was assigned arbitrarily to each knot, is the knot's order in the row (its column number). For example, the knot number of the unknot is 0_1 and the knot number of the trefoil knot is 3_1. We will be using Rolfsen's knot number in our notation for twisted and knotted bands.

1.4 Twisted and Knotted Band Notation

We can now suggest a notation for bands that are half-twisted and that may or may not be knotted. Bands are prefixed using two labels. The first is the standard knot number according to Rolfsen's table, and the second is the total number of half-twists. For example, the unknotted 3-band is denoted by 0_1-3, and its trefoil isomer is denoted by 3_1-3. The trefoil band 3_1-3 is shown in Figure 16.4. Adding more half-twists to a paper strip folded into a trefoil knot will result in a band with a trefoil knot and some additional half-twists. The notation for this band will be 3_1-n, where n is the total number of half-twists (including the three that make up the trefoil knot). Note that the placement of half-twists along the knotted band is easily manipulated by hand and has no effect on the properties discussed in this chapter. This notation is by no means complete, but it is sufficient for the purpose of this chapter. Work on further generalization of this categorization scheme to include bands with links or composite knots, and enantiomorphs is in progress.

Our use of knot notation for surfaces rather than strings is not unreasonable. First and foremost, we are explicitly tying knots and making bands. Second, the boundaries of the surface can be viewed as closed "stringy" knots, and we can mentally shrink our paper bands to its boundaries. Indeed, in topology, some surfaces (e.g., Seifert surfaces) are defined through their knotted or looped boundaries.

1.5 Tricks and Knots

Half-twisted bands have many fascinating characteristics. The following properties are fun to demonstrate and are frequently used by magicians. Cut an (unknotted) 0_1-n-band through its centerline, and the following happens according to the parity of n:

- If n is odd, there will be a knotted band with a total of $2n + 2$ half-twists. For instance, a 0_1-3-band produces the 3_1-8-band.
- If n is even, there will be two bands linked together $n/2$ times, each of which has n half-twists. For example, a 0_1-2-band will create two 0_1-2-bands linked together.

The topology that we get from bisecting a half-twisted band along its centerline can be guessed even before cutting, by observing the boundary of the band. For example, the boundary of the Möbius strip is an unknot after bisection. Similarly, the boundary of the 0_1-2-band is the Hopf link; the 0_1-3-band is the trefoil knot; the 0_1-4-band is Solomon's knot; and so on, in one-to-one correspondence with the topology of the same bands cut through their centers.

2 Flexagons

2.1 Straight-Strip Flexagons

A straight strip of paper has two visible surfaces, front and back. Surprisingly, there is a way to fold a strip into a two-dimensional shape that has more than two visible surfaces. These shapes are known as flexagons [6]. Always popular with children, they have seemingly magical properties [8]. Flexagons can be made from many different paper-strip geometries [9]. An excellent source for learning about flexagons is Scott Sherman's website [13]. In this chapter, we confine our discussion to flexagons made from straight strips, just like the ones used to make half-twisted bands. The strips are methodologically folded along a series of straight lines drawn on the paper, according to a given algorithm, and the ends of the strip are joined. This results in a flat, folded, polygon-shaped structure, visibly divided into equal-area sections: stacks of folded layers of paper. In flexagon terminology, the stacks are called *pats* and the layers are called *leaves*. The magic happens when the user manipulates ("flexes") the flexagon, and hidden surfaces suddenly appear. In flexagon terminology, these surfaces are called *faces*.

There are many types of flexagons, as well as different kinds of flexes. A top-down definition of these flexagons can be found in Pook [10]. Simple flexagons usually have only one or two flexes, but more are being discovered all the time.

Sherman and Schwartz in particular have discovered quite a few flexes and many new, exciting, and intriguing flexagons [12].

Straight-strip flexagons have states and modes. Loosely speaking, a *state* is a combination of the shape of a face, its pat structure, and the ways it can be flexed. A *mode* produces a repeating series of faces. Most flexes iterate the patterns and the shapes in a given mode, but many flexagons have a special flex whose only function is to switch from one mode to another.

2.2 Octa-, Hexa-, Tetra-, and 8-Flexagons

There is still no consistent nomenclature for flexagons of any kind, although many systems have been suggested by various authors. We will follow one that removes an ambiguity. In their seminal work *A Mathematical Tapestry* [5], Hilton and Pedersen categorized regular flexagons by using one or two numbers preceding the word "flexagon." The first—the numeral directly to the left of "flexagon"—represents the number of triangles on the surface of the folded polygon. Therefore, regular 8-flexagons, known in the literature as both octaflexagons or tetraflexagons, are regular polygons with eight triangles per face. Previous systems use the prefixes "octa-," "tetra-," and "hexa-" to stand for the number of the sides of the polygon, and since there are many square and rectangular tetraflexagons not composed of triangles, a regular "8-flexagon" can now specifically refer to the triangle type. Regular 6-flexagons, known in the literature as hexaflexagons, are regular hexagons with six triangles per face.

Hilton and Pedersen added a second number to represent the number of faces. Therefore, the regular 3-6-flexagon is the well-known tri-hexaflexagon, a hexagon with three faces, each made of six triangles. Figure 16.5 shows the tri-hexaflexagon. For straight-strip flexagon templates, like the ones used in this chapter, we redefine this second number as $(n - x)/f$, where n is the number of triangles on both sides of the strip, x is the number of blank triangles at the end of both sides (the triangles that are fastened face-to-face to complete the flexagon, usually 2), and f is the number of triangles on the face of the flexagon.

3 Regular Straight-Strip Flexagons and Their Alter Egos

3.1 Is a Straight-Strip Flexagon a Half-Twisted Band, or Is a Half-Twisted Band a Flexagon?

It might seem that flexagons are exact opposites of half-twisted bands. After all, flexagons have three or more faces, while the bands have less than three sides. But in fact, half-twisted bands and flexagons share the same topology [3].

For example, a 3-6-flexagon is a flattened band with three half-twists. In general, a flexagon with an odd number of faces is topologically equivalent

Figure 16.5. The 3-6-flexagon, slightly spread out. The arrows show three pockets, which make up the three half-twists.

to a band with the same number of half-twists. Similarly, a flexagon with an even number of faces is topologically equivalent to a corresponding band. The number of half-twists in the "band twin" of a flexagon with N faces and n triangles per face is given by Conrad and Hartline [3] as $3N - n$. This implies an important insight: Flexagons constructed by making an even number of half-twists—we call these *even flexagons*—have two independent flexagon cycles, and odd flexagons have only one cycle. We now show how this equivalence is manifested in some classic flexagons and discuss the role of the different flexes. We examine three straight-strip regular flexagons and their band partners.

3.2 The 3-6-Flexagon (Tri-Hexaflexagon)

The strip for this well-known flexagon is shown in Figure 16.6. It is divided into ten equilateral triangles on each side. The band is then folded into a hexagon composed of six triangles by folding the strip along the far side of every third triangle and then taping the two end triangles face to face. Using the formula that relates the number of triangles on one side of the strip to the number of faces—$(20 - 2)/6 = 3$—we see that this is indeed a three-faced flexagon.

Each face is made of six triangles. Two faces are readily apparent: the one on top and the one on the reverse side. A third face emerges only when the

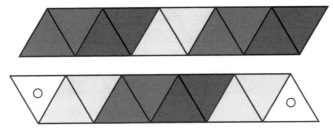

Figure 16.6. Folding a 3-6-flexagon. Shown are the two sides of the strip (rotated about a horizontal axis). To assemble, fold each yellow triangle onto its adjacent "twin." A hexagon will form with red on one side and blue on the other. The white triangles will be positioned face to face and can be fixed with adhesive.

flexagon is flexed. If each face has a unique color, a simple flex reveals all three faces one after the other in a continuous cycle. Hilton and Pedersen [5] call this the *straight flex*, but we use the more descriptive term *pinch flex*. A description of how to perform the pinch flex can be found in Gardner [4, p. 4]. In addition, all the triangles for each face appear at the same time. For instance, the six triangles that make up a blue face stay together, although they reorient themselves relative to one another after a flex.

The 3-6-flexagon has only one pat structure, 1-2: 1-leaf pats alternate with 2-leaf pats regardless of which faces are visible. It also has only one mode. Since there are three 2-leaf pats, a manifestation of the strip's having three half-twists, the flexagon is "odd": the strip has one side (as opposed to three faces) and is topologically equivalent with the 0_1-3-band.

Cutting a 0_1-3-band through its center creates a knotted 3_1-8-band. Making the same cut with the 3-6-flexagon produces the same result. Of course, it is physically difficult to cut through the flexagon. Layers must be pierced one at a time, an awkward move with pats of two layers. But the 3-6-flexagon is really just a flattened 0_1-3-band stretched toward a common center point.

3.3 The 6-6-Flexagon (Hexa-Hexaflexagon)

The strip for this flexagon is shown in Figure 16.7.

Also constructed of equilateral triangles, the 6-6-flexagon is folded in two steps. First, the strip is wound into a spiral: There are triangles on its outside and ones hidden within. The spiral then is folded into a hexagon, with triangles from the outer part of the spiral forming the first visible faces. Together with those from a third face, these faces are made from triangles on the outer spiral. To the frustration of many new to *flexigation*, three additional faces appear one-third as often than those from the first group. Interestingly, all the triangles from the "more friendly" faces appear on one side of the strip, while all the triangles

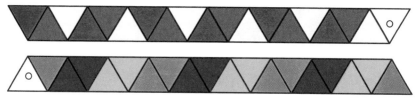

Figure 16.7. Folding a 6-6-flexagon. Shown are the two sides of the strip (rotated about a horizontal axis). To assemble, fold each purple, green, and orange triangle onto its adjacent twin. Then do the same with the now-adjacent pairs of yellow triangles. Position the remaining orange triangles on top of one another. The now-adjacent white triangles can be folded face to face and fixed with adhesive.

from the "more shy" faces are on the reverse—as well as inside the spiral. There are two pat structures: 2-4 and 1-5.

Except after special manipulations, all the triangles for each face in the 6-6-flexagon always appear at the same time. For instance, if each face has its own unique color, the six triangles that make up a red face stay together, although as with the 3-6-flexagon, they reorient themselves after a flex. There is only one way to traverse the 6-6 flexagon to expose all its faces: the pinch flex. Pinch flexing at the same corner until the flexagon can no longer open and then moving to an adjacent corner is known as the Tuckerman traverse. It is the algorithm that is the quickest way to expose all the faces. A description of the Tuckerman traverse, as well as Tuckerman diagrams, appears in Gardner [4, p. 5].

An additional manipulation is the complicated *V-flex*, discovered by T. Bruce McLean [7]. This does not traverse the unicolored faces of the flexagon. Instead, it makes triangles from different faces mingle. Once the V-flex is made, the whole flexagon looks as if it were a completely new hexaflexagon. It has bicolored faces, and the pinch flex only iterates the same three (bicolored) faces, even when we flex from different corners. We regard this flexagon as a *mode* of the original flexagon. Typically, flexagons can have different modes. The mode associated with the flexagon when it is first constructed is called the *normal mode*. In the 6-6-flexagon, the normal mode consists of the set of unicolored faces. The V-flex switches it into another mode that we call the *V-mode*. We can now see the important role of flexes: Each has its own topological function. The pinch flex iterates the different states in each mode, while the V-flex switches the mode of the flexagon. Executed along one diagonal of the hexagon and done successively enough times, the V-flex can even turn the inner spiral completely inside out. Once this happens, the whole flexagon is inverted, as if the "more shy" faces were folded topmost, and behaves exactly like the normal-mode flexagon. The Tuckerman diagrams of the different modes of a flexagon—the diagrams that show the order in which the faces are brought into view—do not

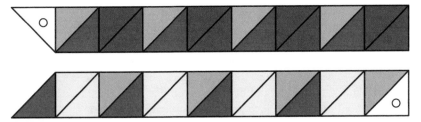

Figure 16.8. Folding a 4-8-flexagon. Shown are the two sides of the strip (rotated about a horizontal axis). To assemble, fold each yellow triangle on top of its adjacent twin. Then fold the pairs of green triangles in the same fashion. A square will form with red on one side and blue on the other. Position the white triangles face to face and fix with adhesive.

necessarily have the same structure. In fact, the V-mode diagram at one stage has the same structure as the one for the 6-3-flexagon.

The 6-6-flexagon is even, with twelve half-twists, and is topologically equivalent to the 0_1-12-band. It therefore has two sides, identified as the inside and outside of the spiral, that are completely different. This explains *why* the V-flex can completely invert the strip, and when this is done, why the resulting mode has the same Tuckerman diagram as the normal mode flexagon.

3.4 The 4-8-Flexagon (Octa- or Tetraflexagon)

The strip for this flexagon is shown in Figure 16.8. The 4-8-flexagon is folded from a strip of eight squares on each side, with each square divided into two right isosceles triangles. It is constructed by making four half-twists on one side along the hypotenuses of the strip's second, sixth, tenth, and fourteenth triangles, and then wrapping the strip around those edges to form a square. Here, four pairs of triangles adjacent along their hypotenuses divide the face into four equal smaller squares. Inspection of the flexagon shows it has eight triangles on each face in a 3-1 pat structure.

There are two readily apparent ways to flex the 8-flexagon. One can pinch flex it along creases running from the midpoints of the sides of the square through the center. This is also called "straight flexing." Alternatively, one can perform the "pass-through flex." Described in *A Mathematical Tapestry* [5], this flex is executed by pinching along the four half-diagonals of the square. The pats then do what looks like a little somersault—and assuming that the model has unicolored faces—showing a glimpse of one color, the pass-through color, before opening flat to another. The straight and the pass-through flexes should really be considered the same pinch flex. The "pinching from the corners" maneuver is the same, but the pass-through flex takes you through an extra intermediate position. Together, they complete the flexes needed to traverse the normal-mode flexagon. According to the $(n - x)/f$ formula, this flexagon should have four faces; however, the pinch and pass-through flexes reveal only three: the ones that appear in the normal mode.

A special "reverse pass-through" flex, also described in *A Mathematical Tapestry*, switches the cycle to the reverse pass-through mode. Analogous to inverting the strip (folding up the flexagon so that the hidden face is visible), it reveals the fourth face. Pinch and pass-through flexing then show only two of the original colors. In each of these two modes, normal and reverse pass-through, only three of the four possible faces appear, correlating to the fact that the original strips have triangles of only three colors on each side. Two of the colors are common to both sides of the strip (and modes), and one color is unique to each.

This flexagon is topologically equivalent to the 0_1-8-band. Since the flexagon is topologically equivalent with an even band, it has two sides, explaining the two modes and the hidden faces.

Note that along with new flexes, the 4-8 flexagon introduces other interesting features. The two visible faces of the flexagon—the top and the reverse—are never identical, although the pat structure is always the same: One side shows eight open folds radiating from the center of the square, while the reverse shows only four. And, curiously, the flexagon must be flipped over to see all the colors in a planar state. Regardless of the mode, one face on each side does not flex out flat. The 4-8-flexagon can also be manipulated into two intermediate shapes that do not flex: a regular hexagon and a rectangle.

4 Summary

Flexagons and half-twisted bands are wonderful objects full of surprises. The flexagon/band connection provides valuable insight into both flexagons and half-twisted bands, and it suggests a broader view of paper-band manipulation. In particular, information about the roles of the different flexes, the number of sides of a flexagon, the existence of new modes, and other properties can be obtained and better understood by examining flexagon/band twins.

In this chapter, we have introduced a classification scheme for half-twisted bands that is akin to the flexagon classification. We then shifted our focus to the connections between bands and 6- and 8-straight-strip flexagons. In general, even-numbered half-twisted bands have a flex or series of flexes that invert its corresponding flexagon. In particular, the V-flex switches the mode of the 6-6-flexagon, eventually turning it inside out; and the reverse pass-through flex inverts the 8-flexagons. Further work is being done to explore the flexagon/band connection for irregular straight-strip bands, which will be followed by work on other flexagon nets. A paper on half-twisted band and paradromic ring notation is in preparation. We are also looking at the possibility of creating twin flexagons from knotted bands. Work along these lines is already in progress.

References

[1] C. Adams. *The Knot Book*. American Mathematical Society, Providence, RI, 2004.

[2] T. F. Banchoff and J. H. White. The behavior of the total twist and self-linking number of a closed space curve under inversions. *Math. Scand.* **36** (1975) 254—262.

[3] A. S. Conrad and D. K. Hartline. Flexagons. Online at http://delta.cs.cinvestav.mx /mcintosh/comun/flexagon/flexagon.html 1962. Last accessed June 2018.

[4] M. Gardner. *Hexaflexagons and Other Mathematical Diversions*. University of Chicago Press, Chicago, 1988.

[5] P. Hilton and J. Pedersen. *A Mathematical Tapestry*. Cambridge University Press, New York, 2012.

[6] P. Jackson. *Flexagons*. British Origami Society, Stratford-upon-Avon, 2013.

[7] T. B. McLean. V-flexing the hexahexaflexagon. *Amer. Math. Monthly* **86** (1979) 457–466.

[8] D. Mitchell. *The Magic of Flexagons*. Tarquin Publications, St. Albans, England, 2008.

[9] L. Pook. *Flexagons Inside Out*. Cambridge University Press, Cambridge, 2003.

[10] L. Pook. *Serious Fun with Flexagons*. Springer, New York, 2009.

[11] D. Rolfsen. *Knots and Links*. AMS Chelsea Publishing, Providence, RI, 2003.

[12] A. Schwartz and J. Rutzky. The hexa-dodeca-flexagon. In *Homage to a Pied Puzzler*, E. Pegg, A. Schoen, and T. Rodgers, eds. A. K. Peters, Wellesley, MA, 2009, 257–268.

[13] S. Sherman. *Flexagons*. Online at http://loki3.com/flex/. Last accessed June 2018.

17

THE SHORTEST CONNECTION PROBLEM
ON TRIANGULAR GRIDS

Jie Mei and Edmund A. Lamagna

The shortest connection network problem studied in this chapter originates from the board game TransAmerica (Figure 17.1). In the game, each player attempts to connect five cities in different regions of the country with as few train tracks as possible. The map on which the players build their railroads is a triangular grid. Since each player can place only two tracks per round, it is advantageous to know the shortest path connecting one's five cities to increase the chances of winning.

On a similar note, Figure 17.2 shows the triangular glass panels at the Dalí Museum in Florida. If lights are installed at the junctions of some panels, their electrical wires need to crawl along the metal frame, which is a triangular grid. How do we design a wire network to minimize the materials used?

We describe our shortest connection problem as follows. Given a set of points on a "regular" triangular grid, with unit length edges, find a path of smallest total length that connects these points. We call such grids simply *triangular grids,* since they are the only type considered here. The points to be connected are referred to as *terminal points.* The paths lie along the grid lines, and additional junction points may be introduced into the network. These junction points are called *Steiner points,* and the resulting connected network is called a *Steiner tree.* A Steiner minimum tree (SMT) has the shortest length among all Steiner trees (Figure 17.3). A vertex in a Steiner tree can be either a terminal point or a Steiner point.

On the Euclidean plane, Melzak [1] first proposed an algorithm to solve the Euclidean Steiner problem. The algorithm constructs Steiner trees for all possible topologies and selects the shortest one to be the SMT. Melzak's algorithm is a brute-force approach and takes exponential time because the number of topologies is large. In fact, the Euclidean Steiner problem was proven to be NP-hard by Garey et al. [2] and Rubinstein et al. [3].

Another geometry where the Steiner problem has been widely studied is the rectilinear grid. This version is important because of its application to

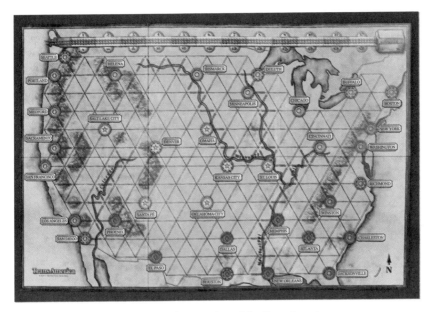

Figure 17.1. The game board for TransAmerica.

Figure 17.2. Triangular glass shell at the Dalí Museum in St. Petersburg, Florida.

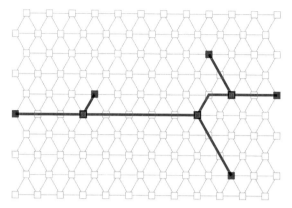

Figure 17.3. A Steiner minimum tree for five terminal points (black). The red points are Steiner points.

integrated-circuit routing design. The problem is also known as the Manhattan distance Steiner problem. In 1966, Hanan [4] showed that the Steiner points can be chosen from a predetermined set of points, and a rectilinear SMT can therefore be computed by an exhaustive search. The rectilinear Steiner tree (RST) problem was shown to be NP-hard by Garey and Johnson in 1977 [5]. They demonstrated that the vertex cover problem, which is known to be NP-complete, can be transformed in polynomial time to the RST problem.

In the Euclidean Steiner tree problem, the edges of a full Steiner minimum tree always meet at 120° angles. Consequently, it seems natural to study the Euclidean SMT using a coordinate system with axes 120° apart instead of the usual Cartesian system. Hwang and Weng [6] develop a calculus for hexagonal coordinates and show how it can be applied to the Euclidean SMT.

λ-geometry provides a general framework for studying Steiner tree problems in a variety of different geometries. In the λ-geometry plane, the feasible orientations of lines connecting two points are fixed and must lie in one of λ possible orientations. Only lines that form angles $k\pi\lambda$ with the positive x-axis for integers $0 \le k < \lambda$ are allowed. We have $\lambda = 2$ for rectilinear coordinates; $\lambda = 3$ for both hexagonal coordinates and the triangular grid, accounting for some of the similarities between them; and $\lambda = \infty$ in the Euclidean plane. The section on λ-geometry and the hexagonal plane in Hwang et al. [7] provides an introduction to Steiner trees in these metric spaces.

To the best of our knowledge, there has been far less research on the Steiner tree problem for a triangular grid, the board geometry in the game of TransAmerica. There are interesting differences between locating Steiner points on a triangular grid and the other geometries mentioned. For example, when connecting three terminal vertices on a triangular grid, there may be multiple choices for the Steiner point, whereas only one choice exists for the Euclidean and rectangular problems. In this chapter, we provide a comprehensive study of the triangular Steiner problem and our approach to its solution.

1 The Triangular Coordinate System

1.1 Triangular Coordinates

A coordinate system based on triangular grid lines is used as the reference system in this research.

Let O be the origin. The axes, x_1, x_2, x_3, are the three directions along the triangular grid lines at O. x_1, x_2, x_3 are $120°$ apart. Each axis separates the plane into a positive half and a negative half. A counterclockwise rotation is defined to be the positive direction. The half plane initially scanned by rotating an axis counterclockwise is the positive half; the other half is the negative half plane (Figure 17.4).

Parallel grid lines are one unit length apart in all three directions. For any point P, let p_1, p_2, p_3 be the distances to the respective axes. Depending on whether P sits on the positive or the negative half plane of x_1, its x_1 coordinate is either $+p_1$ or $-p_1$. The same applies to the other two coordinates of P. For example, in Figure 17.5, the coordinates of P are $(+p_1, -p_2, +p_3) = (3, -5, 2)$. Note that the coordinates of the points and distances discussed throughout this chapter are constrained to be integers. Though only discretized problems are considered here, almost all definitions and theorems also apply to continuous coordinates due to the linearity of the metric system.

We next list some of the basic properties of the triangular coordinate system, providing proofs for those that are not obvious.

Lemma 1.   *The three coordinates of any point add to 0 (constant).*

Proof. Consider a point $P(p_1, p_2, p_3)$ in the triangular coordinate system. Let the polar coordinates of P be (r, θ). Then

$$p_1 = r\sin\theta, \; p_2 = r\sin(\theta - \frac{2\pi}{3}), \; p_3 = r\sin\left(\theta - \frac{4\pi}{3}\right),$$

$$p_1 + p_2 + p_3 = r\sin\theta + r\sin(\theta - \frac{2\pi}{3}) + r\sin(\theta - \frac{4\pi}{3}) = 0. \qquad \square$$

Definition 1 (Distance).   *Consider two points, $A(a_1, a_2, a_3)$ and $B(b_1, b_2, b_3)$. The distance from A to B is the fewest number of steps taken walking from A to B along grid lines.*

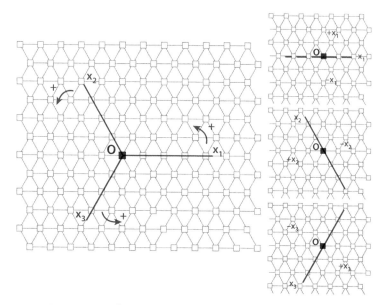

Figure 17.4. The triangular coordinate system and half planes.

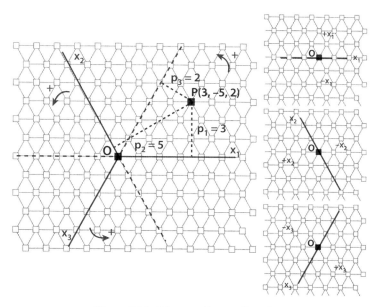

Figure 17.5. The coordinates of a point.

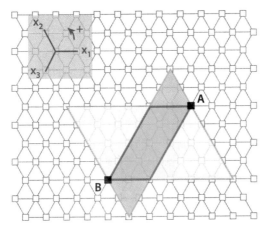

Figure 17.6. Three parallelograms enclosed by the axis lines passing through A and B. The one in the red frame is the distance parallelogram of A and B.

The axis lines passing through A and B enclose three parallelograms having A and B as diagonally opposite vertices. In one of these parallelograms, the angles at both A and B are $60°$. This is the *distance parallelogram* of A and B. The distance between A and B is the sum of the lengths of any two adjacent edges of their distance parallelogram (Figure 17.6).

Lemma 2. *Let* $d_1 = |a_1 - b_1|$, $d_2 = |a_2 - b_2|$, *and* $d_3 = |a_3 - b_3|$. *The distance between A and B is* $\min\{d_1 + d_2, \, d_1 + d_3, \, d_2 + d_3\}$. *It is also* $\max\{d_1, \, d_2, \, d_3\}$, *and equal to* $\dfrac{1}{2}(d_1 + d_2 + d_3)$.

Proof. Without loss of generality, suppose $d_3 = \max\{d_1, d_2, d_3\}$ and $b_3 > a_3$. Because $a_1 + a_2 + a_3 = b_1 + b_2 + b_3$, $(a_1 - b_1) + (a_2 - b_2) = b_3 - a_3 = d_3$, we must have $a_1 - b_1 \geq 0$ (otherwise, $d_2 \geq a_2 - b_2 = d_3 + (b_1 - a_1) > d_3$, which contradicts our assumption). Similarly, $a_2 - b_2 \geq 0$. Thus we have $d_1 + d_2 = d_3$, and $d_1 + d_2 + d_3 = 2d_3$. Therefore, $\min\{d_1 + d_2, d_1 + d_3, d_2 + d_3\} = d_1 + d_2 = d_3 = \frac{1}{2}(d_1 + d_2 + d_3)$ is the smallest sum of any two d_i. □

An observation from studying Figure 17.6 is that all walks from A to B in the distance parallelogram, with directions restricted to being parallel to two edges of the parallelogram, have the same total length.

Definition 2 (Circle). *Given a point P and length l, the set of all points at distance l from P forms a regular hexagon. We call these points the* circle *with center P and radius l on the triangular grid (Figure 17.7).*

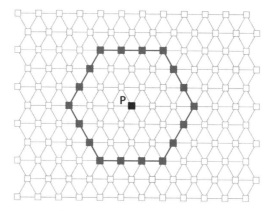

Figure 17.7. A circle on a triangular grid. P is the center; the radius is 3.

Consider a point $P(p_1, p_2, p_3)$. Draw lines through P that are parallel to the three axis lines. These lines divide the plane into three *sectors* around P. Call sector I the infinite region between directions x_2 and x_3, sector II the region between x_1 and x_3, and sector III the region between x_1 and x_2 (Figure 17.8a).

Alternatively, the distance between two points can be considered using the notion of sectors. Consider another point $Q(q_1, q_2, q_3)$ on the grid. If Q is in sector I of P, the distance $|PQ|$ is the difference of their first coordinates: $|PQ| = |p_1 - q_1|$. A similar result holds for points in sectors II and III. To understand this definition visually on the grid, depending on which sector of P the point Q is located, draw lines parallel to the corresponding axis through P and Q. $|PQ|$ is the distance between these parallel lines (Figure 17.8b).

1.2 Equilateral Triangles Aligned to the Grid

Consider three vertices that form an equilateral triangle of side length a with edges along grid lines. Let the vertices be $A(0, 0, 0)$, $B(0, -a, a)$, and $C(a, -a, 0)$ (Figure 17.9).

Lemma 3. *For any point in or on the equilateral triangle, the sum of the distances to each of the three edges adds to a.*

Proof. Let $P(p_1, p_2, p_3)$ be a point in or on $\triangle ABC$. The line equations of the three sides are $x_1 = 0$, $x_2 = -a$ and $x_3 = 0$. Thus, we have $0 \leq p_1 \leq a$, $-a \leq p_2 \leq 0$, $0 \leq p_3 \leq a$. The distance to edge AB ($x_1 = 0$) is p_1; the distance to edge BC ($x_2 = -a$) is $p_2 - (-a) = p_2 + a$; the distance to edge AC ($x_3 = 0$) is p_3. So the sum of the distances is $p_1 + p_2 + a + p_3 = a$ (Figure 17.9). ☐

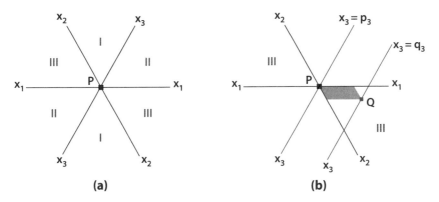

Figure 17.8. (a) Sectors of a vertex P on a triangular grid. (b) Distance between two points based on sectors. Q is in sector III of P. $|PQ|$ is the distance between lines $x_3 = p_3$ and $x_3 = q_3$. The gray shaded parallelogram is the distance parallelogram of P and Q.

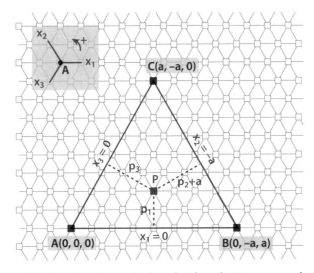

Figure 17.9. Equilateral triangle aligned with grids. P is an internal point.

Lemma 4. *The sum of the distances from any point P in or on the equilateral triangle to the three vertices is 2a (constant).*

Proof.

$$|PA| = \frac{1}{2}(|p_1| + |p_2| + |p_3|) = \frac{1}{2}(p_1 - p_2 + p_3),$$

$$|PB| = \frac{1}{2}(|p_1| + |p_2 + a| + |p_3 - a|) = \frac{1}{2}(p_1 + (p_2 + a) + (a - p_3)),$$

$$|PC| = \frac{1}{2}(|p_1 - a| + |p_2 + a| + |p_3|) = \frac{1}{2}((a - p_1) + (p_2 + a) + p_3),$$

$$|PA| + |PB| + |PC| = 2a. \qquad \square$$

Lemmas 3 and 4 are useful for finding solutions to the triangular Steiner problem for three terminal vertices.

1.3 Steiner Trees on Triangular Grids

We provide some important properties of Steiner trees and Steiner minimum trees (SMTs).

Lemma 5. *For n terminal vertices on a triangular grid, there are at most $n - 2$ Steiner points in the SMT.*

Proof. Let s be the total number of Steiner points and $d_1, d_2, \ldots, d_s$ be the degrees of every Steiner point in the Steiner tree.

The degree of any Steiner point must be greater than 2, that is, $d_i \geq 3$ for any $d_i \in \{d_1, d_2, \ldots, d_s\}$. Otherwise, if a Steiner point had degree 2, it could simply be eliminated. So there is no need to consider cases where a Steiner point connects only two vertices.

Let e be the number of edges in the SMT. Because the sum of the degrees of all vertices in any graph is twice the number of edges,

$$2e = d_1 + d_2 + \cdots + d_s + n = \sum_i^s d_i + n,$$

and for a tree,

$$e = s + n - 1$$

$$\implies \quad 2e = 2(s+n-1) = \sum_i^s d_i + n \geq 3s + n.$$

Thus

$$n - 2 \geq s. \qquad \square$$

Lemma 5 holds not only for triangular grids, but is also shared by the Euclidean SMT, the rectilinear SMT, or that for any other metric space. A Steiner tree that has $n-2$ Steiner points is called a *full Steiner tree*. In a full Steiner tree, the terminals and Steiner points are all distinct; and a terminal point has degree 1. The degree of any Steiner point in an SMT is at most 4.

2 Exact Algorithms

2.1 Three Terminal Vertices and the Median Triangle

Definition 3 (Median triangle). *Consider three terminal vertices $A(a_1, a_2, a_3)$, $B(b_1, b_2, b_3)$, and $C(c_1, c_2, c_3)$ on a triangular grid. Let the median values in each direction be $m_1 = median\{a_1, b_1, c_1\}$, $m_2 = median\{a_2, b_2, c_2\}$, $m_3 = median\{a_3, b_3, c_3\}$. The region enclosed by lines $x_1 = m_1$, $x_2 = m_2$, and $x_3 = m_3$ is an equilateral triangle. This follows naturally because the lines are $120°$ apart. This equilateral triangle is the* median triangle *for the terminal set $\{A, B, C\}$. Lines $x_1 = m_1$, $x_2 = m_2$, and $x_3 = m_3$ are the* median lines *in each direction (Figure 17.10).*

In some cases, the median triangle may degenerate to a single point. These cases will be considered later in Lemma 6.

Based on the definition of the median triangle for three terminal vertices, we are able to provide a theorem that solves the Steiner problem for three terminal vertices.

Theorem 1. *To optimally interconnect three terminal points on a triangular grid, there is one Steiner point, and the possible location of this point may not be unique.*

1. *If the terminal points are collinear (all lie on the same grid line), there is only one possible choice for the Steiner point, and it is the terminal point that lies between the other two (Figure 17.11a).*
2. *If the terminal points are noncollinear, every point in or on the median triangle is a possible choice for the Steiner point, and all yield SMTs of the same shortest length. The region for the Steiner points is called the Steiner region (Figure 17.11b).*

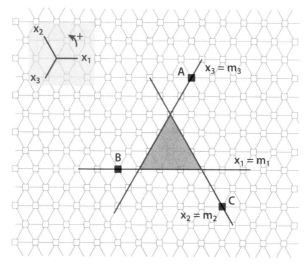

Figure 17.10. The gray shaded equilateral triangle is the median triangle for terminal set $\{A, B, C\}$. The median lines are $x_1 = m_1 = b_1$, $x_2 = m_2 = c_2$ and $x_3 = m_3 = a_3$.

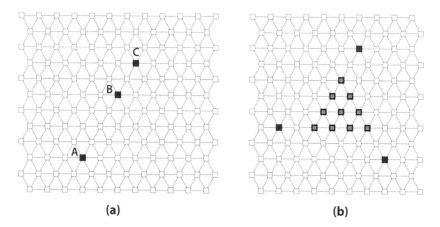

(a)   **(b)**

Figure 17.11. (a) Three collinear terminal points. Terminal B serves as the Steiner point. (b) Three noncollinear terminal points (black). Each blue point can serve as a Steiner point, and the median triangle forms the Steiner region.

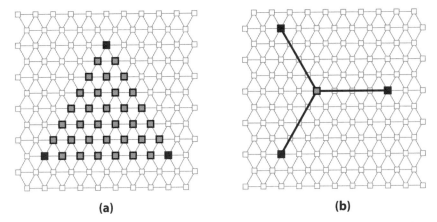

Figure 17.12. (a) The Steiner region is the entire equilateral triangle determined by the terminal points. (b) The spinner set has a unique SMT, the spinner tree. Black denotes terminal points; blue denotes Steiner points.

Lemma 6. *Special cases: (a) For three terminal points that form an equilateral triangle along the grid lines, the Steiner region is coincident with the equilateral triangle itself (Figure 17.12a). (b) For three terminal points whose median lines meet at the same point, the Steiner region degenerates to a single point. There is a unique SMT for this terminal set. Call this terminal set the* spinner set, *and this unique SMT the* spinner tree *(Figure 17.12b).*

2.2 Four Terminal Vertices on a Triangular Grid

Introducing a fourth terminal vertex to the three-point Steiner problem brings uncertainties when we consider the topology of Steiner trees. For three terminal vertices, there is only one Steiner point, and we simply connect it to each terminal to obtain the Steiner tree. The position of the Steiner point and the length of the Steiner tree can be calculated from Theorem 1 without much effort. For four terminal vertices, there are two Steiner points, and a Steiner tree is constructed so that one Steiner point connects two terminal vertices, the other connects the remaining two terminals, and the two Steiner points are connected to each other. The complexity arises because we need to consider all possible pairings among the terminal vertices to find the SMT. The solution to a four-point Steiner problem will provide the foundation for finding a general solution to an n-point triangular Steiner problem.

Consider four terminal points, $A(a_1, a_2, a_3)$, $B(b_1, b_2, b_3)$, $C(c_1, c_2, c_3)$, and $D(d_1, d_2, d_3)$. There will be two Steiner points in a Steiner tree. Steiner point S_1 connects A and B; Steiner point S_2 connects C and D; S_1 is connected with S_2 (Figure 17.13a).

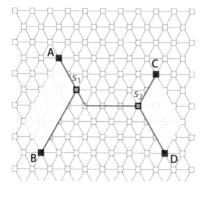

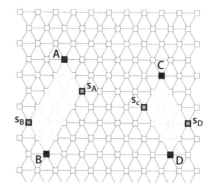

(a) SMT for the terminal set.

(b) $\{A, B, S_A, S_B\}$ is the candidate set for A and B. $\{C, D, S_C, S_D\}$ is the candidate set for C and D.

Figure 17.13. A four terminal set $\{A, B, C, D\}$.

Lemma 7. S_1 *is the Steiner point of* $\triangle ABS_2$, *and* S_2 *is the Steiner point of* $\triangle CDS_1$.

Proof. We prove this by contradiction.

The total length of the tree is $|AS_1| + |BS_1| + |S_1S_2| + |CS_2| + |DS_2|$. Suppose S_1 is not a Steiner point of $\triangle ABS_2$. Then we can find another Steiner point S' for $\triangle ABS_2$ such that $|AS'| + |BS'| + |S'S_2| < |AS_1| + |BS_1| + |S_1S_2|$, a contradiction. Therefore, the satisfying point S_1 must be a Steiner point of $\triangle ABS_2$. The same is true for S_2. □

Lemma 7 is a necessary condition for the Steiner point pairs. From its proof, we also see that by satisfying this condition, we will find the SMT for a particular topology. However, when searching for the Steiner pairs, we need to narrow our choices based on some criteria rather than brute forcing all pairs of points on the grid. We therefore introduce the concept of *candidate points* for a Steiner point.

Definition 4 (Candidate points/set). *Consider two points A and B on the grid. If A and B are noncollinear, the* candidate points *for A and B are the four vertices of their distance parallelogram. Otherwise if A and B are collinear, the* candidate points *are A and B themselves. We shall call the set of candidate points the* candidate set *(Figure 17.13b).*

It can be shown that for a given topology, it is sufficient to search for Steiner pairs in the candidate set. A detailed proof is given in Mei [13]. All topologies must be considered to guarantee obtaining the globally optimal (minimum length) Steiner tree.

2.3 The Recursive Algorithm

The five-point Steiner problem is considered before we tackle the general case with n terminal vertices. Suppose T is an SMT for a five-terminal set $\mathcal{X} = \{A, B, C, D, E\}$. If vertices A and B are connected to Steiner point S_1, then $T - S_1A - S_1B$ must be the SMT for the four-terminal set $\{S_1, C, D, E\}$. Here, $S_1A + S_1B$ is the shortest distance connecting A and B. Thus, we choose S_1 to be in the candidate set of these two terminal vertices, and we say A and B are *merged* to their candidate point S_1.

Choosing the merging point to be in the candidate set can result in the optimal solution. For any other optimal routes where a merging point is not in the candidate set, we can choose it to be a candidate point and obtain an SMT of the same length. The following lemma shows how a Steiner point can be obtained by merging two vertices to their candidate set when we construct an SMT. Additional details can be found in Mei [13].

Lemma 8. *A Steiner point can be found in the candidate set of some vertex pair in a SMT.*

Proof. Suppose in an SMT that, in vertices A and B are joined to Steiner point S, and S is not in the candidate set of (A, B), denoted *Candidate*(A, B). We show by case analysis that we can choose S from the candidate set of some vertex pair to obtain a SMT with the same length.

1. S is a terminal point. Merging the *terminal* vertex pair (S, A) to its candidate set and then merging (S, B) will give the desired topology.
2. S is not a terminal and it has degree 3. Let S connect to A, B, and X. S must be the Steiner point for set $\{A, B, X\}$. In this case, we can choose S from *Candidate*$(A, B) \cup \{X\}$.
3. S is not a terminal, and it has degree 4. Let S connect to A, B, X, and Y. In this case, S can be chosen from *Candidate*$(A, B) \cup$ *Candidate*(X, Y). If $S \notin$ *Candidate*(A, B), then it can be chosen from *Candidate*(X, Y).

Since it follows from Lemma 5 that the degree of a Steiner point can be only 3 or 4, this concludes the proof. □

Now the idea for a recursive algorithm to solve an n-point Steiner problem becomes straightforward. In each round, we choose two terminal points P and Q and merge them into their candidate set $\mathcal{C}$. For every element C in $\mathcal{C}$, we compute the Steiner tree for the $n-1$ terminal set made up of the original terminal set with P and Q replaced by C.

Algorithm 1.   SMT for n points. Input: A set of terminal points $\mathcal{X}$. Output: Length of Steiner minimum tree.

---

    function SMT($\mathcal{X}$):
        n = number of elements in $\mathcal{X}$
        if $n \leq 3$:
            return SMT$_3(\mathcal{X})$ // one-step SMT calculation for 3 points
        for each pair (p_1, p_2) in $\mathcal{X} \times \mathcal{X}, p_1 \neq p_2$:   // $\binom{n}{2}$ pairs
            for each candidate s of (p_1, p_2):
                let $\mathcal{X}' = (\mathcal{X} - \{p_1, p_2\}) \cup \{s\}$
                local_tree_length $= |sp_1| + |sp_2| + $ SMT($\mathcal{X}'$)
                global_min = min(global_min, local_tree_length)
        return global_min

Runtime Analysis. We present a rough estimate of the runtime of the basic recursive algorithm. At every terminal-reducing step, each pair of two points is selected to be merged to its candidate set (maximum size of 4). So the runtime is governed by a recurrence relation of the form $t(n) = 4 \binom{n}{2} t(n-1)$, whose solution is $O(2^n (n!)^2)$, where n is the number of terminal vertices to be connected.

The recursive algorithm was implemented in Maple. Though effective in finding the shortest route, the program is inefficient and can be used to find the SMT for up to only six points within a reasonable amount of time. The results are shown in Table 17.1, presented later in the chapter.

2.4 The Binary Tree Model

The main redundancy of the basic recursive algorithm comes from reconsidering terminal pairs. An example of a Steiner tree for a five-point terminal set $\{1, 2, 3, 4, 5\}$ is shown in Figure 17.14. In the basic recursive program, this topology will be considered at least twice: (1) merge terminal 1 and 2 to $S_1 \rightarrow$ set reduced to $\{S_1, 3, 4, 5\} \rightarrow$ merge terminal 3 and 4 to $S_2 \rightarrow$ set reduced to $\{S_1, S_2, 5\}$; (2) merge terminal 3 and 4 to $S_2 \rightarrow$ set reduced to $\{1, 2, S_2, 5\} \rightarrow$ merge terminal 1 and 2 to $S_1 \rightarrow$ set reduced to $\{S_1, S_2, 5\}$.

To eliminate considering a tree topology multiple times, we use a binary tree model to represent a Steiner tree and consider pairing terminals[1] at the beginning of the program. Figure 17.15 shows the binary tree representation of

---

[1] In our recursive program, terminals not only include those in the original terminal set, but also refer to merged points that are treated as new terminal points in the next recursive step.

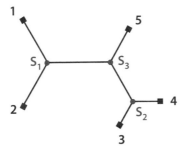

Figure 17.14. A Steiner tree for five terminal vertices. $1, \ldots, 5$ are the terminals; $S_1, S_2,$ and S_3 are Steiner points.

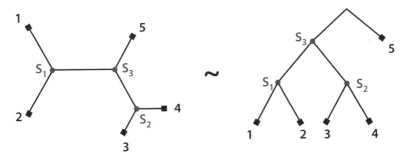

Figure 17.15. A Steiner tree and its binary tree representation.

the example in Figure 17.14. In a binary tree, the leaves correspond to terminal points and all internal vertices excluding the root correspond to Steiner points. Note that the root does not represent any vertex in the Steiner tree, and its children are simply the two components it connects.

Lemma 9. *For n terminal points, there are* $(2n - 3)!!$ *binary trees.*[2]

Proof. Consider how many ways we can add a terminal point to a given topology. The additional point can be merged to either the leaf nodes or the internal vertices (including the root.) So the recurrence relation for the number of binary trees is

$$T(n) = (\underbrace{n-1}_{\text{leaves}} + \underbrace{n-2}_{\text{internal vertices}}) \cdot T(n-1)$$

[2] The *double factorial* (or *semifactorial*) of m, denoted $m!!$, is equal to $m(m-2)(m-4) \cdots$.

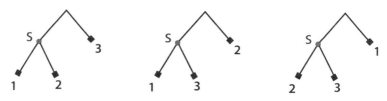

Figure 17.16. Binary trees for three terminal points.

$$\implies \quad T(n) = (2n-3)\,(2n-5)\,\cdots 5\cdot T_3$$
$$= (2n-3)!!, \quad \text{since} \quad T_3 = 3. \qquad \square$$

Implementation and Runtime Analysis. By analogy with the original recursive program, the binary tree model has the runtime recurrence relation $t(n) = 4\,(2n-3)\,t(n-1)$, and

$$t(n) = 4^{n-3}\,(2n-3)!! = 4^{n-3}\,\frac{(2n-2)!}{2^{n-1}\,(n-1)!} = 2^{n-5}\,\frac{(2n-2)!}{(n-1)!}.$$

So asymptotically, the runtime is $O\!\left(2^n\,\dfrac{(2n)!}{n!}\right)$, where n is the size of the initial terminal set. Compared to the basic recursive program, the improvement brought about by the binary tree model is huge, as one can see simply from the two recurrence relations. The basic algorithm is quadratically dependent on the previous term, while the binary tree algorithm is linearly dependent on the previous term.

However, there still remains some redundancy in the binary tree model. As an example, take the simplest Steiner tree, which consists of three terminal points. There are three binary tree representations, while only one topology exists for this terminal set (Figure 17.16). This triple redundancy exists for all Steiner points connecting three components. So, in order to implement the recursive binary tree algorithm efficiently, we used a global hash table to store all previously computed SMTs to avoid any recalculation, at the expense of requiring additional space.

The results of the improved implementation for computing Steiner trees are shown in Table 17.1, including a comparison with the original basic program. The speedup is significant, and the limits of its feasibility have been pushed from six to eight terminal points. However, this performance improvement involves a trade-off between space and time. Part of the factorial cost growth is transferred to space required for the hash table, and how efficiently Maple implements a hash table may also be a limit on computing power.

TABLE 17.1.
Performance comparison of the SMT programs in Maple

Number of terminals	4	5	6	7	8
Basic program runtime	0.14 s	5.5 s	360 s	58875 s	—
Binary tree runtime (BT)	0.13 s	3.2 s	36 s	389 s	5083 s
Basic runtime/BT runtime	1.1	1.7	10	151	—

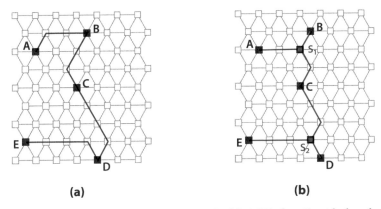

(a)   (b)

Figure 17.17. Terminal set $\mathcal{X} = \{A, B, C, D, E\}$. (a) MST for $\mathcal{X}$ with length 14. $V(MST) = \mathcal{X}$. (b) SMT for $\mathcal{X}$ with length 12. $V(SMT) = \mathcal{X} \bigcup \{S_1, S_2\}$.

3 The Steiner Ratio Conjecture

Given a terminal point set $\mathcal{X}$, a *minimum spanning tree* (MST) is the shortest connection network for $\mathcal{X}$, where all vertices belong to $\mathcal{X}$. A spanning tree differs from a Steiner tree in that no additional vertices are introduced in a spanning tree. An example is given in Figure 17.17. Since there are no known polynomial-time algorithms for computing a Steiner minimum tree (SMT), an MST can conveniently be used to approximate an SMT because there exist fast, polynomial-time algorithms due to Prim and Kruskal (see Prömel and Steger [12]) to compute an MST. Given a terminal set $\mathcal{X}$ on a triangular grid, we can compute the complete graph G for $\mathcal{X}$ and then apply either Prim's or Kruskal's algorithm on G to find an MST for $\mathcal{X}$. With this approach, an MST on a triangular grid can be found in a time that is polynomial in the number of terminal points.

The Steiner ratio has been studied for both the Euclidean and the rectilinear Steiner problems. Although this ratio has been defined in different ways, it is meant to quantify how well we can estimate an SMT with an MST. Our discussion of the Steiner ratio will be based on the following definition.

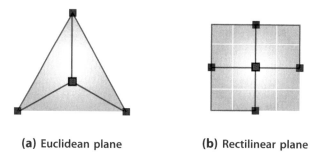

(a) Euclidean plane   **(b)** Rectilinear plane

Figure 17.18. The Steiner ratio achieved for the Euclidean and rectilinear Steiner problems. Black vertices denote terminal points; red vertices denote Steiner points. The red route gives the SMT; the blue route is the MST.

Definition 5 (Steiner ratio).   *Let* $\rho = l_{MST}/l_{SMT}$ *denote the ratio of the length of an MST to that of an SMT for a particular graph. The upper bound of ρ over all graphs is the* Steiner ratio.

In the Euclidean plane, the Steiner ratio is conjectured to be $2/\sqrt{3}$, and this is achieved when three terminal points form an equilateral triangle (Figure 17.18a). This Steiner ratio conjecture on the Euclidean plane was proposed in 1968 by Gilbert and Pollak [8] and allegedly proven in 1990 by Du and Hwang [9]. The proof was shown to be incorrect by Ivanov and Tuzhilin [10] in 2012, and the conjecture remains open. On a rectilinear grid, the Steiner ratio is $3/2$ when four terminals are aligned in a cross shape (Figure 17.18b). The Steiner ratio for the rectilinear Steiner problem was proposed and proved by Hwang in 1976 [11].

Conjecture (Steiner ratio conjecture).   *The Steiner ratio for the triangular Steiner problem is $4/3$ (i.e., $\rho \leq 4/3$). The ratio is achieved when three terminals form a spinner set (defined in Lemma 6), and the three arms of the spinner tree are of equal length (Figure 17.19).*

We will prove our conjecture for three and four point terminal sets.

3.1 The Three-Point Case

Theorem 2.   *The Steiner ratio for three arbitrary points on a triangular grid is $4/3$.*

Proof. Suppose the terminal vertices are $A(a_1, b_1, c_1)$, $B(a_2, b_2, c_2)$, and $C(a_3, b_3, c_3)$. Consider a Steiner tree for $\{A, B, C\}$, where the Steiner point

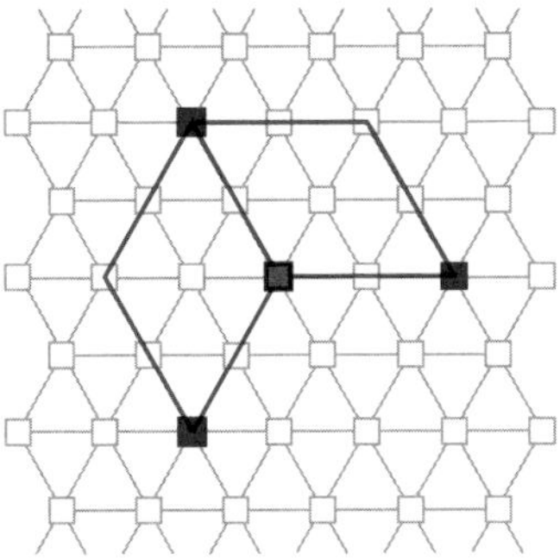

Figure 17.19. A spinner set (black points) that has an SMT with equal-length arms.

is $X(x_1, x_2, x_3)$. Let $M_i = \max\{a_i, b_i, c_i\}$, $m_i = \min\{a_i, b_i, c_i\}$, $d_i = M_i - m_i$, and $md_i = \text{median}\{a_i, b_i, c_i\}$ for $i \in \{1, 2, 3\}$.

The total length of the Steiner tree is

$$l = |XA| + |XB| + |XC| = \frac{1}{2} \sum_{i=1}^{3} (|x_i - md_i| + d_i) = \frac{1}{2}\left(\sum_{i=1}^{3} |x_i - md_i| + \sum_{i=1}^{3} d_i \right),$$

so

$$l_{SMT} \geq \frac{1}{2} \sum_{i=1}^{3} d_i.$$

Now consider the MST:

$$l_{MST} \leq \text{twice the average of } |AB|, |BC|, |AC| = \frac{2}{3}(|AB| + |BC| + |AC|)$$

$$= \frac{2}{3}\left(\frac{1}{2} \sum_{i=1}^{3} |a_i - b_i| + \frac{1}{2} \sum_{i=1}^{3} |b_i - c_i| + \frac{1}{2} \sum_{i=1}^{3} |a_i - c_i| \right)$$

$$= \frac{1}{3} \sum_{i=1}^{3} (|a_i - b_i| + |b_i - c_i| + |a_i - c_i|)$$

$$= \frac{1}{3} \sum_{i=1}^{3} 2(M_i - m_i) = \frac{2}{3} \sum_{i=1}^{3} d_i.$$

Combining the bounds for l_{SMT} and l_{MST}, we have

$$\frac{l_{MST}}{l_{SMT}} \leq \frac{4}{3}. \qquad \square$$

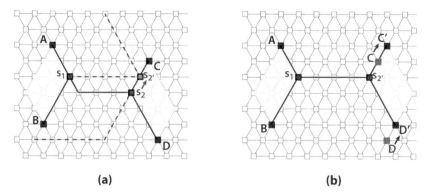

(a)   (b)

Figure 17.20. Terminal set $\{A, B, C, D\}$ transforms to $\{A, B, C', D'\}$ having the same Steiner tree length. The Steiner tree for $\{A, B, C', D'\}$ is a double spinner tree.

3.2 The Four-Point Case

Consider the full Steiner tree for a four-terminal set $\mathcal{X} = \{A, B, C, D\}$ in Figure 17.20a. S_1 and S_2 are two Steiner points. From the Steiner tree construction described in Algorithm 1, assume edges S_1A, S_1B, S_2C, and S_2D all align with grid lines. Edge S_1S_2 may not align with the grid, in which case $\{A, B, S_2\}$ and $\{C, D, S_1\}$ are not spinner sets. We can, however, transform this terminal set into a spinner set without changing the length of the SMT. In this example, we move S_2 along the circle centered at S_1 with radius $|S_1S_2|$ to S'_2. Consequently, terminals C and D are transformed to C' and D' (Figure 17.20b). The resulting terminal set $\mathcal{X}' = \{A, B, C', D'\}$ has the same Steiner tree length as $\mathcal{X}$, and $l_{MST}(\mathcal{X}') \geqslant l_{MST}(\mathcal{X})$. Therefore, $\rho(\mathcal{X}) \leqslant \rho(\mathcal{X}')$. It would be sufficient to prove $\rho(\mathcal{X}') \leqslant 4/3$ for $\mathcal{X}'$, which has a double spinner Steiner tree. For the remainder of this section, a Steiner tree for four terminal points refers to a double spinner Steiner tree for four terminals.

Spanning Trees. Spanning trees consist only of edges between terminal pairs. Given the topology of a double spinner Steiner tree for a terminal set $\mathcal{P} = \{A, B, C, D\}$, the edges joining a pair of terminal vertices can be categorized as one of three types: neighbors, bridges, and crosses (Figure 17.21). A neighbor edge joins two terminals that are connected to the same Steiner point in a Steiner tree. A bridge edge joins two terminals that are directly connected to different Steiner points, and the bridge edge does not intersect the Steiner tree. A cross edge joins two terminals that are directly connected to different Steiner points, but the cross edge intersects the Steiner tree. Examples of these types of spanning tree edges are shown in Figure 17.21 with blue dashed lines, and they reflect the only ways of joining the terminal pairs.

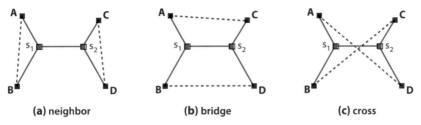

Figure 17.21. Types of edges joining two terminal points: (a) neighbor, (b) bridge, and (c) cross.

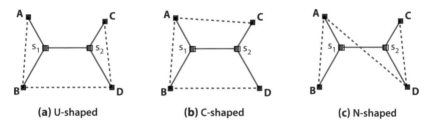

Figure 17.22. Different shaped spanning trees: (a) U-shaped, (b) C-shaped, (c) N-shaped.

Based on the edge types in a spanning tree, there exist a total of seven differently shaped spanning trees for $\mathcal{P}$. We name and describe three that will be useful in our discussion: U-shaped, C-shaped, and N-shaped spanning trees (Figure 17.22). A U-shaped spanning tree contains two neighbor edges and a bridge edge; a C-shaped spanning tree contains two bridge edges and a neighbor edge; an N-shaped spanning tree contains two neighbor edges and a cross edge.

Steiner Ratio.

Lemma 10. *Given a double spinner tree for four terminal points, $l_{MST}/l_{SMT} \leqslant 4/3$.*

Proof. The proof is by contradiction.

Assume $l_{MST} > (4/3)\, l_{SMT}$. This implies that the length of *any* spanning tree must be greater than $(4/3)\, l_{SMT}$. It is sufficient to show a contradiction by choosing any spanning tree(s).

Suppose $\{A, B, C, D\}$ is a four-point terminal set and it has a double spinner Steiner tree with Steiner points S_1 and S_2. Let the lengths of the edges be $|AS_1| = a$, $|BS_1| = b$, $|S_1 S_2| = s$, $|CS_2| = c$, and $|DS_2| = d$. So $l_{SMT} = a + b + c + d + s$.

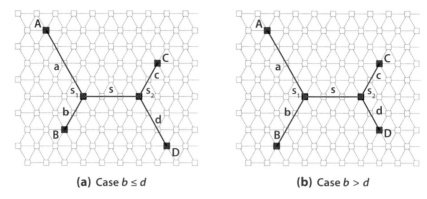

(a) Case $b \leq d$ **(b)** Case $b > d$

Figure 17.23. A double spinner tree for four terminal points. (a) case $b \leq d$; (b) case $b > d$.

Without loss of generality, assume $a = \max\{a, b, c, d\}$. To express the length of the possible components of the spanning tree, we divide the proof into two cases.

1. Case $b \leq d$ (Figure 17.23a).
   The distances between the terminal points are

   $$|AB| = a + b, \; |BD| = s + d, \; |CD| = c + d, \; |BC| = s + b + c, \; |AC| = s + a.$$

   Suppose

   - for a U-shape, $(|AB| + |BD| + |CD|) > \dfrac{4}{3} l_{SMT}$
     $$\implies 3(a + b + s + d + c + d) > 4(a + b + c + d + s)$$
     $$\implies 2d > a + b + c + s; \tag{1}$$
   - for a C-shape, $(|AC| + |CD| + |BD|) > \dfrac{4}{3} l_{SMT}$
     $$\implies 3(s + a + c + d + s + d) > 4(a + b + c + d + s)$$
     $$\implies 2s + 2d > a + 4b + c; \tag{2}$$
   - for an N-shape, $(|AB| + |BC| + |CD|) > \dfrac{4}{3} l_{SMT}$

     $$\implies 3(a + b + s + b + c + c + d) > 4(a + b + c + d + s)$$
     $$\implies 2b + 2c > a + d + s. \tag{3}$$

   Expressions $(1) + (2) + (3) \implies d > a + b$, which contradicts $a > d$.

2. Case $b > d$ (Figure 17.23b).
   The distances between the terminal points are

   $$|AB| = a + b, \; |BD| = s + b, \; |CD| = c + d, \; |AC| = s + a.$$

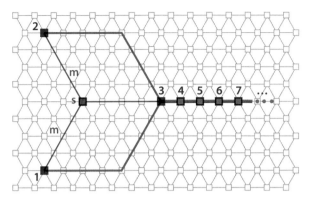

Figure 17.24. A Steiner tree (the red route) of $n \geqslant 4$ terminal points with $\rho \to \frac{4}{3}$. The black and green points denote terminals. The red point S is the only Steiner point in the Steiner tree. The blue route denotes the minimum spanning tree. $m \gg n$.

Suppose

- for a U-shape, $(|AB| + |BD| + |CD|) > \dfrac{4}{3} l_{SMT}$

$$\implies 3(a + b + s + b + c + d) > 4(a + b + c + d + s)$$
$$\implies 2b > a + c + d + s; \tag{4}$$

- for a C-shape, $(|AC| + |CD| + |BD|) > \dfrac{4}{3} l_{SMT}$

$$\implies 3(s + a + c + d + s + b) > 4(a + b + c + d + s)$$
$$\implies 2s > a + b + c + d. \tag{5}$$

Expressions (4) $\times$ 2 + (5) $\implies b > a + c + d$, which contradicts $a > b$.

$\square$

Theorem 3. *The Steiner ratio for four points on a triangular grid is 4/3.*

Proof. Lemma 10 shows that $l_{MST}/l_{SMT} \leq 4/3$. We need only prove that $4/3$ is a tight upper bound.

Construct the Steiner tree shown in Figure 17.23a with $a = b = s = m$, $c = d = 1$. Then

$$l_{SMT} = a + b + c + d + s = 3m + 2, \quad l_{MST} = |AB| + |BD| + |CD| = 4m + 2.$$

The Steiner ratio is $l_{MST}/l_{SMT} = \dfrac{4m + 2}{3m + 2} \to \dfrac{4}{3}$ as $m \to \infty$. $\square$

Note that the $\frac{4}{3}$-bound can be extended to any number of points. For example, in Figure 17.24, from the spinner set $\{1, 2, 3\}$ with an equal-arm spinner tree, add points one unit length apart along ray $S3$ to grow the set. m is the arm length of the spinner tree; n is the total number of terminal points. Let $m \gg n$, then $l_{MST}/l_{SMT} = \frac{4m+(n-3)}{3m+(n-3)} \to \frac{4}{3}$. Generalizing the proof of the Steiner ratio conjecture for four-terminal sets to n points is one goal of our future work.

References

[1] Z. A. Melzak. On the problem of Steiner. *Can. Mathematical Bull.* **4** (1961) 143–148.

[2] M. R. Garey, R. L. Graham, and D. S. Johnson. The complexity of computing Steiner minimal trees. *SIAM J. Appl. Math.* **32** (1977) 835–859.

[3] J. H. Rubinstein, D. A. Thomas, and N. Wormald. Steiner trees for terminals constrained to curves, *SIAM J. Discrete Math.* **10** (1997) 1–17.

[4] M. Hanan. On Steiner's problem with rectilinear distance. *SIAM J. Appl. Math.* **14** (1966) 255–265.

[5] M. R. Garey and D. S. Johnson. The rectilinear Steiner tree problem is NP-complete. *SIAM J. Appl. Math.* **32** (1977) 826–834.

[6] F. K. Hwang and J. F. Weng. Hexagonal coordinate systems and Steiner minimal trees. *Discrete Math.* **62** (1986) 49–57.

[7] F. K. Hwang, D. S. Richards, and P. Winter. *The Steiner Tree Problem.* Annals of Discrete Mathematics **53**. Elsevier North-Holland, Amsterdam, 1992.

[8] E. N. Gilbert and H. O. Pollak. Steiner minimal trees. *SIAM J. Appl. Math.* **16** (1968) 1–29.

[9] D. Z. Du and F. K. Hwang. The Steiner ratio conjecture of Gilbert and Pollak is true. *Proc. National Academy Sci. USA* (1990) 9464–9466.

[10] A. O. Ivanov and A. A. Tuzhilin. The Steiner ratio Gilbert-Pollak conjecture is still open. *Algorithmica* **62** (2012) 630–632.

[11] F. K. Hwang. On Steiner minimal trees with rectilinear distance. *SIAM J. Appl. Math.* **30** (1976) 104–114.

[12] H. J. Prömel and A. Steger. *The Steiner Tree Problem.* Vieweg, Berlin, 2002.

[13] J. Mei. Shortest Connection Networks on Triangular Grids. Master's thesis, University of Rhode Island, Kingston, 2018.

18

ENTROPY OF LEGO JUMPER PLATES

David M. McClendon and Jonathon Wilson

In 2016, the revenue of the LEGO[1] Company was more than \$6.3 billion [11]. One reason LEGO products are so popular might be the seemingly endless number of ways to connect together the small plastic building toys. This leads to an interesting combinatorial question: Exactly how many different ways can n LEGO bricks of the same size, color, and shape be interlocked? If n is small, then this number can be counted exactly, if one has enough computing power. Begfinnur Durhuus and Søren Eilers studied this question for 2×4 rectangular LEGO bricks and were able to determine that there are

$$8,274,075,616,387$$

different ways to connect eight 2×4 LEGO bricks [6]. To put this number into perspective, suppose that you could build one of these constructions every 5 seconds. It would take you 1.31 million years to run through all these constructions!

Unfortunately, once n becomes large (for 2×4 bricks, "large" means ten [6]), the exact number of different configurations is still not known—no closed mathematical formula exists, and the run time for any known computer algorithm is too large. The good news, however, is that if one defines $T_B(n)$ to be the number of different configurations that can be built from n LEGO bricks from some particular class B of brick, then in many cases, one can show that $T_B(n)$ grows exponentially in n, and upper and lower bounds on the exponential growth rate of this function can be obtained. Indeed, Durhuus and Eilers [6] compute upper and lower bounds on this growth rate for standard rectangular LEGO bricks.

In this chapter, we study a different type of LEGO brick, called a *jumper plate*. A jumper plate is a 1×2 LEGO element that has only one stud on its top, and three locations on its bottom into that studs can be inserted (see Figure 18.1).

---

[1] LEGO is a trademark of the LEGO Company.

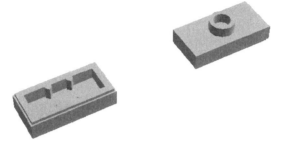

Figure 18.1. The bottom and top of a LEGO jumper plate. To attach two jumper plates, the stud on the top of the child can be inserted into any of the three "slots" on the bottom of the parent.

When two jumper plates are attached in this way, we can arrange them so that their studs point upward, and refer to the plate on the top of the connection as the *parent* and the plate underneath as a *child* (we use the term *grandchild* in the obvious way).

In this chapter, we study the function $T_{\mathcal{J}}(n)$, which counts the number of contiguous LEGO buildings that can be made from n jumper plates. We are especially interested in determining how fast this function grows: Is it exponential? Superexponential? Or something else? We will prove that $T_{\mathcal{J}}$ grows at an exponential rate and give bounds on the rate of its exponential growth.

1 Counting Small Buildings

1.1 Why Study Jumper Plates?

For some classes $\mathcal{B}$ of LEGO bricks, $T_{\mathcal{B}}(n)$ is very easy to compute. For example, for a standard 1×1 LEGO brick, the only way to connect n such bricks together is to make a 1×1 tower of height n, so $T_{1 \times 1}(n)$ is the constant function $T_{1 \times 1}(n) = 1$.

LEGO also produces a "double jumper" plate (denote this class of plate by $\mathcal{D}$), which is a 2×2 plate with a single stud in the center of the top (see Figure 18.2). There are five ways to attach one double jumper plate to another (by placing the stud of the child in either the center or one of the four corners of the parent). Since a double jumper plate can have at most one child, specifying which of the five connections is used to attach each child to its parent completely describes a building made from n double jumper plates. Since there are $n-1$ plates in such a building that are children, $T_{\mathcal{D}}(n) = 5^{n-1}$.

One reason these two classes of bricks have easy-to-describe functions T is that the buildings one can make from them lack three "dimensions" of freedom, in that the number of pieces being used completely determines the building's height. Jumper plates are the simplest LEGO elements that allow for bonafide

Figure 18.2. Top and bottom view of a LEGO double jumper plate.

three-dimensional constructions, in which one can build outward in nontrivial ways as well as directly up and down, and that is why we choose to study them.

Jumper plates are popular with LEGO aficionados, because unlike standard rectangular LEGO bricks, jumper plates allow for creations that have a "half-stud" offset; this "jumping" of a half-stud gives the piece its name.

1.2 What Makes Two Buildings "Different"?

We said earlier that $T_J(n)$ is the number of contiguous LEGO buildings that can be made from n jumper plates. To clarify this definition, we need to describe exactly what makes one building "different" from another. First, since each jumper plate has only one stud on its top, the building has to have a unique jumper plate on its topmost level; call this jumper plate the *root* of the building. To account for translational symmetry, we specify that $T_J(n)$ is the number of buildings that can be made from n jumper plates, where the root occupies a fixed position.

If one thinks of buildings as being identified up to rotational symmetry, then each building is counted twice in our computation of $T_J(n)$ (because when the root is rotated by 180° about its center, it occupies the same position). However, the exponential growth rate of $T_J(n)$ would be the same whether such buildings are identified or not, so we will not bother with identifying buildings that are rotationally symmetric. As an example, in Figure 18.3, we treat buildings 1 and 3 as two separate buildings (each made from two jumper plates), even though rotating building 1 by 180° produces building 3. In particular, this means $T_J(2) = 6$. Notice that the "half-stud" offset permitted in some of the attachments shown in Figure 18.3 means that jumper plates will not form the same kinds of buildings as the standard rectangular LEGO bricks studied in Durhuus and Eilers [6].

1.3 Configurations Made From a Small Number of Jumper Plates

To get an idea of how the function T_J behaves, let's actually compute some values of $T_J(n)$. When $n = 3$, we can just build each of the constructions and count them (see Figure 18.4). In particular, buildings made from three jumper

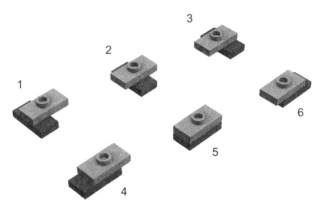

Figure 18.3. The six different ways to attach two jumper plates. In each building, the light gray plate is the parent, and the dark gray plate is the child. In the top three connections pictured, we say that the child is perpendicular to the parent; in the bottom three connections, the child is parallel to the parent.

plates of height 3 can be cataloged by first choosing one of the six connections described in Figure 18.3 to specify how to attach the middle plate to the root, and then choosing one of the six connections in Figure 18.3 for how the bottom plate attaches to the middle plate. This gives $6(6) = 36$ buildings of height 3 made from three jumper plates (in general, there are 6^{n-1} buildings of height n made from n jumper plates). There is one building of height 2 made from three jumper plates (shown at the bottom of Figure 18.4), making a total of thirty-seven buildings made from three jumper plates.

Furthermore, if n is small enough, we can count $T_{\mathcal{J}}(n)$ by hand (see Table 18.1 for the values when $n \leq 8$). To get an idea of how these values are obtained, we'll go through the case $n = 5$. Buildings made from five jumper plates must have height 3, 4, or 5, so we can count the number of buildings of each height separately and add:

If the building is five plates high, every plate (other than the bottom one) has exactly one child; since there are six ways to attach each nonroot plate to its parent, we obtain a total of $6^4 = 1296$ such buildings.

If the building is four plates high, exactly one of the five plates must have two children.

- If the root has two children, then the grandchild of the root must be attached to one of the two children in one of six ways (so there are $2 \cdot 6 = 12$ ways to attach the grandchild to the bottom of the building); then there are six ways to attach the last plate underneath the grandchild. So there are $12 \cdot 6 = 72$ buildings when only the root has two children.

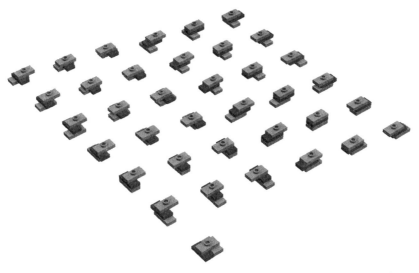

Figure 18.4. The thirty-seven buildings that can be made from three jumper plates. Notice that the top-most plate in each construction (the root) occupies a fixed position. The bottom-most building in this picture is the only building of height 2 that can be made from three jumper plates; the other buildings all have height 3.

TABLE 18.1.
Values of $T_{\mathcal{J}}(n)$ for $n \leq 8$, computed by hand

n	$T_{\mathcal{J}}(n)$
1	1
2	6
3	37
4	234
5	1489
6	9534
7	61169
8	393314

- If the child of the root is the only plate with two children, then a similar argument yields seventy-two buildings in this case as well.
- If the grandchild of the root is the plate with two children, then there are thirty-six buildings (six ways to attach the child to the root, six ways to attach the grandchild of the root underneath the child, and one way to attach the last two plates under the grandchild of the root).

If the building is three plates high, the building consists of the root, the two children of the root, and two plates attached under the children of the

TABLE 18.2.
Values of $T_{\mathcal{J}}(n)$ for $9 \leq n \leq 14$, computed via computer calculations by S. Eilers [7]

n	$T_{\mathcal{J}}(n)$
9	2531777
10	16316262
11	105237737
12	679336650
13	4388301841
14	28366361206

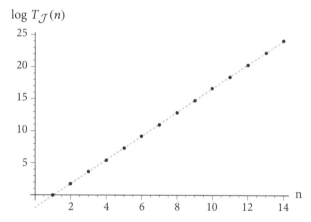

Figure 18.5. The graph of $\log T_{\mathcal{J}}(n)$ versus n for $n \in \{1, ..., 14\}$. The least-squares line derived from these points (shown by the dashed line) has equation $y \approx 1.85531x - 1.93902$, suggesting that $T_{\mathcal{J}}(n) \approx e^{1.85531n - 1.93902}$.

root. The only freedom in such a building is in attaching the bottom-most plates; there are thirteen ways to do this (nine ways in which the plates in the bottom level are parallel to the plates in the second level, and four ways in which the plates in the bottom level are perpendicular to those in the middle level).

Combining these cases, we obtain $T_{\mathcal{J}}(5) = 1296 + 72 + 72 + 36 + 13 = 1489$.

Sadly, once $n \geq 9$, this method begins to break down, because there are too many cases to efficiently count. However, S. Eilers [7] recently communicated to us the values of $T_{\mathcal{J}}(n)$ for $9 \leq n \leq 14$, obtained by computer calculations (see Table 18.2); unfortunately, even computer calculations do not help in computing $T_{\mathcal{J}}(n)$ for larger n—the run time for known algorithms becomes prohibitively large.

Plotting the values of $\log T_{\mathcal{J}}(n)$ against n for $n \leq 14$ (Figure 18.5), note that $T_{\mathcal{J}}$ appears to have exponential growth (in Section 3, we prove that $T_{\mathcal{J}}(n)$

does in fact grow exponentially). Furthermore, the least-squares linear equation for $\log T_{\mathcal{J}}(n)$ plotted against n, derived from the values of $T_{\mathcal{J}}(n)$ for $n \leq 14$, suggests that $T_{\mathcal{J}}(n) \approx e^{1.85531n - 1.93902}$, that is, $T_{\mathcal{J}}(n)$ has exponential growth rate $1.85531 \approx \log 6.39368$. (We show rigorously in Section 4 that this numerical approximation underestimates, at least slightly, the actual exponential growth rate of $T_{\mathcal{J}}$.)

2 Entropy

We saw from the numerics in the previous section that there is good reason to believe $T_{\mathcal{J}}(n)$ grows exponentially; that is, $T_{\mathcal{J}}(n)$ grows like $C \exp(h_{\mathcal{J}} n)$ for suitable constants C and $h_{\mathcal{J}}$. We are interested in studying the value of $h_{\mathcal{J}}$; assuming $T_{\mathcal{J}}(n) \approx C \exp(h_{\mathcal{J}} n)$, we can solve for $h_{\mathcal{J}}$ to obtain

$$h_{\mathcal{J}} \approx \frac{1}{n} \left(\log T_{\mathcal{J}}(n) - \log C \right).$$

As $n \to \infty$, the $\frac{1}{n} \log C$ term goes to zero, leaving $h_{\mathcal{J}} \approx \frac{1}{n} \log T_{\mathcal{J}}(n)$. With this idea in mind, we define the *entropy* of a jumper plate to be

$$h_{\mathcal{J}} := \lim_{n \to \infty} \frac{1}{n} \log T_{\mathcal{J}}(n),$$

provided this limit exists (we will show that it does in Section 3).

We use the word "entropy," because the formula used to define $h_{\mathcal{J}}$ resembles a formula used in information science to compute a quantity called entropy: Consider a stationary, ergodic process (as an example of such a process, think of a ticker-tape printing out 0s and 1s randomly according to some probability law). Now take the words of length n coming from this process (in the ticker-tape example, a "word" of length 6 would be something like 001011 or 110111), and order these words by their probabilities. After fixing $\lambda \in (0, 1)$, select words of length n in the above order, one at a time, starting with the most likely word, until the probabilities of the selected words sum to at least λ. Defining $N_n(\lambda)$ to be the number of words it takes to do this, it turns out that

$$\lim_{n \to \infty} \frac{1}{n} \log N_n(\lambda) = h, \tag{1}$$

where h is a number, independent of λ, called the *entropy* of the stochastic process. The quantity h measures the amount of "randomness" or "chaos" in the process. The entropy is an important invariant of stationary processes that has applications in data compression and ergodic theory. Essentially, the fact in equation (1) is a corollary of what is known in information theory as the Shannon-McMillan-Breiman Theorem, or the "asymptotic equipartition property" [15, 13, 4, 2].

Remark: The base of the logarithm is irrelevant in the definition of entropy, as choosing two different bases yields the same answer up to a constant multiple. We choose base e, but the only time in our chapter that this matters is in the proof of Theorem 2.

2.1 History and Summary of the Main Results

Several mathematicians and computer scientists have done work on counting numbers of various LEGO configurations made from rectangular bricks [1, 5, 6, 9, 10, 12]. As mentioned earlier, Durhuus and Eilers [6] showed that the number $T_{b \times w}(n)$ of buildings that can be made from n standard rectangular $b \times w$ LEGO bricks grows exponentially in n, and they described upper and lower bounds on the entropy of 2×4 bricks.

We investigate the entropy of LEGO jumper plates, using some methods borrowed from Durhuus and Eilers [6] and other methods involving the combinatorics of objects we call "labeled binary graphs." In the next two sections, we show that the limit defining the entropy exists and is at least $\log 6.44947$. The techniques in these sections are borrowed heavily from Durhuus and Eilers, who studied the entropy of (nonjumper) rectangular LEGO bricks [6]. In Section 5, we prove that the entropy is at most $\log(6 + \sqrt{2})$, using a new method of associating a "labeled binary tree" to each building and counting the number of such labeled trees. The method of associating a graph to a LEGO construction was used in Durhuus and Eilers [5], but the idea of labeling the graphs (and the associated combinatorics) is, as far as we know, new.

Section 6 contains an outline of how our methods might be further improved, and the last section outlines how our methods can be applied to a different type of LEGO element called a "roof tile."

3 Existence of Entropy

Durhuus and Eilers [6] establish the existence of the entropy for configurations of rectangular $b \times w$ bricks; we mimic their argument to explain why $h_{\mathcal{J}}$ exists.

Theorem 1. *Let $T_{\mathcal{J}}(n)$ be the number of buildings made from n 1×2 jumper plates. Then*

$$h_{\mathcal{J}} = \lim_{n \to \infty} \frac{1}{n} \log T_{\mathcal{J}}(n)$$

exists in $[0, \infty]$.

Proof. Denote by $B_{\mathcal{J}}(n)$ the set of buildings that can be made from n jumper plates. Then let $A_{\mathcal{J}}(n)$ be the subset of $B_{\mathcal{J}}(n)$ consisting of buildings whose bottom-most layer contains exactly one jumper plate; let $a_n = \#(A_{\mathcal{J}}(n))$. Observe that

$$T_{\mathcal{J}}(n-1) \leq a_n \leq T_{\mathcal{J}}(n). \tag{2}$$

To see the left-hand inequality, notice that by removing the bottom plate from each member of $A_{\mathcal{J}}(n)$, we obtain a member of $B_{\mathcal{J}}(n-1)$, and every member of $B_{\mathcal{J}}(n-1)$ can be obtained in this fashion. The right-hand inequality follows from the fact that $A_{\mathcal{J}}(n) \subseteq B_{\mathcal{J}}(n)$.

From Expression (2), we see that

$$h_{\mathcal{J}} = \lim_{n\to\infty} \frac{1}{n} \log T_{\mathcal{J}}(n) = \lim_{n\to\infty} \frac{1}{n} \log a_n.$$

Next, notice that $a_{n+m} \geq a_n a_m$. To see why, observe that attaching the root of any element of $A_{\mathcal{J}}(n)$ to the underside of the plate on the bottom level of any building in $A_{\mathcal{J}}(m)$ produces a building in $A_{\mathcal{J}}(m+n)$. This procedure yields an injection $A_{\mathcal{J}}(n) \times A_{\mathcal{J}}(m) \hookrightarrow A_{\mathcal{J}}(m+n)$, giving the desired inequality. Therefore, for all m and n, $\log a_{m+n} \geq \log a_m + \log a_n$, so by Fekete's subadditive lemma, $\left\{ \frac{1}{n} \log a_n \right\}$ converges as $n \to \infty$ to $\sup \left\{ \frac{1}{n} \log a_n \right\} \in [0, \infty]$. $\qquad\square$

4 A Lower Bound on the Entropy

As there are six choices for how an only child can be attached to its parent, the number of buildings of height n that can be made from n jumper plates is 6^{n-1}, thus producing the trivial lower bound

$$h_{\mathcal{J}} \geq \lim_{n\to\infty} \frac{1}{n} \log 6^{n-1} = \log 6.$$

We tighten this bound by applying a technique developed in Durhuus and Eilers [6] that counts the number of buildings with a fixed number of "bottlenecks."

Theorem 2. *Let $h_{\mathcal{J}}$ be the entropy of a 1×2 LEGO jumper plate. Then*

$$h_{\mathcal{J}} \geq \log 6.44947.$$

Proof. Recall $A_{\mathcal{J}}(n+1)$ is the set of LEGO buildings made from $n+1$ jumper plates such that the top and bottom layers of the building each consists of a single plate. We say that a building in $A_{\mathcal{J}}(n+1)$ has a *bottleneck* if the building

Figure 18.6. A LEGO building with two bottlenecks, located at the black jumper plates. Each of the black jumper plates is the only plate in its level of the building.

has a layer (other than the top and/or bottom layer) that has only a single brick in it (see Figure 18.6).

For $n \geq 0$, let c_n denote the number of buildings in $A_{\mathcal{J}}(n+1)$ that have no bottlenecks. Durhuus and Eilers [6] show using generating functions that for any n,

$$\sum_{j=1}^{n} c_j \left(e^{h_{\mathcal{J}}}\right)^{-j} \leq 1, \tag{3}$$

and their argument carries over to our context. By explicitly computing values of c_j for some j, we obtain a lower bound for $h_{\mathcal{J}}$. First, it is clear that $c_1 = 6$, since any building made from two plates has no bottleneck. Note also that $c_2 = 0$, because any building with three plates arranged in three layers must have a bottleneck in the middle layer.

To determine c_3, note that any building in $A_{\mathcal{J}}(4)$ without bottlenecks must have one plate in its bottom layer, and two in the middle layer that are both children of the root. There is therefore only one way to hook the top three plates together, so c_3 is equal to the number of ways to attach a single jumper plate to the bottom of one of two parallel jumper plates. There are six ways to attach this last plate to its parent, and two choices of parent, so $c_4 = 6(2) = 12$.

Next, $c_4 = 0$, because any building made from five jumper plates where the top and bottom layers consist of a single plate must have a bottleneck in it.

For c_5, observe that any building in $A_{\mathcal{J}}(6)$ with no bottlenecks must have one plate in the top layer (call this layer 0), one plate in the bottom layer (layer 3), and two plates in each of the two intermediate layers (layers 1 and 2). Therefore, there is one way to configure the three jumper plates in layers 0 and 1. Once those three are attached, two more jumper plates need to be attached underneath layer 1 to form layer 2. There are nine ways to do this so that the plates in layer 2 are parallel to the plates in layer 1, and four ways to do this so that the plates in layer 2 are perpendicular to the plates in layer 1. Once layer

2 is made, the last plate (which comprises layer 3) needs to be attached to the bottom of one of the two plates in layer 2 to finish the building; there are twelve ways to do this. Altogether, $c_5 = (4+9)12 = 156$.

At this point we know from expression (3) that

$$6\left(e^{h_{\mathcal{J}}}\right)^{-1} + 12\left(e^{h_{\mathcal{J}}}\right)^{-3} + 156\left(e^{h_{\mathcal{J}}}\right)^{-5} \leq 1,$$

from which it follows that $h_{\mathcal{J}} \geq \log 6.3877$.

S. Eilers [7] relayed to us computer-generated computations of $c_7 = 2652$, $c_8 = 144$, $c_9 = 59100$, $c_{10} = 18192$, and $c_{11} = 1,615,740$; applying these values, the lower bound improves to $h_{\mathcal{J}} \geq \log 6.44947$. $\square$

Notice that this lower bound is greater than the value of $h_{\mathcal{J}}$ suggested by the least-squares computation in Section 1 (which was $\log 6.39368$).

5 An Upper Bound on the Entropy

In this section, we look for an upper bound on $h_{\mathcal{J}}$. Remember from the proof of Theorem 1 that $B_{\mathcal{J}}(n)$ denotes the set of LEGO buildings made from n jumper plates.

5.1 A Crude Upper Bound

We obtain $h_{\mathcal{J}} \leq \log 8$ by applying a method described in Durhuus and Eilers [6] that associates to each LEGO building a string of characters taken from a finite alphabet. More specifically, let $b \in B_{\mathcal{J}}(n)$ be a building. Number the plates in b from 1 to $n-1$ as follows: Call the root of the building "plate 1," then number the child(ren) of the root "plate 2" (and "plate 3," if the root has two children), then continue inductively, numbering the children of plate 2, then any children of plate 3, and so forth. Any time that a plate has two children, choose a standard way to order the children (for example, choose a compass direction to represent north, and when a plate has two children, give the smaller number to the plate that is either further south or further west, depending on its orientation).

Next, number the different ways to connect two jumper plates 1 to 6, as shown in Figure 18.3, and define $\mathcal{A} = \{0, \bowtie, 1, 2, 3, 4, 5, 6\}$.

Then define $f : B_{\mathcal{J}}(n) \hookrightarrow \mathcal{A}^{n-1}$ as follows: $f(b) = (x_1, ..., x_{n-1})$ if, for every $j \in \{1, ..., n-1\}$,

$$x_j = \begin{cases} 0 & \text{if plate } j \text{ of the building has no children} \\ \bowtie & \text{if plate } j \text{ of the building has two children} \\ z \in \{1, ..., 6\} & \text{if plate } j \text{ of the building has exactly one child, which is} \\ & \quad \text{attached to its parent via connection } z \text{ as shown in} \\ & \quad \text{Figure 18.3.} \end{cases}$$

In the last case above, to distinguish between connections like those numbered 1 and 3 in Figure 18.3, one can decree that if the child is attached to the southernmost or westernmost slot of the parent, then the connection is type 1; otherwise it is type 3.

Essentially, the symbol in the jth position of $f(b)$ tells you how to attach children to the jth plate in building b. As such, a string of $n-1$ symbols provides directions to construct at most one building, so f is $1-1$. Thus

$$T_{\mathcal{J}}(n) = \#(B_{\mathcal{J}}(n)) \leq \#(\mathcal{A}^{n-1}) = 8^{n-1},$$

and it follows that

$$h_{\mathcal{J}} \leq \lim_{n \to \infty} \frac{1}{n} \log 8^{n-1} = \log 8.$$

5.2 Bounding the Entropy by Counting Labeled Binary Trees

In this section, we improve the upper bound to $h_{\mathcal{J}} \leq \log\left(6 + \sqrt{2}\right)$ by a new method that associates, with each LEGO building, a "labeled binary tree," counting the number of such labeled trees with specific properties, and counting the maximum number of buildings that can be associated to each such labeled tree.

First, by a *binary tree* $\mathcal{T}$, we mean a tree where every node has either zero or two children, and where the left and right children at each node are distinguished. More formally, we decree a binary tree to be a rooted tree that is also an ordered tree, where every node has either zero or two children. Given such a binary tree, a *branching* of the tree is a node that has two children. We denote the set of nodes of binary tree $\mathcal{T}$ by $V(\mathcal{T})$.

Next, a *labeled binary tree* is a pair $(\mathcal{T}, f)$, where $\mathcal{T}$ is a binary tree, and f is a function that assigns to each node in $\mathcal{T}$ a positive integer. For each $n \in \{1, 2, 3, ...\}$, let L_n be the set of labeled binary trees $(\mathcal{T}, f)$ such that

$$\sum_{v \in V(\mathcal{T})} f(v) = n;$$

denote the cardinality of L_n by $Q(n)$. For each $n \in \{1, 2, 3, ...\}$ and $k \in \{0, 1, 2, ...\}$, define $L_{n,k}$ to be the set of labeled binary trees in L_n that have exactly k branchings. See Figure 18.7 for an example.

The first key observation related to our counting of LEGO structures is the following theorem.

Theorem 3. *Let $T_{\mathcal{J}}(n)$ be the number of LEGO buildings that can be made from n 1×2 jumper plates. Then*

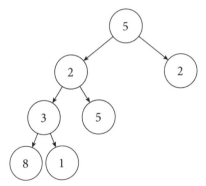

Figure 18.7. A labeled binary tree belonging to the set $L_{26,3}$. The values of the labeling function f are written inside each circle; 26 is the sum of the labels on the nodes; 3 is the number of branchings.

$$T_{\mathcal{J}}(n) \leq \sum_{k=0}^{\lfloor \frac{n-1}{2} \rfloor} 6^{n-1-2k} \#(L_{n,k}).$$

Proof. We begin by associating to each building $b \in B_{\mathcal{J}}(n)$ a labeled binary graph $\Theta(b) \in L_{n,k}$. The idea of how this association works is shown in Figures 18.8 and 18.9. The concept is that the nodes of tree $\Theta(b)$ indicate the parents in the original building that have two children, and the labels on the nodes (i.e., the values of f) indicate how many generations one needs to pass through before seeing the next plate with two children. Now for the details.

First, given $b \in B_{\mathcal{J}}(n)$, associate a graph to b by thinking of the individual jumper plates comprising b as nodes and saying that the nodes are related if the corresponding plates are attached, similar to what was done in Durhuus and Eilers [5]. This produces a tree $\theta(b)$, whose root corresponds to the root of the building, where each node in $\theta(b)$ has at most two children.

To order the tree $\theta(b)$, we need to consider the situation where a plate in the building has two children. To do this, choose a compass direction to represent north. If a plate has two children, either one child is south of the other, or one child is west of the other. In the first case, decree the left branch in $\theta(b)$ to correspond to the southernmost child, and in the second case, decree the left branch in $\theta(b)$ to correspond to the westernmost child.

To obtain the labeled binary tree $\Theta(b)$, we next define an equivalence relation on the nodes of $\theta(b)$. Given nodes v and w in $\theta(b)$, say that $v \preceq w$ if there is a chain of nodes $v = v_0, v_1, v_2, ..., v_n = w$ such that for each $j \in \{1, 2, 3, ..., n\}$, v_j is the *only* child of v_{j-1}. (Notice that for all nodes v, $v \preceq v$ by setting $n = 0$.) Then declare nodes v and w to be equivalent if $v \preceq w$ or $w \preceq v$.

Figure 18.8. A LEGO building b made from sixteen jumper plates.

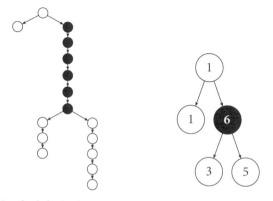

Figure 18.9. On the left, the binary tree $\theta(b)$ is shown for the building b in Figure 18.8. This binary tree essentially "forgets" the orientation of the plates and simply records how the individual pieces are attached. In particular, the nodes colored black in this tree come from the plates colored black in the building pictured in Figure 18.8. On the right, the labeled binary tree $\Theta(b) \in L_{16,2}$ is pictured; note that the six individual black nodes in $\theta(b)$ have been collapsed to a single node labeled "6" in $\Theta(b)$.

Denoting the equivalence class of a node v under this relation by $[v]$, we obtain a labeled binary tree $\Theta(b)$

1. whose vertices are the equivalence classes $[v]$,
2. whose edge relations are defined by saying $[w]$ is the child of $[v]$ if and only if some member of $[w]$ is the child of some member of $[v]$ in tree $\theta(b)$, and
3. whose labeling function f is defined by $f([a]) = \#([a])$. This completes the formal definition of Θ.

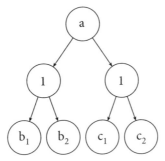

Figure 18.10. No labeled graph of this type (where a, b_1, b_2, c_1, and c_2 are positive integers) can be obtained as $\Theta(b)$ for any building $b \in B_{\mathcal{J}}(n)$, because the plate at the top-most branching would have four grandchildren.

For the second part of the proof, we count the maximum number of preimages a labeled binary tree has under Θ. If $\Theta(b) \in L_{n,k}$, then b must be a building made of n jumper plates, of which k have exactly two children; these plates are at locations specified by the labels of $\Theta(b)$. To describe such a building, one therefore needs only to specify how to attach the jumper plates that are only children. There are $n - 1 - 2k$ such plates that need to be attached to make such a building, and for each such plate, there are ≤ 6 ways to attach the plate to its parent. Thus there are at most 6^{n-1-2k} buildings b for which $\Theta(b)$ is a fixed labeled tree in $L_{n,k}$. Summing this count from the minimum number of branchings (zero) to the maximum number of branchings in a building made from n jumper plates ($\lfloor \frac{n-1}{2} \rfloor$) gives the inequality of the theorem. □

In light of Theorem 3, one way we could find an upper bound on $T_{\mathcal{J}}(n)$ would be to study the growth rate of the sequence $\#(L_{n,k})$. But in fact, we can do better by observing that the function Θ defined in Theorem 3 is very far from surjective. As an example, suppose that some jumper plate in a building has two children. Because these two children must be parallel and share a common boundary of length 2, it is impossible for either of those two children to themselves have two children unless their sibling is childless. So in a building made from jumper plates, no plate can have more than two grandchildren, meaning, for example, that a labeled graph such as the one shown in Figure 18.10 cannot be $\Theta(b)$ for any $b \in B_{\mathcal{J}}(n)$.

Defining $L^*_{n,k}$ to be the set of labeled binary graphs in $L_{n,k}$ that are actually obtained as $\Theta(b)$ for at least one building $b \in B_{\mathcal{J}}(n)$, and denoting the cardinality of the set $L^*_{n,k}$ as $Q(n, k)$, it follows from the reasoning in the last paragraph of the proof of Theorem 2 that

$$T_{\mathcal{J}}(n) \leq \sum_{k=0}^{\lfloor \frac{n-1}{2} \rfloor} 6^{n-1-2k} Q(n, k). \qquad (4)$$

Instead of studying the growth rate of $\#(L_{n,k})$, we instead will find an effective upper bound on the size of $Q(n, k)$. The next three lemmas work toward this goal. In Lemma 1, we lay out some preliminary properties of $Q(n, k)$. Lemma 2 establishes a recursively defined upper bound for $Q(n, k)$, and Lemma 3 uses the preceding two lemmas to establish a closed formula for a nice upper bound on $Q(n, k)$.

Lemma 1 (Properties of $Q(n, k)$).   *Let $Q(n, k)$ be defined as above. Then*

     1. *If $n < 2k + 1$, then $Q(n, k) = 0$.*
     2. *For any $n \in \{1, 2, 3, ...\}$, $Q(n, 0) = 1$.*
     3. *For any $k \in \{1, 2, ...\}$, $Q(2k + 1, k) = 2^{k-1}$.*

Proof. For property 1, observe that a (full) binary tree $\mathcal{T}$ with k branchings must have exactly $2k + 1$ nodes. Thus, for any function $f : V(\mathcal{T}) \to \{1, 2, 3, ...\}$, we have that

$$\sum_{v \in V(\mathcal{T})} f(v) \geq 2k + 1,$$

so no pair $(\mathcal{T}, f)$ can exist in $L_{n,k}$ if $n < 2k + 1$.

For property 2, note that a tree with zero branchings consists of a single node. The only element of $L_{n,0}$ is therefore this single node, together with the function assigning n to that node.

Last, to show property 3, notice that a labeled binary tree belongs to $L_{2k+1,k}$ if and only if the tree has k branchings and $f(v) = 1$ for every node in $V(\mathcal{T})$. For such a tree to come from a building made from jumper plates, there must be only one branching at each level of the tree (otherwise, there would be a plate in the building with four grandchildren, which is impossible). Thus at each level other than the root, there are two choices for which child in the tree has a branching (the left or the right). Since there are $k - 1$ branchings other than the one at the root, we obtain $Q(2k + 1, k) = 2^{k-1}$, as wanted.   ☐

Lemma 2 (Recursive upper bound for $Q(n, k)$).   *For any $n \in \{1, 2, 3, ...\}$ and any $k \in \{0, 1, 2, ...\}$, we have*

$$Q(n, k) \leq Q(n - 1, k) + \sum_{j=0}^{n-1} \sum_{s=0}^{k-1} Q(j, s) Q(n - j - 1, k - s - 1).$$

Proof. Let $L_{n,k}^{(1)}$ be the set of labeled binary trees in $L_{n,k}^{*}$ such that f assigns 1 to the root vertex.

First, we count the complement of $L_{n,k}^{(1)}$. To do this, observe that any such tree can be associated to a tree in $L_{n-1,k}^{*}$ by subtracting 1 from the label on the root. More precisely, for any $(\mathcal{T}, f) \in L_{n,k}^{*} - L_{n,k}^{(1)}$, define $g : V(\mathcal{T}) \to \{1, 2, 3, ...\}$ by

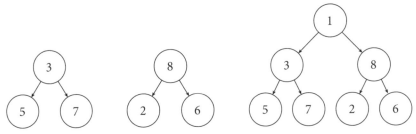

Figure 18.11. An example of how we join labeled trees. Count the number of labeled trees in $L_{n,k}^{(1)}$ by observing that every labeled tree in that set comes from joining a "left subtree" (T_l, f_l) to a "right subtree" (T_r, f_r). This figure shows an example in which we start with two trees: $(T_l, f_l) \in L_{15,1}^*$ (shown at left) and $(T_r, f_r) \in L_{16,1}^*$ (shown in the middle); we join these trees together to make $J = J((T_l, f_l), (T_r, f_r)) \in L_{15+16,1+1+1}^{(1)} = L_{31,3}^{(1)}$, the tree pictured at right.

$$g(v) = \begin{cases} f(v) & \text{if } v \text{ is not the root of } \mathcal{T}, \\ f(v) - 1 & \text{if } v \text{ is the root of } \mathcal{T}. \end{cases}$$

The mapping $(\mathcal{T}, f) \mapsto (\mathcal{T}, g)$ therefore gives a bijection between $L_{n,k}^* - L_{n,k}^{(1)}$ and $L_{n-1,k}^*$, so

$$\# \left(L_{n,k}^* - L_{n,k}^{(1)} \right) = Q(n-1, k).$$

Next we count $L_{n,k}^{(1)}$. Let $j \in \{0, 1, ..., n-1\}$, and let $s \in \{0, 1, ..., k-1\}$. Given any two labeled trees $(T_l, f_l) \in L_{j,s}^*$ and $(T_r, f_r) \in L_{n-j-1,k-s-1}^*$, we can "join" those trees together to create a tree in $L_{n,k}^{(1)}$ as shown in Figure 18.11. More precisely, we build a labeled tree

$$J = J\left((T_l, f_l), (T_r, f_r)\right)$$

as follows: the tree $\mathcal{T}$ of J is formed by taking trees T_l and T_r, adding one more node to serve as the root of the new tree, and decreeing the roots of T_l and T_r to be, respectively, the left and right children of the new root. The function $f : V(\mathcal{T}) \to \{1, 2, 3, ...\}$ assigns 1 to the root of $\mathcal{T}$, and agrees with f_l and f_r on T_l and T_r. Notice that

$$\sum_{v \in V(\mathcal{T})} f(v) = f(\text{root}(\mathcal{T})) + \sum_{v \in V(T_l)} f_l(v) + \sum_{v \in V(T_r)} f_r(v)$$

$$= 1 + j + (n - j - 1) = n,$$

and the number of branchings in J is 1 (from the new root) plus s (the number of branchings in $\mathcal{T}_l$) plus $k - s - 1$ (the number of branchings in $\mathcal{T}_r$). Therefore, $J \in L_{n,k}$. Of course, J might not be in the image of Θ, but every building in $L_{n,k}^{(1)}$ can be obtained in this way, so we know that

$$\# \left(L_{n,k}^{(1)} \right) \leq \sum_{j=0}^{n-1} \sum_{s=0}^{k-1} Q(j, s) Q(n - j - 1, k - s - 1).$$

Adding together the counts of $L_{n,k}^{(1)}$ and $L_{n,k}^* - L_{n,k}^{(1)}$ gives the inequality in the statement of the lemma. □

The "initial values" of $Q(n, k)$ given in Lemma 1 and the recursive formula of Lemma 2 can be combined to obtain the following upper bound on $Q(n, k)$:

Lemma 3 (Upper bound on $Q(n, k)$). *Let $Q(n, k)$ be defined as above. Then*

$$Q(n, k) \leq \binom{n - 1}{2k} 2^{k-1}.$$

Proof. For $n \geq 1$ and $k \geq 0$, define numbers $R(n, k)$ by using the information from Lemma 1 and pretending that the inequality given in Lemma 2 is actually an equality: more formally, set

$$R(n, k) = 0 \text{ for any } n < 2k + 1, \tag{5}$$

$$R(n, 0) = 1 \text{ for any } n \geq 0, \tag{6}$$

$$R(2k + 1, k) = 2^{k-1} \text{ for any } k \geq 1, \tag{7}$$

and recursively define, for any $n > 2k + 1$,

$$R(n, k) = R(n - 1, k) + \sum_{j=0}^{n-1} \sum_{s=0}^{k-1} R(j, s) R(n - j - 1, k - s - 1). \tag{8}$$

In light of Lemma 2, $Q(n, k) \leq R(n, k)$ for all n and k. We will prove the lemma by showing $R(n, k) = \binom{n-1}{2k} 2^{k-1}$.

The key to this lemma is to see that for each $n \in \{1, 2, 3, ...\}$ and each $k \in \{0, 1, 2, ...\}$, $R(n, k)$ is a polynomial of degree $2k$ in the variable n. To prove this claim, we use induction on k. The base case $k = 0$ is obvious, since $R(n, 0) = 1$ for all $n \geq 1$.

For the induction step, fix $k > 0$, and assume that for all $j < k$, $R(n, j)$ is a degree $2j$ polynomial in the variable n. Now define $R(0, k) = 0$ and for $n \in \{1, 2, 3, ...\}$, set

$$D(n, k) = R(n, k) - R(n - 1, k).$$

By formula (8) above, we see that

$$D(n, k) = \sum_{j=0}^{n-1} \sum_{s=0}^{k-1} R(j, s) R(n - j - 1, k - s - 1).$$

By the induction hypothesis, $R(j, s)$ is a polynomial of degree $2s$ in the variable j, and $R(n - j - 1, k - s - 1)$ is a polynomial of degree $2(k - s - 1)$ in the variable $n - j - 1$. Therefore, for each j and s, the expression

$$R(j, s) R(n - j - 1, k - s - 1)$$

is a polynomial in two variables n and j (degree $2(k - 1)$ in the variable j and degree $2(k - s - 1)$ in the variable n). When these polynomials are summed from $s = 0$ to $k - 1$, we obtain a polynomial that is degree $2(k - 1) = 2k - 2$ in the variable j and degree $2(k - 1)$ in the variable n. Therefore, $D(n, k)$, being the sum from $j = 0$ to $n - 1$ of such polynomials, is a polynomial of degree $2k - 1$ in the variable n (the highest degree coming from the polynomials in variable j being added together). Finally,

$$R(n, k) = \sum_{j=0}^{n} D(j, k)$$

is the sum of $n + 1$ polynomials of degree $2k - 1$ in the variable n, which is a polynomial of degree $2k$. This establishes the claim.

At this point, we know that $R(n, k)$ is a polynomial of degree $2k$ that has roots when $n = 1, 2, 3, 4, ..., 2k$. Therefore, for some constant C depending on k,

$$R(n, k) = C(n - 1)(n - 2) \cdots (n - 2k),$$

and in particular, $R(2k + 1, k) = C(2k)!$. But we know $R(2k + 1, k) = 2^{k-1}$ by equation (7), and it follows that $C = \frac{2^{k-1}}{(2k)!}$. Therefore

$$R(n, k) = \frac{2^{k-1}}{(2k)!}(n - 1)(n - 2) \cdots (n - 2k)$$

$$= \frac{2^{k-1}}{(2k)!} \frac{(n - 1)!}{(n - 2k - 1)!}.$$

$$= 2^{k-1} \binom{n-1}{2k},$$

as needed. □

To summarize, at this point we know by combining Theorem 3 and Lemma 3 that the number $T_{\mathcal{J}}(n)$ of buildings made from n LEGO jumper plates satisfies

$$T_{\mathcal{J}}(n) \le \sum_{k=0}^{\lfloor \frac{n-1}{2} \rfloor} Q(n,k)6^{n-1-2k}$$

$$\le \sum_{k=0}^{\lfloor \frac{n-1}{2} \rfloor} \binom{n-1}{2k} 2^{k-1} \cdot 6^{n-1} \left(\frac{1}{36} \right)^k.$$

Using the convention that $\binom{n}{k} = 0$ when $n < k$, this inequality can be rewritten as

$$T_{\mathcal{J}}(n) \le \frac{1}{2} \cdot 6^{n-1} \sum_{k=0}^{\infty} \binom{n-1}{2k} \left(\frac{1}{18} \right)^k. \tag{9}$$

We have obtained a series on the right-hand side expression (9) that fortunately, can be summed using the binomial theorem.

Lemma 4. *For any $r \in (0,1)$,*

$$\sum_{k=0}^{\infty} \binom{n-1}{2k} r^k = \frac{(1 + \sqrt{r})^{n-1} + (1 - \sqrt{r})^{n-1}}{2}.$$

Proof. Let $r \in (0,1)$. From the binomial theorem,

$$(1 + \sqrt{r})^{n-1} = \sum_{k=0}^{\infty} \binom{n-1}{k} (\sqrt{r})^k (1)^{n-1-k}$$

$$= \binom{n-1}{0} + \binom{n-1}{1} r^{1/2} + \binom{n-1}{2} r + \binom{n-1}{3} r^{3/2} + \cdots,$$

and also

$$(1 - \sqrt{r})^{n-1} = \sum_{k=0}^{\infty} \binom{n-1}{k} (-\sqrt{r})^k (1)^{n-1-k}$$

$$= \binom{n-1}{0} - \binom{n-1}{1} r^{1/2} + \binom{n-1}{2} r - \binom{n-1}{3} r^{3/2} + \cdots .$$

When the two preceding series are added together, the noninteger powers of r cancel; dividing the sum by 2 gives the formula in the claim.   □

Finally, we are able to put all the work of this section together and deduce the upper bound on $h_{\mathcal{J}}$.

Theorem 4. *The entropy* $h_{\mathcal{J}}$ *of a* 1×2 *jumper plate satisfies* $h_{\mathcal{J}} \le \log(6 + \sqrt{2})$.

Proof. Applying the formula in Lemma 4 to equation (9), we get

$$T_{\mathcal{J}}(n) \le 6^{n-1} \left(\frac{1}{4}\right) \left[\left(1 + \sqrt{\frac{1}{18}}\right)^{n-1} + \left(1 - \sqrt{\frac{1}{18}}\right)^{n-1} \right]$$

$$= \frac{1}{4} \left[\left(6 + \sqrt{2}\right)^{n-1} + \left(6 - \sqrt{2}\right)^{n-1} \right],$$

and therefore

$$h_{\mathcal{J}} = \lim_{n \to \infty} \frac{1}{n} \log T_{\mathcal{J}}(n)$$

$$\le \lim_{n \to \infty} \frac{1}{n} \log \frac{1}{4} \left[\left(6 + \sqrt{2}\right)^{n-1} + \left(6 - \sqrt{2}\right)^{n-1} \right].$$

The dominant exponential term inside the brackets comes from the $\left(6 + \sqrt{2}\right)^{n-1}$ term, so we obtain $h_{\mathcal{J}} \le \log(6 + \sqrt{2})$, as desired. (Alternatively, this limit can be evaluated rigorously using logarithm rules and L'Hopital's Rule.)   □

6 Further Improvements to the Upper Bound

Recall from the discussion in Section 5 that many labeled binary graphs in $L_{n,k}$ cannot be obtained as $\Theta(b)$ for any building b made from n jumper plates. We obtained the upper bound $h_{\mathcal{J}} \le \log(6 + \sqrt{2})$ by throwing out some labeled binary graphs for which some nodes have four grandchildren. However, this technique does not come close to discarding all the labeled binary graphs which are not in the range of Θ. In particular, any labeled binary graph containing a subgraph like either of those in Figures 18.10 or 18.12 cannot be $\Theta(b)$ for any building b.

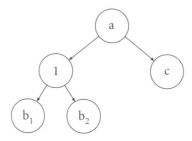

Figure 18.12. If $c > 1$, then no matter the values of positive integers a, b_1 and b_2, this labeled graph cannot be $\Theta(b)$ for any $b \in B_{\mathcal{J}}(n)$ for any n, because the plate corresponding to the top-most branching would have three grandchildren.

We propose a program to further improve our upper bound as follows. As before, for each n and k, let

$$L^*_{n,k} = \text{Range}(\Theta) \cap L_{n,k},$$

and let $Q^*(n, k)$ be any sequence satisfying $Q^*(n, k) \geq \#(L^*_{n,k})$. By the same argument as the one given in Section 5, we have

$$h_{\mathcal{J}} \leq \sum_{k=0}^{\lfloor \frac{n-1}{2} \rfloor} Q^*(n, k) 6^{n-1-2k},$$

so any $Q^*(n, k)$ that grows more slowly than ours (in this language, what we used for "$Q^*(n, k)$" in Section 4 was $R(n, k) = \binom{n-1}{2k} 2^k$) would improve the upper bound. One way to do this is would be to count the number of labeled binary graphs coming from binary graphs in which no node has more than two grandchildren. By itself, however, this improvement would not give the exact value of $Q^*(n, k)$, as there are other restrictions on the kinds of labeled binary graphs that lie in the range of Θ.

7 Remarks on the Entropy of LEGO Roof Tiles

S. Eilers asked whether our methods can be adapted to study 2×1 LEGO roof tiles (such a piece, which we denote by class $\mathcal{R}$, slopes inward at a 45° angle from a base that measures 1×2 to a top that measures 1×1; see Figure 18.13). Unlike jumper plates, the bottom of a roof tile allows only two slots for the top stud of a child to be inserted into its parent.

First, the argument given in Section 3 for jumper plates carries over directly to show that $h_{\mathcal{R}} = \lim_{n \to \infty} \frac{1}{n} \log T_{\mathcal{R}}(n)$ exists. Second, it is easy to see that

Figure 18.13. The top and bottom of a 2 × 1 LEGO roof tile.

for 2 × 1 roof tiles, $T_{\mathcal{R}}(2) = 8$, because to connect two such pieces together, one needs to choose a slot to insert the child (two options) and independently choose a direction for the slope of the child to point (four options). Thus, a crude lower bound on the entropy is $h_{\mathcal{R}} \geq \log 8$, and this bound could be improved by methods akin to what we executed for jumper plates in Section 4 (for this class of brick, $c_1 = 8$ and $c_3 = 144$, producing a lower bound of $h_{\mathcal{R}} \geq \log 9.57174$).

As for an upper bound on $h_{\mathcal{R}}$, by associating a labeled graph $\Theta(b)$ to any building b made from n 2 × 1 roof plates in the same way we did for jumper plates in Section 5, one obtains

$$T_{\mathcal{R}}(n) \leq \sum_{k=0}^{\lfloor \frac{n-1}{2} \rfloor} 8^{n-1-2k} 9^k \#(L_{n,k}).$$

The extra 9^k term in this expression comes from the fact that there are nine distinct ways to attach two roof tiles underneath a particular roof tile, and a building associated to a graph in $L_{n,k}$ has exactly k of these branchings.

Unfortunately, at this point, the problem becomes more difficult, because unlike jumper plates, a roof plate may have four grandchildren. Our methods do provide an upper bound of $h_{\mathcal{R}} \leq \log 14$, but with far more complicated analysis using special mathematical functions, which we briefly outline in the next paragraph, leaving the details for the interested reader to work out.

Using methods similar to what we did in Lemmas 1–3, one can show that

$$\#(L_{n,k}) = \binom{n-1}{2k} \binom{2k}{k} \frac{1}{k+1},$$

and therefore,

$$T_{\mathcal{R}}(n) \leq 8^{n-1} \sum_{k=0}^{\lfloor \frac{n-1}{2} \rfloor} \binom{n-1}{2k} \binom{2k}{k} \frac{1}{k+1} \left(\frac{9}{64} \right)^k$$

$$= 8^{n-1} \sum_{k=0}^{\infty} \frac{(n-1)!}{(n-2k-1)!(k+1)!} \frac{(9/64)^k}{k!}$$

$$= 8^{n-1} \, {}_2F_1 \left(\frac{1}{2} - \frac{n}{2}, 1 - \frac{n}{2}, 2; \frac{9}{16} \right),$$

where $_2F_1$ is the Gauss hypergeometric function [14, 16]. Using a symmetry of the hypergeometric function called the "Pfaff transformation" (see, for example, Khawaja and Daalhuis [8]) and the definition of the Jacobi polynomial $P_m^{(a,b)}$ (see chapter 4 of Szegö [17]), this upper bound can be rewritten as

$$8^{n-1} \left(\frac{7}{16} \right)^m \frac{1}{m+1} P_m^{(1, \frac{-1}{2})} \left(\frac{25}{7} \right),$$

where $m = \frac{1}{2}(n-1)$. For large m, this Jacobi polynomial can subsequently be approximated using Darboux's formula (see formula (1.2) of Bosbach and Gawronski [3]), as

$$P_m^{(1, \frac{-1}{2})} \left(\frac{25}{7} \right) = \frac{K}{\sqrt{m}} 7^{m+1} (1 + o(1))$$

for a suitable constant K. Putting all this together gives

$$T_{\mathcal{R}}(n) \le 8^{n-1} \left(\frac{7}{16} \right)^{\frac{n-1}{2}} 7^{\frac{n+1}{2}} \frac{K}{(m+1)\sqrt{m}} (1 + o(1)),$$

and by taking the logarithm of this expression, dividing by n, and letting $n \to \infty$, one obtains the upper bound

$$h_{\mathcal{R}} \le \log 8 + \log \frac{\sqrt{7}}{4} + \log \sqrt{7} = \log 14.$$

It would be interesting to study other LEGO bricks for which an upper bound on their entropy can be computed using our methods.

Acknowledgments

The authors thank the organizers of the 2017 MOVES conference, especially Tim Chartier, who recommended that the second author give a talk on related work at the conference. We also thank Søren Eilers for providing computer computations of some constants in this chapter, suggesting the application of our technique to roof tiles, and for other helpful discussions.

References

[1] M. Abrahamsen and S. Eilers. On the asymptotic enumeration of LEGO structures. *Exp. Math.* **20** (2011) 145–152.

[2] P. H. Algoet and T. Cover. A sandwich proof of the Shannon-McMillan-Breiman theorem. *Ann. Prob.* **16** (1988) 899–909.

[3] C. Bosbach and W. Gawronski. Strong asymptotics for Jacobi polynomials with varying weights. *Methods Appl. Anal.* **6** (1999) 39–54.

[4] L. Breiman. The individual ergodic theorem of information theory. *Ann. Math. Statist.* **28** (1957) 809–811; correction **31** (1960) 809–810.

[5] B. Durhuus and S. Eilers. Combinatorial aspects of pyramids of one-dimensional pieces of arbitrary fixed integer length. In *21st International Meeting on Probabilistic, Combinatorial and Asymptotic Methods in the Analysis of Algorithms, pp.* 143–158, *Discrete Math. Theor. Comput. Sci. Proc., Am. Assoc. Discrete Math. Theor. Comput. Sci., Nancy,* 2010.

[6] B. Durhuus and S. Eilers. On the entropy of LEGO. *J. Appl. Math. Comput.* **45** (2014) 433–448.

[7] S. Eilers. Personal communication (2018).

[8] S. F. Khawaja and A. B. Olde Daalhuis. Uniform asymptotic expansions for hypergeometric functions with large parameters IV. *Anal. Appl. (Singapore.)* **12** (2014) 667–710.

[9] J. Kirk Christiansen. Taljonglering med klodser—eller talrige klodser [in Danish]. Klodshans. LEGO Company Newsletter (1974).

[10] K. Kirk Christiansen. *The Ultimate LEGO Book.* Dorling Kindersley, London, 1999.

[11] Lego Group. Annual Report 2016. Available at https://www.lego.com/en-us /aboutus/news-room/2017/march/annual-results-2016.

[12] D. Lipkowitz. *The LEGO Book.* Dorling Kindersley, London, 2012.

[13] B. McMillan. The basic theorems of information theory. *Ann. Math. Statist.* **24** (1953) 196–219.

[14] A. B. Olde Daalhuis. Hypergeometric function. In *NIST Handbook of Mathematical Functions.* US Department of Commerce, Washington, DC, 2010, 383–401.

[15] C. Shannon. A mathematical theory of communication. *Bell System Tech. J.* **27** (1948) 379–423, 623–656.

[16] L. J. Slater. *Generalized Hypergeometric Functions.* Cambridge University Press, Cambridge, 1966.

[17] G. Szegö. *Orthogonal Polynomials.* American Mathematical Society, Providence, RI, 1975.

About the Editors

Jennifer Beineke is a professor of mathematics at Western New England University, Springfield, MA. She earned undergraduate degrees in mathematics and French from Purdue University, West Lafayette, IN, and obtained her PhD from the University of California, Los Angeles. She held a visiting position at Trinity College, Hartford, CT, where she received the Arthur H. Hughes Award for Outstanding Teaching Achievement. Her research in the area of analytic number theory has most recently focused on moments of the Riemann zeta function. She enjoys sharing her love of mathematics, especially number theory and recreational mathematics, with others, usually traveling to math conferences with some combination of her husband, parents, and three children.

Jason Rosenhouse is a professor of mathematics at James Madison University, Harrisonburg, VA, specializing in algebraic graph thoery. He received his PhD from Dartmouth College, Hanover, NH, in 2000 and has previously taught at Kansas State University, Manhattan. He is the author of the books *The Monty Hall Problem: The Remarkable Story of Math's Most Contentious Brainteaser* and *Among the Creationists: Dispatches from the Anti-Evolutionist Front Line*. With Laura Taalman, he is the coauthor of *Taking Sudoku Seriously: The Math Behind the World's Most Popular Pencil Puzzle*, which won the 2012 PROSE Award from the Association of American Publishers in the category "Popular Science and Popular Mathematics." He is the editor-elect of *Mathematics Magazine*, published by the Mathematical Association of America. Currently, he is working on a book about logic puzzles, forthcoming from Princeton University Press.

Beineke and Rosenhouse are the editors of *The Mathematics of Various Entertaining Subjects,* Vols. I and II, published by Princeton University Press in association with the National Museum of Mathematics. Volume I was named a *Choice* Outstanding Academic Title for 2016. *Choice* is a publication of the American Library Association.

Max A. Alekseyev is an associate professor of mathematics and computational biology at the George Washington University, Washington, DC. He holds an MS in mathematics from Lobachevsky State University, Russia (1999) and a PhD in computer science from the University of California, San Diego (2007). He is a recipient of the CAREER award from the National Science Foundation (2013), and the John Riordan prize (2015) from the OEIS Foundation. His research interests range from discrete mathematics (particularly, combinatorics and graph theory) to computational biology (particularly, comparative genomics and genome assembly).

Michael P. Allocca is an associate professor of mathematics at Muhlenberg College, Allentown, PA. He earned his PhD from North Carolina State University, Raleigh. He has published in multiple areas of mathematics, ranging from homotopy algebras to the intersection of group theory and molecular biology. He considers himself a "coffee shop mathematician" who loves to talk about any interesting mathematics over a warm cup of white mocha. When not teaching and researching mathematics, he enjoys woodworking and completing other projects around his home.

Walker Anderson is a high school student at Central Bucks High School, West, in Doylestown, PA. He is a current member of the US Puzzle Team and has competed in the 2016 and 2017 World Puzzle Championships in Slovakia and India. He runs a puzzle blog at www.wa1729.blogspot.com and has submitted puzzles to Grandmaster Puzzles and Logic Masters India. He recently wrote the iBook *A Beginner's Guide to Logic Puzzles*.

Barry A. Balof is a professor of mathematics at Whitman College, Walla Walla, WA. He earned his BA from Colorado College, Colorado Springs, and his PhD from Dartmouth College, Hanover, NH. His research interests lie in ordered set theory, graph theory, and enumerative combinatorics. When not in the classroom, you are likely to find him at the bridge table.

Arthur T. Benjamin is the Smallwood Family Professor of Mathematics at Harvey Mudd College, Claremeont, CA. With Jennifer Quinn, he wrote *Proofs That Really Count: The Art of Combinatorial Proof*, which received the Beckenbach Book Prize from the Mathematical Association of America. Professors Benjamin and Quinn were editors of *Math Horizons* magazine from 2004 through 2008. Dr. Benjamin is also a professional magician who performs his mixture of math and magic to audiences all over the world. He has given several TED talks, which have been viewed more than 20 million times. In

2017, he received the Communications Award from the Joint Policy Board of Mathematics.

Jeffrey Bosboom received a BS from the University of California, Irvine, in 2011, and an MS from the Massachusetts Institute of Technology, Cambridge, in 2014. He is currently a PhD student at the Massachusetts Institute of Technology, Cambridge, researching compilers and programming languages for performance engineering, combinatorial search and enumeration, and the computational complexity of games and puzzles.

Spencer Congero was born in Hartford, CT. He received a BS in electrical engineering with a minor in music recording from the University of Southern California, Los Angeles, in 2016. He is currently a PhD student in electrical and computer engineering at the University of California, San Diego.

Jay Cordes is a data scientist who enjoys tackling challenging problems, including how to guide future data scientists away from the common pitfalls he saw in the corporate world. He is a recent graduate from the University of California, Berkeley's Master of Information and Data Science (MIDS) program and graduated from Pomona College, Claremont, CA, with a mathematics major. He has worked as a software developer and a data analyst and was also a strategic advisor and sparring partner for the winning pokerbot in the 2007 AAAI Computer Poker Competition world championship.

Erik D. Demaine is a professor of computer science at the Massachusetts Institute of Technology, Cambridge. He received a MacArthur Fellowship (2003) as a "computational geometer tackling and solving difficult problems related to folding and bending—moving readily between the theoretical and the playful, with a keen eye to revealing the former in the latter." He has cowritten two books (*Geometric Folding Algorithms*, 2007, and *Games, Puzzles, and Computation*, 2009). With his father Martin, his interests span mathematics and art, including curved origami sculptures in the permanent collections of the Museum of Modern Art in New York, and the Renwick Gallery in the Smithsonian, Washington, DC.

Martin L. Demaine is the Angelika and Barton Weller Artist-in-Residence at the Massachusetts Institute of Technology, Cambridge. He started the first private hot glass studio in Canada and has been called the father of Canadian glass. Martin works together with his son Erik, using sculpture to help visualize and understand unsolved problems in mathematics, and their scientific abilities to inspire new art forms. Their artistic work includes curved origami sculptures in the permanent collections of the Museum of Modern Art in New York, and the Renwick Gallery in the Smithsonian, Washington, DC. Their scientific work includes more than 120 joint papers, many about combining art and mathematics.

Persi Diaconis is the Mary V. Sunseri Professor of Statistics and Professor of Mathematics at Stanford University, CA. With Ron Graham, he is the author of *Magical Mathematics: The Mathematical Ideas That Animate Great Magic Tricks*, which received the Euler Prize from the Mathematical Association of America.

Steven T. Dougherty is a professor of mathematics at the University of Scranton, PA, and has been published more than 100 times. He has lectured in twelve countries and is the recipient of the 2005 Hasse Prize from the Mathematical Association of America.

Yossi Elran heads the Innovation Center at the Davidson Institute of Science Education, the educational arm of the Weizmann Institute of Science in Israel. He leads the Institute's recreational math activities, including math circles, festivals, conferences, the K-6 Math-by-Mail program, and other online courses. He is a member of the Gathering for Gardner's Celebration of Mind Committee, which he chaired from 2014 to 2016. He holds a PhD in mathematics from Bar-Ilan University, Israel, has done post-doctoral research in theoretical quantum chemistry, and has written many papers on quantum mechanics, technology in education, and recreational mathematics. He is a co-author of *The Paper Puzzle Book*.

Darren B. Glass received his BA in mathematics from Rice University, Houston, TX, in 1997 and his PhD from the University of Pennsylvania in 2002. He now teaches at Gettysburg College, PA, where he also directs the First-Year Seminar Program. His mathematical interests range from combinatorics to algebraic geometry to cryptography, and he is always looking for fun ways to introduce all of these subjects to undergraduate students.

Ron Graham is Distinguished Research Professor of Mathematics, Computer Science and Engineering at the University of California, San Diego. With Persi Diaconis, he is the author of *Magical Mathematics: The Mathematical Ideas That Animate Great Magic Tricks*, which received the Euler Prize from the Mathematical Association of America.

Brian Hopkins is a professor of mathematics at Saint Peter's University, Jersey City, NJ. He served as editor of *The College Mathematics Journal* from 2014 to 2018 and has received two other recognitions from the Mathematical Association of America: a 2005 George Pólya Award for writing, and a 2015 Deborah and Franklin Tepper Haimo Award for teaching. Hopkins is also active in teacher professional development, primarily through the Institute for Advanced Study/Park City Mathematics Institute.

Tanya Khovanova is a Lecturer at the Massachusetts Institute of Technology, Cambridge, and likes to entertain people with mathematics. She received her PhD in mathematics from the Moscow State University in 1988. Her current research interests lie in recreational mathematics, including puzzles, magic tricks, combinatorics, number theory, geometry, and probability theory. Her website is located at tanyakhovanova.com, her highly popular math blog at blog.tanyakhovanova.com, and her Number Gossip website at numbergossip.com.

Joseph Kisenwether is a mathematics and game design consultant to the casino industry and the founder of Craftsman Gaming. He likes to describe his job as "the reason you cannot win."

Edmund A. Lamagna is a professor of computer science at the University of Rhode Island, Kingston. He earned a PhD from Brown University, Providence, RI, where he completed both his undergraduate and graduate studies. His professional interests lie at the intersection of computer science and mathematics. He has contributed to the fields of computer algebra and the design and analysis of algorithms, and is the author of *Computer Algebra: Concepts and Techniques*. He has also developed innovative approaches for teaching mathematics, including the use of computer algebra to enhance learning in calculus, and the use of puzzles and games to develop analytical thinking skills.

Anany Levitin holds a PhD in mathematics from the Hebrew University, Jerusalem, and an MS degree in computer science from the University of Kentucky, Lexington. He currently is a professor of computing sciences at Villanova University, PA. From 1990 to 1995, he also worked as a consultant at AT&T Bell Laboratories. In addition to several dozen papers, he has authored two books: *Introduction to the Design and Analysis of Algorithms*, which has been translated into five languages, and, jointly with Maria Levitin, *Algorithmic Puzzles*, translated into Chinese, Japanese, Korean, and Russian.

Stephen K. Lucas received his Bachelor of Mathematics from the University of Wollongong, New South Wales, Australia, in 1989 and his PhD from the University of Sydney, Australia, in 1994. In 2002 he received the Michell Medal for Outstanding New Researchers from Australian and New Zealand Industrial and Applied Mathematics (ANZIAM), Australia. He is currently a professor at James Madison University, Harrisonburg, VA, after a postdoc at Harvard University, Cambridge, MA, and a faculty position at the University of South Australia, Adelaide. His research interests span a wide range of topics in applied and pure mathematics, usually with a numerical bent. He enjoys the fact that there is still something interesting to say about Chutes and Ladders!

Jayson Lynch received a BS in physics and computer science and an Master of Engineering. in computer science from the Massachusetts Institute of Technology, Cambridge, in 2013 and 2015, respectively. Currently Jayson is a PhD student at the Massachusetts Institute of Technology under Professor Erik Demaine. Research interests include computational geometry, I/O efficient algorithms, reversible computing, and the computational complexity of games and puzzles.

David McClendon is an associate professor of mathematics at Ferris State University, Big Rapids, MI. Born and raised in Florida, he received his BS in mathematics from the University of North Carolina, Chapel Hill, and his PhD from the University of Maryland, College Park. His mathematical research is in ergodic theory and topological dynamics, and his pedagogical interests center on the use of technology as a tool for inquiry-based learning. In his spare time, he enjoys board games, skiing, and building models from his collection of more than 700,000 LEGO bricks.

Jie Mei received an MS in computer science from the University of Rhode Island, Kingston, in May 2018. Her contribution to in this volume, coauthored with Edmund Lamagna, represents a portion of her thesis research. She did her undergraduate work at, and earned an MS in materials chemistry from, Peking University in China.

David Molnar received his PhD from the University of Connecticut, Storrs, in 2010. His mathematical interests include number theory; dynamical systems; graph theory; and, evidently, games. Keeping in mind his own undergraduate experience, including a term with the Budapest Semesters in mathematics program, he strives to broaden students' awareness of what constitutes Mathematics. One of his outlets for doing so is competitions; since 2009, he has been involved with the New Jersey Undergraduate Mathematics Competition. He is currently assistant teaching professor of mathematics at Rutgers University, New Brunswick, NJ.

David Nacin is a professor at William Paterson University, Wayne, NJ. He enjoys designing and studying puzzles that involve finite groups, the motion of chess pieces, partition identities, and other mathematical structures. He regularly contributes puzzles to a variety of Mathematical Association of America publications and is the author of the puzzle blog, Quadrata (http://quadratablog.blogspot.com/).

Yusra Naqvi is an assistant professor at the University of Sydney, Australia. She earned her PhD from Rutgers University, New Brunswick, NJ. Her research interests focus on Lie theory and related combinatorics.

Jonathan S. Needleman received his BA in math from Oberlin College, OH, in 2003 and his PhD from Cornell University, Ithaca, NY, in 2009. He is currently a professor at Le Moyne College, Syracuse, NY, where he enjoys working on mathematical problems related to puzzles and games. His favorite thing is when he can use recreational math to convince students to get involved in mathematics research. He also has two small children and is always interested in ways to shorten Chutes and Ladders.

Ann Schwartz has discovered more than a dozen flexagons and presented many at seven of the biennial Gathering for Gardner conferences. In 2015, she was the guest speaker at the Recreational Math, Puzzles and Games Conference at the Davidson Institute in Israel, and in 2016, she led a flexagon workshop at MOMATH in New York City. She coauthored "The Hexa-Dodeca-Flexagon," a chapter in *Homage to a Pied Puzzler* (2009). Martin Gardner mentioned her by name in *Hexaflexagons, Probability Paradoxes, and the Tower of Hanoi* (2008). She lives with her husband in New York City.

James D. Stein graduated from Yale University, New Haven, CT, with a BA in mathematics in 1962, and received his MA and PhD degrees from the University of California, Berkeley, in 1967. He retired from the Department of Mathematics at California State University, Long Beach, in 2013, having published more than thirty-five research articles on Banach spaces and fixed-point theory, and having acquired an Erdös number of two. He is the author of several books on mathematics and science, including two selected for the *Scientific American* Book Club. He lives in Redondo Beach, CA, with his wife, Linda.

Jennifer F. Vasquez earned her PhD in 2007 from Indiana University, Bloomington, and is professor of mathematics at the University of Scranton, PA. She was a 2007-08 Project NExT fellow (Sun Dot). Her main research interests lie in geometric topology and quantum computing. Recently, she has begun exploring applications of topology to biology.

Leonard M. Wapner has taught mathematics at El Camino College, Torrance, CA, since 1973. He received his BA and MAT degrees in mathematics from the University of California, Los Angeles. He is the author of *The Pea and the Sun: A Mathematical Paradox* and *Unexpected Expectations: The Curiosities of a Mathematical Crystal Ball*. His writings on mathematics education have appeared in *The Mathematics Teacher* (published by the National Council of Teachers of Mathematics) and *The AMATYC Review* (The American Mathematical Association of Two Year Colleges). He lives in Seal Beach, CA, with his wife Mona and their labrador retriever Toby.

Jonathon Wilson is an undergraduate student at Ferris State University, Big Rapids, MI. He will receive a BS in applied mathematics with a computer science concentration in Spring 2019, after which, he hopes to attend graduate school and become a professor. He enjoys tutoring and helping other students succeed in their math and computer science courses. Mathematically, Jonathon is interested in prime numbers, combinatorics, and other branches of discrete mathematics. His other interests include computer programming and professional wrestling.

Peter Winkler is William Morrill Professor of Mathematics and Computer Science at Dartmouth College, Hanover, NH. He is the author of about 150 research papers and holds a dozen patents in marine navigation, cryptolography, holography, gaming, optical networking, and distributed computing. His research is primarily in combinatorics, probability, and the theory of computing, with forays into statistical physics. He is a winner of the Mathematical Association of America's Lester R. Ford and David P. Robbins prizes.

Index